SILICON PROCESSING
FOR PHOTOVOLTAICS
II

MATERIALS PROCESSING – THEORY AND PRACTICES

VOLUME 6

Series editor

F.F.Y. WANG

NORTH-HOLLAND
AMSTERDAM · OXFORD · NEW YORK · TOKYO

SILICON PROCESSING FOR PHOTOVOLTAICS II

Edited by

CHANDRA P. KHATTAK
Crystal Systems, Inc.
35 Congress Street
Salem, Massachusetts, USA

K.V. RAVI
Mobil Solar Energy Corp.
16 Hickory Drive
Waltham, Massachusetts, USA

1987

NORTH-HOLLAND
AMSTERDAM · OXFORD · NEW YORK · TOKYO

ISBN: 0444 87024 5

Published by:
North-Holland Physics Publishing
a division of
Elsevier Science Publishers B.V.
P.O. Box 103
1000 AC Amsterdam
The Netherlands

Sole distributors for the USA and Canada:
Elsevier Science Publishing Company, Inc.
52 Vanderbilt Avenue
New York, NY 10017
USA

Library of Congress Cataloging-in-Publication Data

Silicon processing for photovoltaics II.
(Materials processing, theory and practices ; v. 6)
Bibliography: p.
Includes index.
1. Solar cells – Materials. 2. Silicon.
I. Khattak, Chandra P. II. Ravi, K. V., 1940– . III. Series.
TK2960.S52 1987 621.31'244 86-23660
ISBN 0-444-87024-5

Printed in The Netherlands

INTRODUCTION TO THE SERIES

Modern technological advances place demanding requirements for the designs and applications of materials. In many instances, the processing of materials becomes the critical step in the manufacturing processes. However, within the vast realm of technical literature, materials processing has not received its proper attention. It is primarily due to the lack of a proper communication forum. Since the materials processing is intimately concerned with specific products, those who are experts have no need to communicate. On the other hand, those who are involved with a different product will develop, in time, the technology of materials processing when required.

It is the objective of this series, Materials Processing – Theory and Practices, to promote the dissemination of technical information about the materials processing. It provides a broad prospective about the theory and practices concerning a particular process of material processing. A material process, intended for one technology, may have an applicability in another. This series provides a bridge between the materials engineering community and the processing engineering community. It is a proper forum of dialogues between the academic and the industrial communities.

Materials processing is a fast-moving field. Within the constraints of time and printed spaces, this series does not aim to be encyclopedic, and all-inclusive. Rather, it supplies an examination of material processes by the active workers. The view will be, by necessity, subjective. But the view will include both near-term and long-term prospectives. It is the fondest hope of this general editor that the volumes in this series can serve as first reference books in the field of materials processing.

Franklin F.Y. WANG
Stony Brook, New York.

PREVIOUS VOLUMES IN THE SERIES

1. Fine line lithography
   R. Newman, volume editor
2. Impurity doping processes in silicon
   F.F.Y. Wang, volume editor
3. Laser materials processing
   M. Bass, volume editor
4. Dry etching for microelectronics
   R.A. Powell, volume editor
5. Silicon processing for photovoltaics I
   C.P. Khattak and K.V. Ravi, volume editors
   (contents on page xii)

In preparation

7. III–V Semiconductor materials and devices
   R.J. Malik, volume editor

PREFACE TO VOLUME 6

The processing of semiconductor silicon for manufacturing low-cost photovoltaic products has been a field of increasing activity over the past decade. A number of papers have been published in the technical literature. A previous volume (volume 5) in this series on Materials Processing – Theory and Practices presented comprehensive reviews on some of the key technologies developed for processing silicon for photovoltaic applications. This volume is complementary to the earlier volume and adds to the list of the important technologies.

The technological mix for production of photovoltaic modules has been changing and it is now possible to select out of a menu of technologies to suit the applications as an alternate to single-crystal silicon wafer processing. One of the areas of increasing importance is the use of multicrystalline silicon wafers. The understanding, production, as well as processing of silicon wafers with grain boundaries require variations from the conventional processes. Further, depending on the technology used, the material shows different characteristics. This volume presents comprehensive reviews of seven technologies for processing silicon for low-cost photovoltaic applications. All these reviews are confined to silicon and no papers on non-silicon semiconductors are included.

The volume starts with a review on the effect of introducing grain boundaries in silicon. This chapter by De Pauw, Mertens and Van Overstraeten examines this problem from a theoretical point of view, correlates the mathematical formulation with experimental evidence, evaluates the influence of grain boundaries and microdefects, and suggests methods to reduce their influence. The next two chapters describe two different processes in commercial production for multicrystalline silicon ingots. Helmreich discusses the Wacker ingot casting process, the largest producer of multicrystalline silicon. Khattak and Schmid describe the adaptation of the heat exchanger method (HEM) of directional solidification and its use as a purification process. The EFG process review by Kalejs discusses the ribbon process which is widely gaining commercial status.

In contrast to the semiconductor industry, the silicon meltstock costs constitute a large portion of the final photovoltaic product costs. D'Aiello, Robinson and Miller discuss the epitaxial solar cell approach in order to minimize the use of high-purity silicon. Dietl, in his chapter, presents a comprehensive review of metallurgical ways of producing silicon meltstocks for photovoltaic applications. Both these approaches are alternatives to reducing costs of silicon starting material.

The final chapter describes an innovative type of solar cell in which light is collected on both sides of the solar cell. This fitting conclusion suggests that even though a number of technologies have been utilized for photovoltaics there is still room for newer approaches.

The editors of this volume have had the unique privilege of working with leaders in the fascinating field of photovoltaics and would like to express their gratitude to the contributors for their thoroughness of discussions and high quality of their contributors.

Chandra P. KHATTAK
Crystal Systems, Inc.
Salem, MA, USA

K.V. RAVI
Mobil Solar Energy Corp.
Waltham, MA, USA

ADVISORY BOARD

CONTENTS

CONTENTS OF SILICON PROCESSING IN PHOTOVOLTAICS I
(Materials Processing – Theory and Practices, volume 5)

CHAPTER 1

EFFECTS OF GRAIN BOUNDARIES AND INTRAGRAIN DEFECTS IN SILICON FOR PHOTOVOLTAIC APPLICATIONS

P. DE PAUW, R. MERTENS and R. VAN OVERSTRAETEN

IMEC v.z.w.
Kapeldreef 75
B-3030 Leuven
Belgium

Silicon Processing for Photovoltaics II, edited by C.P. Khattak and K.V. Ravi

Contents

List of symbols

C_{GB}	Grain boundary capacitance
C_d	Depletion layer capacitance
D	Minority carrier diffusion constant
d	Grain size
E_F	Fermi-level energy
E_{MAX}	Maximum electric field at a grain boundary
FF	Fill factor of a solar cell
G	Half the average interdefect distance
g	Generation rate
g_{AM1}	Average AM1 generation rate
I_{EBIC}	Electron beam induced current
I_{LBIC}	Light beam induced current
I_{S1}	Saturation current of the first exponential
I_{S2}	Saturation current of the second exponential
J_r	Recombination current density at the defect surface
J_{sat}	Saturation current in a semiconductor
J_{sc}	Short-circuit current density of a solar cell
k	Boltzmann constant
L	Minority carrier diffusion length in the crystal between the defects
L_{eff}	Effective minority carrier diffusion length in a defected crystal
L_{AM1}	Effective minority carrier diffusion length at AM1 illumination level
m	Ideality factor of the second exponential
$n(r)$	Minority carrier concentration
$n_0(r)$	Equilibrium minority carrier concentration
n_i	Intrinsic carrier concentration
N_A	Acceptor dopant concentration
N_{is}	Total midgap surface state density at the defect surface
N^A_{is}	Midgap surface state density for acceptor levels
N^D_{is}	Midgap surface state density for donor levels

Q	Integrated excess minority carrier concentration in the volume around the defect
Q_{defect}	Total charge in the defect
$s(n(r_d))$	Surface recombination velocity at the defect surface
$s_{\text{eff}}(n(W))$	Effective surface recombination velocity at $r=W$
T	Temperature
V_{oc}	Open-circuit voltage of a solar cell
v_{th}	Thermal velocity of the carriers
w	Effective width of a grain boundary
W	Depletion layer edge
$\alpha(\lambda)$	Light absorption coefficient at wavelenght λ
ε_{Si}	Relative dielectric permittivity of silicon
ε_0	Absolute dielectric permittivity of silicon
$\varphi_n(r)$	Electron quasi Fermi level at r
$\varphi_p(r)$	Hole quasi Fermi level at r
$\psi(r)$	Electrostatic potential at r
ψ_B	Barrier height at the defect surface
σ_C	Carrier capture cross-section for a coulombic attractive recombination center
σ_N	Carrier capture cross-section for a neutral recombination center
τ_B	Bulk minority carrier lifetime in a crystal with intragrain defects
τ_{eff}	Effective lifetime in a defected crystal

Introduction

During the last decade a number of potentially low-cost solidification techniques for silicon materials useful for photovoltaic applications has been developed. These materials are solidified from very pure silicon. The cost reduction lies in the solidification technique, which proceeds at a higher rate and with lower cost equipment and materials relative to single crystalline silicon. The grain size of this material is very large. In order to distinguish this silicon from single crystalline and polycrystalline materials, the material is called semicrystalline or plesio* single crystalline.

Thermal stresses occur during cooling because of the high solidification rate, which results in non-uniform temperatures, and the adhesion of the solidified silicon to the crucible walls or the supporting substrates, which have a thermal expansion coefficient different from silicon. Also, due to the high reactivity of molten silicon, contamination from the crucible wall material occurs. As a consequence, these materials contain crystal defects like grain boundaries, dislocations and precipitates. The influence of these defects on the solar cell characteristics is the subject of this work.

An overview of the different silicon materials for photovoltaic applications has been given by Dietl et al. (1981) and by Ciszek (1982). Only the three most important, and at the same time the only commercially available, materials will be considered here: Silso (from Wacker), HEM (from Crystal Systems) and UCP (from Semix). These three materials have been solidified in a graphite mold or crucible from very pure silicon.

1. Modeling of grain boundaries and microdefects in defected silicon

In low-cost silicon materials flaws are found that are not present in good quality Czochralski (CZ) or Floating Zone (FZ) single crystalline silicon. These flaws affect both the majority and the minority carrier current flow.

* Plesio is a Greek word, which means 'nearly'.

In a solar cell, minority carrier current flow controls the device behaviour; therefore, the influence of defects on the minority-carrier recombination current is of prime importance. In this section we present a mathematical model completely describing the defect properties.

1.1. Grain boundaries and microdefects in defected silicon

Crystal defects in silicon may have various shapes such as spherical defects (precipitates, dislocation clusters), line defects (different types of dislocations), or planar defects (stacking faults, grain and subgrain boundaries). In order to model these defects it is not necessary to know their exact structure. Experiments indicate that these randomly distributed crystal defects constitute a large density of trapping centers for both types of carriers. The distribution of these traps throughout the energy gap can be U-shaped, constant, or may have other shapes (see section 2.3). Generally, both acceptor- and donor-type traps are present. Since, in the dark, majority carriers are present in a much higher concentration than minority carriers, the defect states predominantly trap the majority carriers. The majority-carrier charge in the defect repels the majority carriers, while it at-

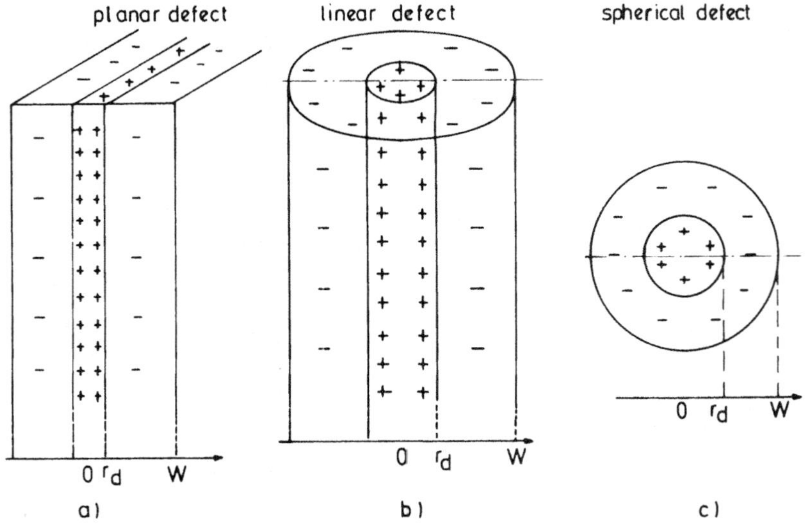

Fig. 1. Different types of defects: (a) planar (grain and subgrain boundaries), (b) cylindrical (different types of dislocations), (c) spherical (precipitates, dislocation clusters).

tracts the minority carriers. This corresponds to a space-charge layer, depleted from minority carriers, around each crystal defect and also gives rise to an enhanced recombination of excess minority carriers. The presence of the space-charge layer also results in a barrier height at the defect. In the case of a planar grain boundary a space-charge layer will be induced on both sides of the boundary (fig. 1a); for a linear defect (fig. 1b) or for a spherical defect (fig. 1c) the space-charge layer surrounds the defect surface.

In order to determine the influence of these defects on the recombination current one has to evaluate the boundary condition imposed at the space-charge layer edge W (fig. 1). The presence of the defect introduces an effective recombination velocity s_{eff} at this boundary. It will be shown in the next sections that s_{eff} depends on the minority carrier concentration $n(W)$ at the space-charge layer edge. To determine the recombination current it is necessary to know the relationship between s_{eff} and $n(W)$. The knowledge of this non-linear boundary condition $s_{eff}(n(W))$ allows the solution of the minority-carrier transport equations in the crystal from which the recombination current can be derived. These solutions allow to determine the influence of defects on the effective lifetime, on the recombination in the space-charge region, on the short-circuit current and on the open-circuit voltage. Therefore, sections 1.2 and 1.3. only consider the defect and the space-charge surrounding it. From this model, $s_{eff}(n(W))$ is derived and the result will be used in sections 3 and 4.

1.2. Mathematical formulation

Let us assume p-type material. (A completely similar calculation can be done for n-type material.) The problem is the calculation of the recombination current density J_r at the defect surface (fig. 2) for a given minority-carrier concentration at the depletion layer edge, $n(W)$. From this result $s_{eff}(n(W))$ is readily derived. For the sake of simplicity, we assume the majority-carrier quasi-Fermi level to remain constant throughout the depletion layer. It is further assumed that all the defect-trapping states are situated on the defect outer surface. The defect characteristics, i.e., the defect surface shape, the density, the distribution through the energy gap, and the capture cross-sections of the acceptor and donor traps are assumed to be known. The following three regions can now be distinguished (fig. 2):

(I) The defect itself, for which the Shockley–Read–Hall theory of recombination is assumed to hold.

(II) The space-charge region between the defect surface and the deple-

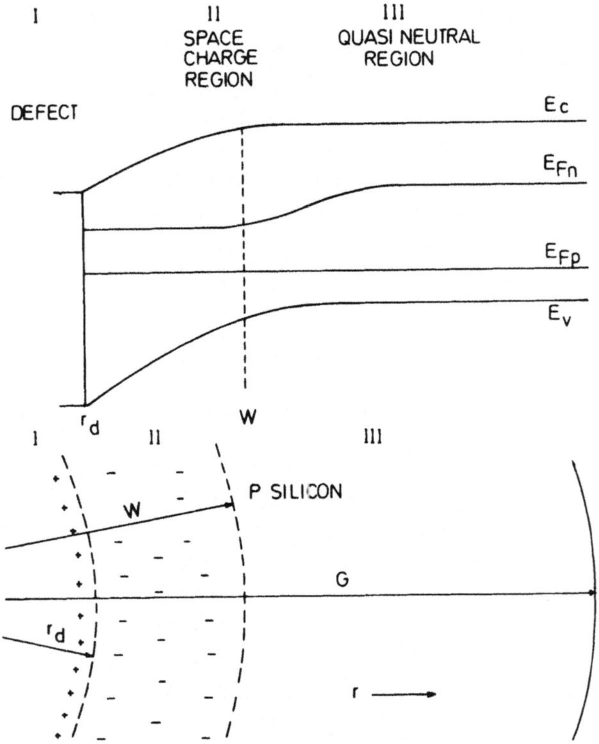

Fig. 2. Band bending and charges around a defect. Three regions can be distinguished: (I) the defect; (II) the space-charge region, and (III) the quasi-neutral region.

tion layer edge. This space charge is caused by the trapping of majority carriers in the defect. Recombination will be neglected in this region.

(III) The quasi-neutral region. In this region there is a diffusion current towards the edge of the space-charge layer and bulk recombination. The characteristics of the bulk recombination centers are not necessarily equal to those of the defect.

Figure 2, in which p-type silicon is considered, indicates schematically the quasi-Fermi levels and the band edges along a cross-section. In order to calculate $s_{\mathrm{eff}}(n(W))$, defined as $J_n(W)/q(n(W)-n_0(W)) \equiv J_r/q[n(W)-n_0(W)]$, the recombination current J_r flowing from W towards the defect must be evaluated. This necessitates the solution of the transport equations in region II. The boundary conditions at the defect surface, where $r=r_d$, require that the minority carrier concentration $n(r_d)$ is the same in the space-charge layer at $r=r_d$ and at the defect surface, and that the minority-carrier current J_r flowing out of region II must be equal to the recombination current at the defect surface.

For each value of $n(W)$, the calculation of the minority-carrier current proceeds as follows:

The carrier densities at the defect surface, $n(r_d)$ and $p(r_d)$, can be expressed in terms of the band bending ψ_B, defined as $\psi(r_d) - \psi(W)$, by the following equations,

$$p(r_d) = n_i \exp\left\{\frac{q[\psi(r_d) - \varphi_p(r_d)]}{kT}\right\} = n_i \exp\left\{\frac{q(\psi_B + \psi(W) - \varphi_p(r_d))}{kT}\right\}, \tag{1.1}$$

$$n(r_d) = n_i \exp\left\{\frac{q[\varphi_n(r_d) - \psi(r_d)]}{kT}\right\} = n_i \exp\left\{\frac{q[\varphi_n(r_d) - \psi_B - \psi(W)]}{kT}\right\}, \tag{1.2}$$

where $\varphi_p(r_d)$ and $\varphi_n(r_d)$ are the hole and electron quasi-Fermi levels at $r = r_d$, and $\psi(r_d)$ and $\psi(W)$ are electrostatic potentials. Under steady-state conditions, the following charge neutrality equation holds

$$Q_{\text{defect}} = Q_{\text{depletion layer}}, \tag{1.3}$$

with Q_{defect} being the charge in the recombination centers at the defect surface. Note that for a grain boundary this charge is for one side of the grain boundary. The total charge in the grain boundary is twice the value Q_{defect}.

First, Q_{defect} and $Q_{\text{depletion layer}}$ are calculated, assuming constant quasi-Fermi levels for majority and minority carriers. This implies negligible recombination and a low current density in the depletion region. We also assume that the transport of carriers across the depletion layer is the same as when the surfaces at $r = r_d$ and $r = W$ are parallel planes. This last assumption is always valid for planar grain boundaries. In the case of other defect shapes, it is also justified for r_d larger than 1 μm because the depletion layer width is then small compared to r_d.

The charge in the defect, Q_{defect}, is obtained by integrating the density of states, multiplied by their occupancy over the energy gap. As stated before, one can assume, without loss of generality, that the trap density is localized at the defect surface. The total number of traps in the defect is then given by the surface trap density multiplied by the area of the defect.

It is further assumed that the Shockley–Read–Hall (SRH) theory is valid. This is not trivial, because the SRH theory assumes that each imperfection center is separated from its neighbour by a distance that is large compared to the extent of the bound-state wave function, localized at the center. The transition rate (by tunneling or hopping on a two-dimensional surface) of an electron (or hole) bound to one center to the adjacent centers is normally much smaller than the rate of transition to the conduction- or valence-band edge states, where many quantum states exist, available for the transition. In this case the defects can be considered as independent and non-interacting. A more in-depth discussion of this subject has been given by Lindholm and Fossum (1981) and Mataré (1971). As a conclusion, for a grain boundary or a dislocation with a high concentration of defect states, the SRH theory may not always be applicable.

We assume here that the distribution of the acceptor and donor states over the energy gap is uniform: the simulations indicate that the shape of this distribution is not very critical. At each energy level, we assume that half of the traps are acceptor, the other half donor-like, and all degeneracy factors are assumed to be unity. These two assumptions are not critical. For the capture cross-sections a distinction has been made between neutral states, characterized by a capture cross-section σ_N, and coulombic attractive centers, with a capture cross-section σ_C (Panayotatos and Card 1980). Little is known about the precise values of the capture cross-sections; σ_C is taken 100 × larger than σ_N. However, this large difference is not a necessary element of this model.

Using the assumptions listed above, Q_{defect} can be expressed as

$$Q_{\text{defect}} = -q \int_{E_V}^{E_C} \frac{\left[\sigma_N n(r_d) + \sigma_C n_i \exp\left(\frac{E_i - E_t}{kT}\right)\right] N_{is}^{A}(E_t)\,\mathrm{d}E_t}{\sigma_N\left[n(r_d) + n_i \exp\left(\frac{E_t - E_i}{kT}\right)\right] + \sigma_C\left[p(r_d) + n_i \exp\left(\frac{E_i - E_t}{kT}\right)\right]}$$
$$+ q \int_{E_V}^{E_C} \frac{\left[\sigma_N p(r_d) + \sigma_C n_i \exp\left(\frac{E_t - E_i}{kT}\right)\right] N_{is}^{D}(E_t)\,\mathrm{d}E_t}{\sigma_C\left[n(r_d) + n_i \exp\left(\frac{E_t - E_i}{kT}\right)\right] + \sigma_N\left[p(r_d) + n_i \exp\left(\frac{E_i - E_t}{kT}\right)\right]} \tag{1.4}$$

In eq.(1.4), $N_{is}^{A}(E_t)$ and $N_{is}^{D}(E_t)$ are the acceptor and donor trap densities at the energy E_t in the energy gap. Equation (1.4) is obtained as follows: the SRH theory gives the expression for the occupancy of a recombination center at an energy E_t in the energy gap. This formula is multiplied by the density of centers at this energy, and integrated over the energy gap, for both acceptor and donor states. $Q_{\text{depletion layer}}$ has been calculated, taking the minority-carrier charge in the depletion layer into account, using the formula derived by Grove and Fitzgerald (1966),

$$Q_{\text{depletion layer}} = -2qn_i L_s[F(\psi(r_d), \varphi_p, \varphi_n)]^{1/2}, \tag{1.5}$$

in which $L_s = (kT/q)^{1/2}(\varepsilon_{si}\varepsilon_0/2qn_i)^{1/2}$ and F follows from

$$F(\psi, \varphi_p, \varphi_n) = \exp\left[\frac{q(\varphi_p - \psi)}{kT}\right] - \exp\left[\frac{q\varphi_p}{kT}\right] + \exp\left[\frac{q(\psi - \varphi_n)}{kT}\right] - \exp\left[\frac{q\varphi_n}{kT}\right] + \frac{q\psi}{kT}\frac{N_A - N_D}{n_i}. \tag{1.6}$$

Equations (1.4), (1.5) and (1.6) contain $n(r_d)$ and $p(r_d)$, due to the assumption of the constant quasi-Fermi levels,

$$n(r_d) = n(W)\exp\left(\frac{q}{kT}\psi_B\right), \tag{1.7}$$

$$p(r_d) = p(W)\exp\left(-\frac{q}{kT}\psi_B\right) = N_A\exp\left(-\frac{q}{kT}\psi_B\right), \tag{1.8}$$

where N_A is the acceptor doping level. Combination of eqs. (1.1)–(1.8) allows the calculation of ψ_B as a function of $n(W)$. The recombination current, J_r, can now be found from the Shockley–Read–Hall theory (Shockley and Read 1952) as follows

$$J_r(n(r_d), p(r_d)) = [n(r_d)p(r_d) - n_i^2]\frac{qv_{th}\sigma_N\sigma_C}{n_i} \times \left(\int_{E_V}^{E_C} dE_t\, N_{is}^A(E_t)\left\{\sigma_N\left[\frac{n(r_d)}{n_i} + \exp\left(\frac{E_t - E_i}{kT}\right)\right] + \sigma_C\left[\frac{p(r_d)}{n_i} + \exp\left(\frac{E_i - E_t}{kT}\right)\right]\right\}^{-1} + \int_{E_V}^{E_C} dE_t\, N_{is}^D(E_t)\left\{\sigma_C\left[\frac{n(r_d)}{n_i} + \exp\left(\frac{E_t - E_i}{kT}\right)\right] + \sigma_N\left[\frac{p(r_d)}{n_i} + \exp\left(\frac{E_i - E_t}{kT}\right)\right]\right\}^{-1}\right), \tag{1.9}$$

where v_{th} is the kinetic limiting velocity. Using eq. (1.9), J_r can be expressed as a function of $n(W)$. In the case of large trap densities at the defect surface, J_r becomes large and the flat quasi-Fermi level approximation for the minority carriers becomes poor. The calculated values of J_r and ψ_B can be improved, using the Gummel–Poon relation (Gummel and Poon 1970). This relation results from the integration of the current continuity equation, neglecting recombination in the space-charge layer,

$$\exp\left[-\frac{q\varphi_n(r_d)}{kT}\right] = \frac{-|J_r|}{qn_iD_n}\int_{r_c}^{W}\exp\left[-\frac{q\psi(r)}{kT}\right]dr + \exp\left[-\frac{q\varphi_n(W)}{kT}\right]. \tag{1.10}$$

To find an improved value of $\varphi_n(r_d)$, using eq. (1.10), one starts from the values of ψ_B and J_r calculated from the constant quasi-Fermi level approach. Using eqs. (1.1)–(1.9), a new calculation of ψ_B and J_r for this value of $\varphi_n(r_d)$ is performed. Using eq. (1.10) and the new values of ψ_B and J_r, the new value of $\varphi_n(r_d)$ is found. This process of iteration is repeated until self-consistency to the required degree of accuracy is obtained.

The only limit imposed on J_r is that it may not exceed the minority or majority carrier saturation current at any place in the crystal. This saturation is attained when all possible minority or majority carriers are used to participate in the current at their maximum allowed speed. This situation may occur at low illumination levels, when there are still only few minority carriers, or when J_r is very high. This saturation current can now be calculated.

The carrier flux F in one direction is given by (McKelvey 1966)

$$F = \tfrac{1}{4}(N/V)v_{th}, \tag{1.11}$$

with N the total number of carriers (holes or electrons) in the volume V.

Under thermal equilibrium, the carrier flux to the defect surface equals the flux coming from the surface. Under non-equilibrium conditions, the flux coming from the defect surface can be assumed to remain the same. This results in a net carrier transport towards the surface, given by

$$F - F' = \tfrac{1}{4}\left(\frac{N}{V} - \frac{N_0}{V}\right)v_{th}, \tag{1.12}$$

with F' the flux from the surface and F the flux to the surface. The net electron current density is then

$$|J_{n\,sat}| = q|F' - F| = \tfrac{1}{4}q(n - n_0)v_{th}, \tag{1.13a}$$

and similarly for the hole current we have

$$|J_{p\,sat}| = q|F' - F| = \tfrac{1}{4}q(p - p_0)v_{th}, \tag{1.13b}$$

For the minority carriers, the minority carrier concentration is smallest at the depletion layer edge ($r = W$). Minority-carrier saturation first occurs there, and the corresponding saturation current is given by

$$|J_{n\,sat}| = \tfrac{1}{4}q\,[n(W) - n_0(W)]\,v_{th} \tag{1.14}$$

In this particular case, $\varphi_n(r_d)$ can be calculated from the equation

$$|J_r| = |J_{n\,sat}|. \tag{1.15}$$

Once $\varphi_n(r_d)$ is known, ψ_B is calculated using the charge neutrality equation, eq. (1.3).

For the majority carriers, the lowest majority-carrier concentration occurs at the defect surface. Majority carrier saturation first occurs there, and the corresponding saturation current is given by

$$\begin{aligned}|J_{p\,sat}| &= \tfrac{1}{4}q[p(r_d) - p_0(r_d)]v_{th}\\ &= \tfrac{1}{4}q[p(W)\exp(-q\psi_B/kT) - p_0(W)\exp(-q\psi_{0B})]v_{th}\end{aligned} \tag{1.16}$$

Simulations reveal, however, that majority-carrier saturation is never a limitation. This can be understood because the majority carriers are present in a much higher concentration and (as will be seen in the next section) the barrier height at room temperature is always lower than 0.5 V. This result indicates that the majority carrier transport does not determine the recombination at the grain boundary, as stated by Seager (1982). Therefore, to model the grain boundary behaviour, it is better to assume that the recombination of minority carriers at the defect surface determines the grain boundary recombination behaviour. The situation can be described as follows: when no bias is applied across the grain boundary, the total minority plus majority current has to be zero. Therefore, the net majority carrier flow has to adapt itself so that it becomes equal to the minority carrier flow. The current value given by eq. (1.13b) is the maximum allowed majority current over the space-charge layer towards the grain boundary.

1.3. Barrier height and effective recombination velocity at a defect

The iterative analysis, outlined in the previous section, allows to calculate the properties of microdefects and grain boundaries. The results of the computer calculations will be shown in this section.

As already mentioned, the most important quantity for majority carrier flow is the grain boundary barrier height, ψ_B. The most important quantity for minority carrier flow is the effective recombination velocity at the defect surface, $s_{eff}(n(W))$.

Although the dark barrier height ψ_B can be calculated using a simplified treatment, it is treated in section 1.2 as a special case of the more general analysis under illumination.

The calculated values of the dark barrier height ψ_B are given in fig. 3 for different doping levels N_A and interface state densities $N_{is} = N_{is}^A + N_{is}^D$. One clearly observes that for each value of N_{is} the ψ_B versus N_A curve has a maximum. This can be explained as follows: at low doping levels the bulk Fermi level position lies close to the middle of the gap. Due to the symmetry of the assumed interface parameters, this results in a low interface charge, and a small band bending. Experimental data indicate such a behaviour (E.S. Yang et al. 1981, Seager et al. 1979, 1980, Seager and Pike 1979). As the doping level increases, the bulk Fermi level moves towards the band edge and the barrier height at the defect will increase. At high doping levels, a second effect becomes important. For an exponentially increasing bulk dopant concentration, the Fermi level (and as a consequence also the charge in the traps) increases only linearly. A higher space-charge density results in a smaller band bending for the same total charge in the depletion layer. Therefore, at high doping levels the band bending reduces, although the charge at the defect surface still increases with increasing dopant concentration. The net result is the existence of a

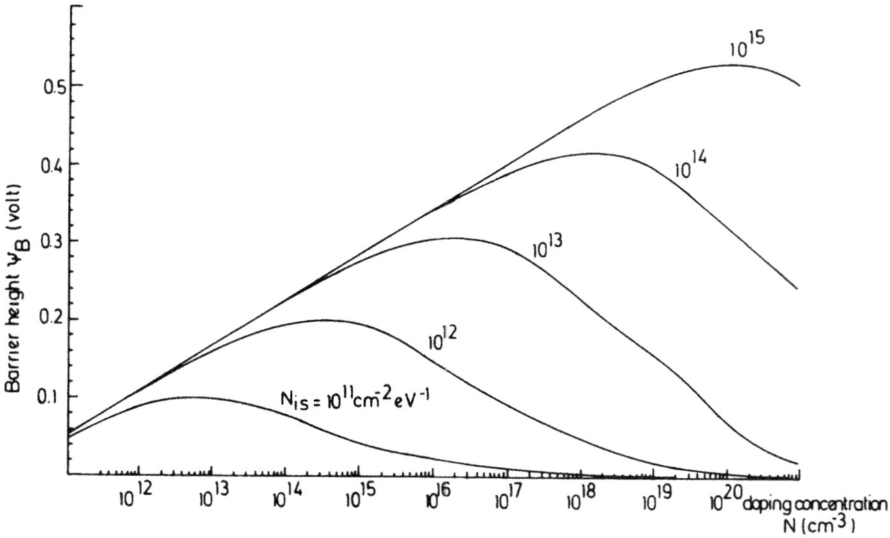

Fig. 3. Calculated grain boundary or microdefect barrier height ψ_B in the dark as a function of doping level, for different values of the surface trapping state density $N_{is} = N_{is}^A + N_{is}^D$.

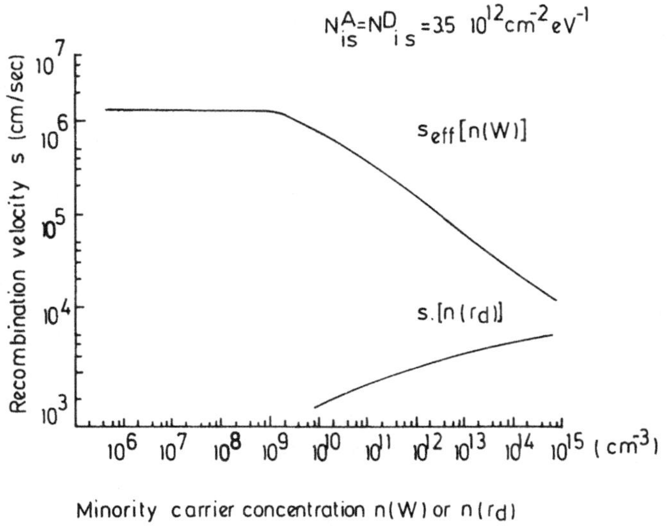

Fig. 4. Calculated dependence of the effective defect recombination velocity $s_{eff}(n(W))$ on $n(W)$, and of $s(n(r_d))$ on $n(r_d)$ for $N^A_{is} = N^D_{is} = 3.5 \times 10^{12} cm^{-2} eV^{-1}$, and $\sigma_C = 1 \times 10^{-14} cm^2$ and $\sigma_N = 1 \times 10^{-16} cm^2$.

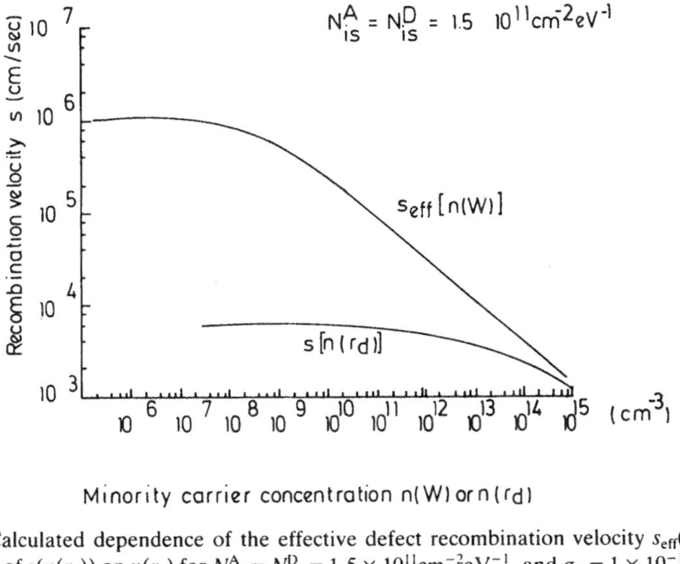

Fig. 5. Calculated dependence of the effective defect recombination velocity $s_{eff}(n(W)$ on $n(W)$, and of $s(n(r_d))$ on $n(r_d)$ for $N^A_{is} = N^D_{is} = 1.5 \times 10^{11} cm^{-2} eV^{-1}$, and $\sigma_C = 1 \times 10^{-14} cm^2$ and $\sigma_N = 1 \times 10^{-16} cm^2$.

doping level for which ψ_B has a maximum. This is in accordance with the measurements of Seto (1975) and Baccarani et al. (1978) on fine grained CVD polysilicon.

The recombination velocity at a semiconductor surface is defined by the recombination rate J_r/q divided by the excess minority carrier concentration at the surface in the absence of barrier bending. However, in reality, band bending always occurs. In this case one has to define an effective minority carrier recombination velocity s_{eff}. For recombination at a surface, s_{eff} is defined as (see section 1.4)

$$s_{eff}(n(W)) = J_r/q[n(W) - n_0(W)] \quad (1.17)$$

$$\cong J_n(W)/q[n(W) - n_0(W)]. \quad (1.18)$$

In eqs. (1.17) and (1.18), $n(W)$ denotes the minority carrier concentration and $n_0(W)$ the equilibrium minority carrier concentration at the depletion layer edge W. Definition (1.17) is also valid for a grain boundary. But $J_n(W)$ is, in general, different from J_r, because minority carriers are transmitted to the other grain.

When minority carrier saturation occurs, J_r equals $J_{n\,sat}$ and by comparing eq. (1.18) with eq. (1.11) it is clear that the maximum value of s_{eff} is $\frac{1}{4}v_{th}$.

An example of the dependence of s_{eff} is given in fig. 4. This curve has been calculated using the following defect parameter values: $\sigma_N = 1 \times 10^{-16}\text{cm}^2$, $\sigma_C = 1 \times 10^{-14}\text{cm}^2$, and $N^D_{is} = 3.5 \times 10^{12}\text{cm}^{-2}\,\text{eV}^{-1}$. A second example is shown in fig. 5, where a lower defect state density is assumed: $N^A_{is} = N^D_{is} = 1.5 \times 10^{11}\text{cm}^{-2}\text{eV}^{-1}$. Both figures indicate that the minority carrier saturation occurs at very low injection levels. When minority carrier saturation occurs, the effective recombination velocity $s_{eff}(n(W))$ equals $\frac{1}{4}v_{th} \cong 2.5 \times 10^6\text{cm sec}^{-1}$. It is important to note that outside the saturation region, $s_{eff}(n(W))$ decreases for increasing $n(W)$ or with injection level. This effect can be explained as follows: the barrier bending at the defect enhances the excess minority carrier concentration at the defect surface $\Delta n(r_d)$ by a factor $\exp(q\psi_B/kT)$, compared to the bulk. Due to the much larger excess minority carrier concentration at the defect surface under illumination, the defect states trap minority carriers in large quantities even at low illumination levels. This minority carrier trapping lowers the net charge in the defect and also the barrier height ψ_B. The barrier lowering reduces the excess minority carrier concentration at the defect surface and as a consequence J_r, being roughly proportional to $\Delta n(r_d)$, is also reduced. As a result the ratio $J_r/q[n(W) - n_0(W)] = s_{eff}(n(W))$ decreases. As will be discussed in section 3, the decrease of $s_{eff}(n(W))$ with illumination is responsible for the existence of a non-linear boundary condition characterizing the defects.

Although the recombination velocity at the defect surface $s(n(r_d))$, defined as $J_r/q[n(r_d) - n_0(r_d)]$, has no physical meaning and cannot be measured, it is used by several authors. A very similar situation holds for a semiconductor surface; also here the surface is in many cases depleted and the effective surface recombination velocity is defined at the depletion layer edge (Van Overstraeten and Van De Wiele 1969). $s(n(r_d))$ is shown in fig. 4 and 5 as a function of $n(r_d)$. In fig. 4, $s(n(r_d))$ increases with $n(r_d)$ although $s_{eff}(n(W))$ decreases with $n(W)$. The increase of $s(n(r_d))$ with $n(r_d)$ is caused by J_r and $n(r_d)$, both increasing less than $n(W)$. In fig. 5, which shows an example for a lower defect state density, $s(n(r_d))$ and $s_{eff}(n(W))$ both decrease with increasing injection level. The decrease of both parameters is caused by $n(r_d)$ and $n(W)$ both increasing more rapidly than J_r.

Further insight into the dependence of $s_{eff}(n(W))$ on the injection level can be gained by looking at the change in barrier height with increasing injection level or illumination. Fig. 6 shows ψ_B versus $\Delta E_F(r_d) = E_F p(r_d) - E_F n(r_d)$ with the doping concentration N as a parameter. These results clearly

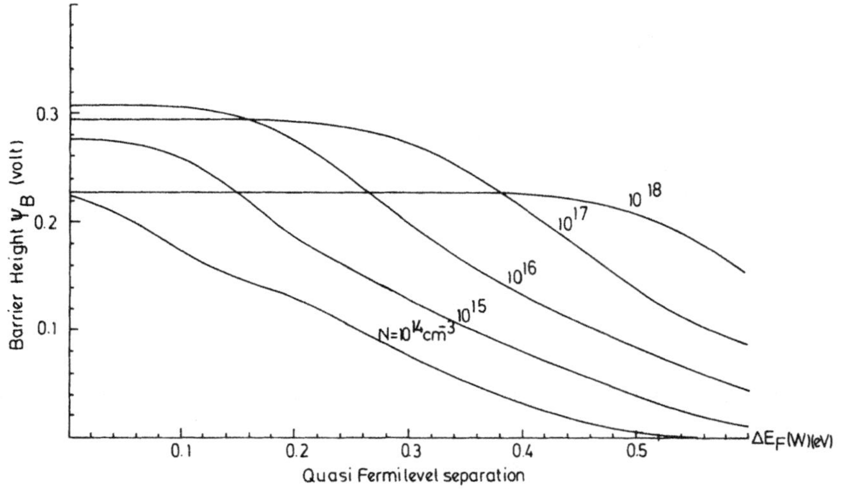

Fig. 6. Calculated grain boundary or microdefect barrier height ψ_B as a function of injection level for different substrate doping levels N. The injection level is characterized by the quasi-Fermi level separation $\Delta E_F(W)$ at the depletion layer edge. In the calculation $N_{is} = A^A_{is} + N^D_{is}$ was assumed to be $1 \times 10^{13} cm^{-2}eV^{-1}$.

indicate that the barrier height decreases with increasing illumination. This causes the decrease of $s_{\rm eff}(n(W))$ with increasing injection level. At high doping levels, the minority carrier current saturates at low injection levels. When current saturation occurs, the barrier height cannot increase anymore since the recombination at the defect surface is limited to its saturation value; any further increase in barrier height would result in a larger J_r, which is physically impossible.

1.4. Current transport through a grain boundary

When a voltage is applied across the grain boundary, the current through the grain boundary is limited by the majority carrier saturation in the depletion layer, and by the scattering taking place at the grain boundary.

If the scattering in the depletion layer and at the grain boundary is neglected, the current through the grain boundary is determined by the majority carrier saturation current in the depletion region. The majority carrier flux is given by eq. (1.11)

$$qF = \tfrac{1}{4} q p(r_{\rm d}) v_{\rm th}. \tag{1.19}$$

With $p(r_{\rm d}) = p(W)\exp(q\psi_{\rm B}/kT)$ we obtain

$$qF = \tfrac{1}{4} q p(W)\exp(q\psi_{\rm B}/kT) v_{\rm th}. \tag{1.20}$$

With $p(W) = N_{\rm V}\exp\{-[q\varphi_{\rm p}(W) - E_{\rm V}]/kT\}$ we obtain

$$N_{\rm V} = 2(2\pi kTm_{\rm h}^*/h^2)^{3/2},$$

$$v_{\rm th} = (16kT/2\pi m_{\rm h}^*)^{1/2}$$

$$\cong 1\times 10^7\ \text{cm/s} \quad \text{(for silicon)}$$

we can rewrite eq. (1.19) as

$$qF = A^* T^2 \exp(-q\Phi_{\rm B}/kT), \tag{1.21}$$

with $A^* = 4\pi q m_{\rm h}^* k^2/h^3$ and $\Phi_{\rm B} = \psi_{\rm B} + \varphi_{\rm p} - E_V/q$. The constant A^* is called the Richardson constant. A semiconductor for which eq. (1.21) is valid is called a thermionic emitter, for which the thermionic emission theory is valid.

Equation (1.21) is the majority carrier flux going from the neutral bulk through the space-charge region to the grain boundary. The electric field in the first space-charge layer repels the majority carriers and is a barrier to majority carrier flow. The electric field in the second space-charge layer helps the majority carriers at the grain boundary to the other side. By increasing the temperature, the majority carriers get more kinetic energy and their probability of crossing the barrier increases.

For a symmetrical grain boundary, a positive applied voltage V reduces the first (retarding) barrier height and increases the second (accelerating) barrier. For low voltages, we assume that the voltage is equally distributed over both barriers. The majority carriers (holes in this case) coming from the positive electrode encounter a reduced barrier height $\psi_B - \frac{1}{2}V$. The number of majority carriers able to surmount the first barrier and which are then swept across the second depletion layer, increases exponentially with the applied bias, as

$$F = A^* T^2 \exp(-q\Phi_B / kT) \exp(qV / 2kT). \tag{1.22a}$$

The majority carriers coming from the other side, encounter an increased barrier height $\psi_B + \frac{1}{2}V$. Their flux is given by

$$F' = A^* T^2 \exp(-q\Phi_B / kT) \exp(-qV / 2kT). \tag{1.22b}$$

The total current over the grain boundary, J_{GB}, is the difference between those two fluxes,

$$\begin{aligned} J_{GB} &= q(F - F') \\ &= A^* T^2 \exp(-q\Phi_B / kT) [\exp(qV / 2kT) - \exp(-qV / 2kT)] \\ &= 2A^* T^2 \exp(-q\Phi_B / kT) \sinh(qV / 2kT). \end{aligned} \tag{1.23}$$

To get an agreement between the measured and the calculated I–V characteristic of a grain boundary, an attenuation factor c must be introduced in the thermionic emission model to account for the much smaller experimental measured currents. For a metal–silicon barrier, c equals one, but for a grain boundary, c can be very small. Measured transmission factors for majority carriers range from 0.06–0.2 (Seto 1975, Mandurah et al. 1979, Lu et al. 1980, 1981a). This means that the current through the grain boundary is much smaller than predicted by thermionic emission theory.

The origin of the low value of the transmission factor is still a point of discussion. Several mechanisms are important: scattering in the depletion layer, Coulombic scattering at the fixed charge in the grain boundary, and scattering at the grain boundary structure itself.

The diffusion theory describes the scattering in the depletion layer. According to the theory of Crowell and Sze (1966), the diffusion theory is more important than the thermionic emission theory when

$$\mu E_{max} > A^* T^2 / q N_V, \tag{1.24}$$

with μ the majority carrier mobility, and E_{max} the maximum electric field in the depletion layer. Using

$$E_{max} = \left(\frac{q N_A 2 \psi_B}{\varepsilon_{Si} \varepsilon_0} \right)^{1/2}, \tag{1.25}$$

we can rewrite eq. (1.23) as

$$\psi_B > \frac{\varepsilon_{Si} \varepsilon_0}{2 q N_A} \left(\frac{A^* T^2}{\mu q N_V} \right)^2. \tag{1.26}$$

With $\mu = 450\,\text{cm}^2/\text{Vs}$, $N_V = 1.04 \times 10^{19}\text{cm}^{-3}$, $A^* = 0.66 \times 120\,\text{A}/\text{cm}^2/\text{K}^2$, $T = 300\,\text{K}$, and $\varepsilon_{Si} = 11.9$, we obtain

$$\psi_B > 3 \times 10^{14} / N_A, \tag{1.27}$$

where N_A is expressed in cm^{-3}. Equation (1.27) is only fulfilled for extremely low barrier heights and doping levels. Therefore, in most practical cases, the diffusion theory is not an important limitation to the transport of carriers through a silicon grain boundary.

Wu and E.S. Yang (1982a,b) calculated the transmission factor caused by Coulombic scattering at the grain boundary. The larger the charge in the grain boundary states, the smaller the transmission factor. The grain boundary charge increases with the grain boundary trapping state density and with the doping level.

Since undoped polycrystalline silicon has a higher resistivity than the corresponding single crystalline silicon, another mechanism is also active. This mechanism is found in the scattering at the grain boundary structure itself, which is independent of the grain boundary charge. Therefore, this scattering mechanism is independent of the doping concentration and the injection level. The scattering at the grain boundary structure itself can be modelled by assuming that the grain boundary is a very thin layer of amorphous silicon. This was suggested by Taniguchi et al. (1980) and Kim et al. (1984).

The larger the grain boundary charge, and the more disordered the grain boundary Si–Si bonds, the smaller the grain boundary transmission factor. Since the grain boundary barrier height increases with the grain boundary charge, and since the grain boundary charge increases with the degree of disorder in the Si–Si bonds, the transmission factor will decrease with increasing grain boundary barrier height. Therefore, large grain boundaries can be characterized by a very small value of the transmission factor, while small grain boundaries barriers can be characterized by a transmission factor close to one.

As a result, the maximum current through a grain boundary is given by

$$J_{GB} = cA^*T^2 \exp(-q\Phi_B/kT)\sinh(qV/2kT). \tag{1.28}$$

Equation (1.28) is the expression for the current through a grain boundary when a small voltage is applied across it. This equation is slightly different from the expression for the thermionic current over a Schottky barrier, which, in most publications (e.g., Pike and Seager 1979, Card and Hwang 1980), is also assumed to be valid for a grain boundary.

It is also known that a grain boundary with a large barrier height can withstand large voltages of about a hundred volt without breakdown (Taylor et al. 1952). In this case the depletion layer width of the barrier, which is reduced by the applied voltage, is reduced to a small value and remains constant, while the depletion layer width of the barrier, which is increased by the applied voltage, is increased and withstands nearly the whole voltage. For large voltages, the current through the grain boundary is given by

$$J_{GB} = A^*T^2\exp(-q\Phi_B/kT)\,[\exp(qV_1/2kT) - \exp(-qV_2/2kT)] \tag{1.29}$$

where V_1 is the voltage drop over the barrier which is reduced, V_2 is the voltage drop over the barrier which is increased, and $V_1 + V_2 = V$.

2. Experimental data

In this section the observed properties of defects are compared with the theoretical model presented in section 1.

We show experimentally that, in semicrystalline silicon, not only grain boundaries are important for the minority carrier recombination, but also that intragrain defects in large grained silicon have significant importance.

Experimental evidence for a space-charge layer surrounding the defects is presented through the measurements of grain boundary I–V characteristics, and the grain boundary capacitance. Measurements of the grain boundary zero-bias resistance versus $1/T$ make it possible to derive the grain boundary barrier height. All measurements on Silso grain boundaries agree with the theoretical model developed in section 1.

We also give an overview of the measured fundamental electronic properties of grain boundaries, like trap state density, trap state distribution and capture cross-sections. A large variation in the measured trapping state densities is observed.

Finally, an overview is given of the methods to determine the effective grain boundary recombination velocity, s_{eff}. s_{eff} is measured on several grain boundaries on Silso silicon, using the EBIC as well as the LBIC method. Values of 6×10^3–4×10^4cm/s have been measured. Also the change of grain boundary characteristics under different illumination levels is discussed. The decrease of the grain boundary resistance is observed. Furthermore, it is shown experimentally that s_{eff} decreases with injection level.

2.1. Recombination in defected silicon

Originally, most defect-related research in polycrystalline silicon was devoted to grain boundaries. Since only fine grained polysilicon was used for integrated circuits and since only two-dimensional defects cause an important barrier to majority carrier flow, the earlier work was mainly concentrated on grain boundaries. In solar cells, on the contrary, minority carrier recombination is most important; the transport through the grain boundaries is only of secondary importance because of the large grain sizes in these silicon wafers. For the minority carrier current flow, all intragrain defects (spherical, linear or planar) are very important as is indicated by our experimental data discussed in this section.

A typical example of the importance of defects is the dependence of the effective minority carrier diffusion length, L_{eff}, on the defect density. A consequence of this is the variation of L_{eff} with the injection level, even at low illumination levels (De Pauw et al. 1984a). Several defected silicon materials are examined: semicrystalline p- and n-type Silso (Wacker), p-type HEM (Crystal Systems) – two typical potentially low-cost substrates for the production of solar cells – and p-type CZ-pulled defected single-crystal wafers from Metallurgie Hoboken Overpelt (MHO). A defect-free single-crystal wafer is added as reference. This control wafer has a high antimony doping level (5×10^{18}cm^{-3}) to assure a uniform bulk diffusion length smaller than the bulk thickness. In all cases the silicon feedstock material for the crystal growth is of a semiconductor grade purity. Also a semicrystalline upgraded metallurgical silicon wafer (UMG), obtained by CZ-pulling of commercially available MG silicon is added. On these substrates solar cells with an area of 3.24 cm^2 are fabricated using a screen printing process (Frisson et al. 1978). The Silso, HEM and UMG have millimeter sized grains. The MHO material is single crystalline with a dislocation density between 5×10^3 and 5×10^4cm^{-2}, as specified by the manufacturer. The effective, or averaged, diffussion length in the material,

L_{eff}, is determined by fitting of the diode long-wavelength quantum efficiency, in the wavelength region from 0.9–1.0 μm. To measure the dependence of L_{eff} on the illumination level, a DC white-bias light is used, on which a chopped monochromatic light beam is superimposed. The results are presented in fig. 7 and 8. Fig. 7 clearly indicates that many samples of the defected silicon substrates (Silso, HEM and MHO) show a strong increase of the effective diffusion length at low illumination levels. Similar curves also have been published in the literature by Fabre et al. (1975) and Fabre and Mautref (1975), and on CZ silicon, by Makram-Ebeid (1979) on RAD silicon, by Pogany (1980) on CAST silicon, and by Ho et al. (1977), Hari Rao et al. (1978), Ho and Mathias (1981), Shimokawa and Hayashi (1982, 1983) on EFG ribbons. The possible reasons why L_{eff} does not increase in some of the defected materials of fig. 8 and increases in others will be discussed in section 3. The L_{eff} increase is not observed in the defect-free single-crystal control wafer (figure 8) and also not in the UMG Si wafer. The result on the control wafer is not new. Although the UMG

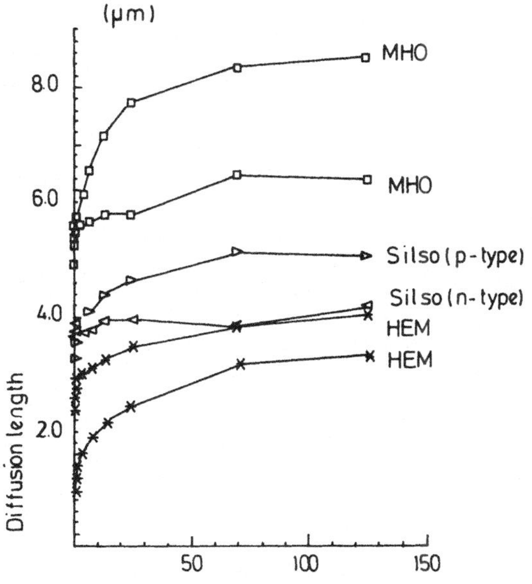

Fig. 7. Measured variation of the effective diffusion length in defected silicon. This figure clearly indicates that some of the defected substrates show a strong increase of the effective diffusion length with illumination at low injection levels.

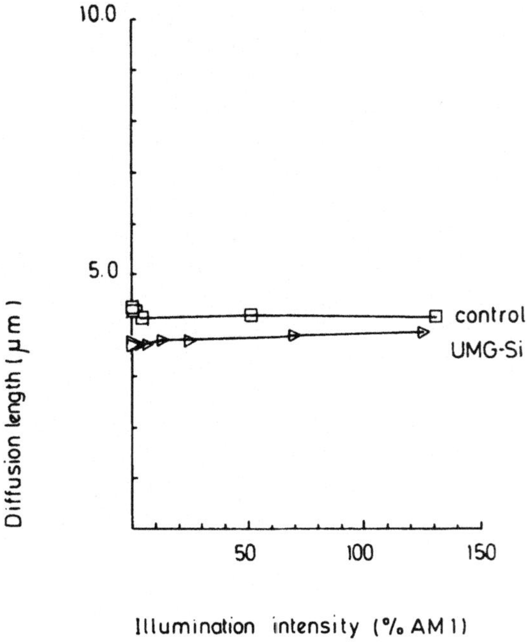

Fig. 8. The increase of L_{eff} is not observed in the defect-free single-crystal control wafer and also not in the UMG Si wafer.

Si wafer contains a lot of crystal defects such as grain boundaries and intragrain defects, this can not explain the low value (about 4 μm) of the effective diffusion length in this material. Probably, the high impurity concentration in the bulk of the dopant species boron and aluminum, and other harmful elements such as iron and titanium, explain this low value of L_{eff}. This may also explain why L_{eff} in UMG Si does not change with injection level. An important conclusion from these measurements is that the increase of L_{eff} seems to be related to the material quality. In all cases, the saturated effective diffusion length is much smaller than the average grain size. Therefore, grain boundaries alone cannot be responsible for the small diffusion lengths commonly observed in these defected materials.

In order to further investigate these effects, the intragrain material is examined utilizing the electron beam induced current (EBIC) technique. Figure 9 shows the EBIC response of Silso material in a region of high intragrain activity. Clearly visible is the variation of the EBIC in the intragrain material of these grains. In other grains a very low intragrain electrical

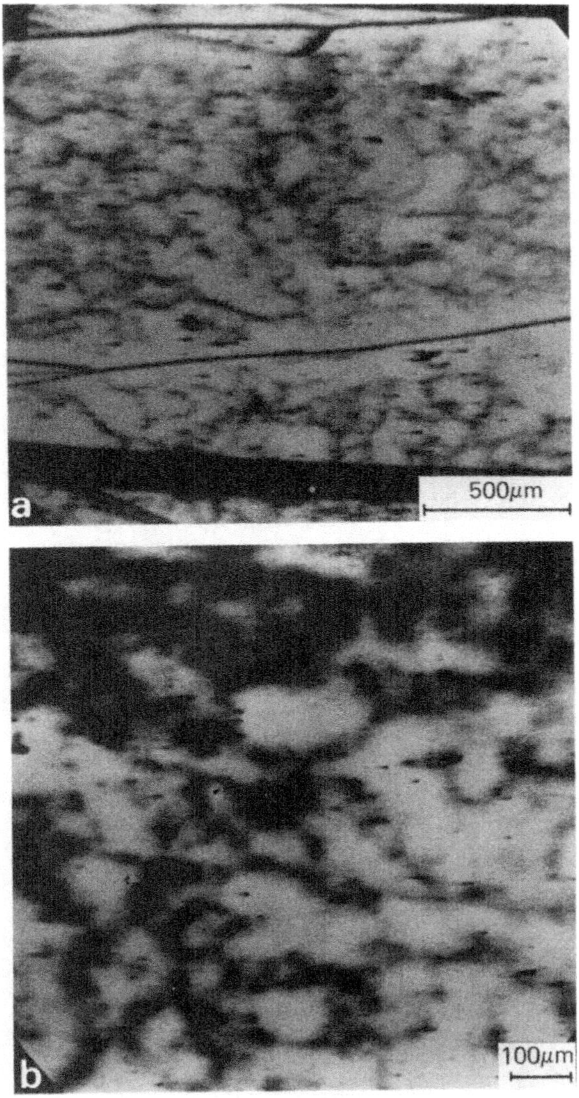

Fig. 9. EBIC response of regions of high intragrain electrical activity in Silso.

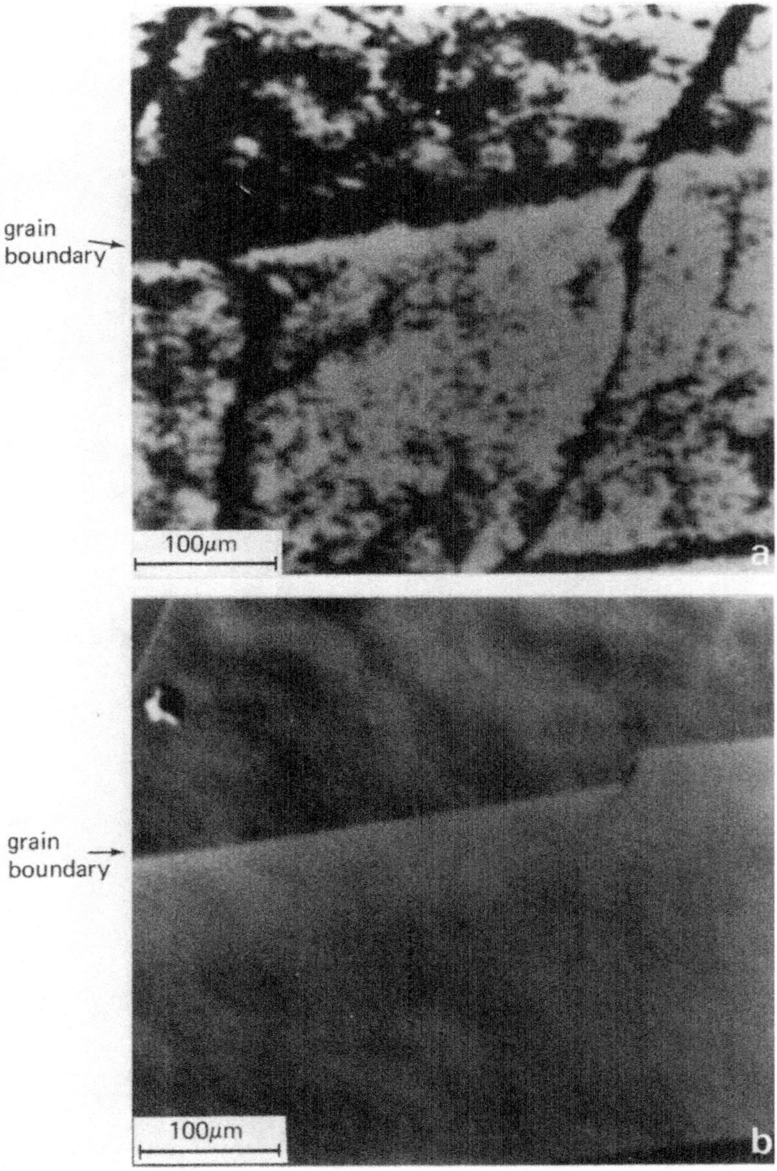

Fig. 10. (a) EBIC response of a region of high intragrain electrical activity in HEM, (b) Secondary electron image showing the sample surface at the same location.

activity is found. Figure 10 shows that the EBIC response of HEM material. Here also a strong intragrain defect electrical activity is observed. These measurements agree with earlier reported EBIC measurements (Hanoka 1979/80, Huber et al. 1980, Texier et al. 1980, Inoue et al. 1981, Cunnigham et al. 1982) and with results obtained by the laser scanning method (Inoue et al. 1981, Hyland et al. 1982) on similar materials.

For the single-crystal control material no spatial variation of the EBIC signal could be detected, even when the EBIC contrast is set to the maximum. The observed spatial variation of the EBIC signal indicates that the intragrain material contains active defects. From the EBIC images one can derive the interdefect distance to be of the order of a few tens of microns.

Additional evidence for the existence of the intragrain defects comes from defect etching experiments. Figures 11a,b,c show the intragrain etch pit density on the p-type Silso, HEM and MHO materials, respectively. The samples are first Secco etched [a mixture of HF, $K_2Cr_2O_3$ and water–Secco d'Aragona (1972)] to reveal the intragrain defects. On the materials Silso and HEM, which show a large increase of L_{eff}, large etch pit densities are measured, ranging from 5×10^4 to $1 \times 10^6\,cm^{-2}$. Large differences in densities are found from one grain to another. Similar densities have been reported on the same materials (see for example Finetti et al. 1980) and on similar materials (see, for example, Hyland et al. 1982, S.M. Johnson and Culik 1982). If one assumes all the etch pits to be in the surface plane, the corresponding interdefect distance is from 10 to 30 micron. The defected single crystalline MHO materials, which also show a large increase in L_{eff}, have etch pit densities between 5×10^3 and $5 \times 10^4\,cm^{-2}$. Because the increase in L_{eff} is not seen in the defect-free wafer, the increase is expected to be dependent on the interdefect distance, as will be shown in section 3.

The influence of grain boundaries on the minority carrier recombination is well known. The unexpected result of these measurements is the sometimes important intragrain defect density in large grained silicon. We review here some other published, but less known, data emphasizing the importance of intragrain defects.

Recent results on an ingot fabricated by a fast variant of the Bridgman method indicated a poor correlation between the intragrain diffusion length and the average grain size (Oualid et al. 1982a). However, an excellent correlation between the intragrain diffusion length and the intragrain defect density has been found. Similar conclusions have been drawn on other potentially low-cost large-grained silicon substrates: zone refined

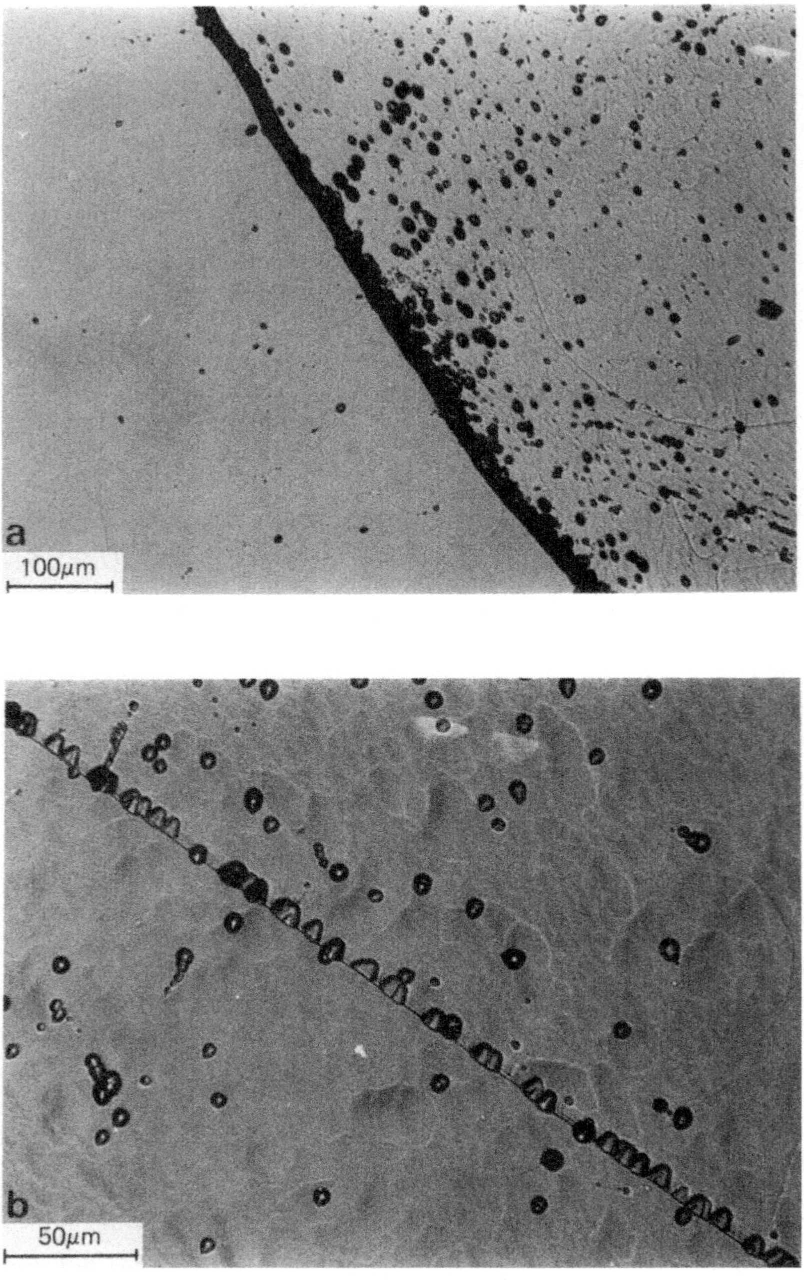
a
100μm
b
50μm

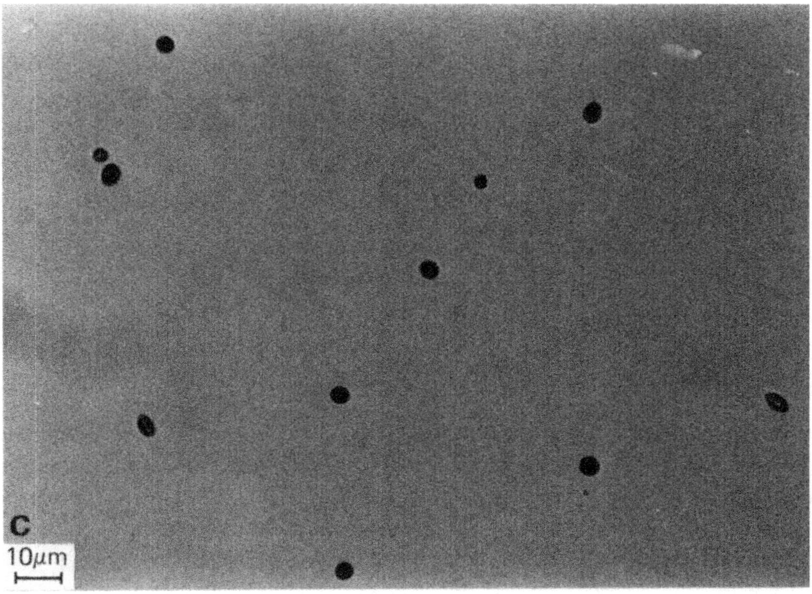

Fig. 11. Defect etching by Secco etch of intragrain defects: (a) Silso, (b) HEM, and (c) MHO-silicon.

silicon (Inoue et al. 1981), CAST silicon ribbons (K. Yang et al. 1980), UCP silicon (S.M. Johnson and Culik 1982), Silso silicon (Finetti et al. 1980), and RTR silicon ribbons (Sopori et al. 1982). This suggests that when L_{eff} increases with injection level, the diffusion length is defect limited. Since many of the potential low-cost substrates have a high defect density, one expects that most of these materials will have an illumination-dependent L_{eff}.

Finetti et al. (1980) on Silso, Wohlgemuth et al. (1983) on HEM, and Culik et al. (1983) found that the V_{oc} and J_{sc} of their cells is limited by the intragrain lifetime and not by the grain boundaries.

As will be treated in section 5.3, diffusion occurs much faster along grain boundaries and defects, than in single crystalline silicon. Lindmayer (1978) measured the diode capacitance per unit cell surface and found that this capacitance is much larger for junctions on semicrystalline silicon, than on single-crystal good quality silicon. The difference in capacitance can be explained by preferential diffusion along the grain boundaries, which increases the junction area and so the capacitance per unit cell

Table 1
Electrical activity of crystal defects in CAST ribbons (K. Yang et al. 1980).

Group I strongly active	Group II moderately active	Group III nonactive
SiC dendrites and inclusions	Multiple overlapping stacking faults	First order (coherent twins)
Grain boundaries	Grown-in dislocations	Stacking faults
Degenerated twins, Low-angle boundaries		
Dislocation bands		

surface. Lindmayer also found that this is also true for diodes entirely made in intragrain material, giving strong evidence for a high intragrain defect density, together with preferential diffusion along these defects.

Finally, in section 5.2, it will be shown that the influence of passivation treatments will depend largely on the intragrain defect density.

One question that remains unanswered concerns the shape and the nature of these intragrain defects. K. Yang et al. (1980) made a classification of the defects found in silicon ribbons as shown in table 1. A large variation exists. For the cases shown in figs. 9 and 10, it is likely that they are dislocations or precipitates. Inoue et al. (1981) found many etch pits corresponding to a dip in the EBIC signal, which indicates a relation between the etch pits and the EBIC dips. Defect etching reveals preferential orientations in the etch pits, which indicates that many of them are dislocations. In order to reduce costs, many of these materials have been solidified and cooled down at a relatively high speed. Therefore, relatively large thermal gradients exist, which give rise to stressed regions and dislocation generation and multiplication (see, for example, Indenbom 1979). In future materials the defect density can be much lower. An example is the Oscillating Crucible Technique (Daud et al. 1982), which is currently under development and seems to result in a reduced intragrain defect density.

2.2. *Experimental evidence for a space-charge layer surrounding defects and grain boundaries.*

The first experimental studies concerning grain boundaries in semiconductors date from more than 30 years ago (Taylor et al. 1952). The original work focussed on germanium. To explain grain boundary I–V characteristics and grain boundary capacitances, Taylor et al. were the first to propose the double depletion layer model in which the grain boundary is assumed to be surrounded by a space-charge layer at each side (fig. 2).

As shown in section 1.4, the double depletion layer model leads to eq. (1.28) to describe the grain boundary *I–V* characteristics. Typical *I–V* curves of a Silso grain boundary for different temperatures are given in figure 12. These characteristics can be explained by the double depletion layer model, since they show an increasing current with increasing bias voltage and temperature.

The double depletion layer model also explains the capacitance of grain boundaries. The total grain boundary capacitance C_{GB} is the series connection of the two depletion layer capacitances C_d. The high-frequency grain boundary capacitance C_{GB} is related to the barrier height ψ_B, as given by eq. (2.1),

$$C_{GB} = \frac{C_d}{2} = \left(\frac{q\varepsilon_{Si}\varepsilon_0 N}{8\psi_B}\right)^{1/2}, \tag{2.1}$$

where N is the impurity concentration and ε_{Si} and ε_0 are the relative and the absolute dielectric permittivity, respectively. Figure 13 gives the barrier height for a Silso grain boundary for different temperatures, obtained from high-frequency capacitance measurements.

Other experimental support for the double depletion layer model comes from electron beam induced current (EBIC) measurements on a grain

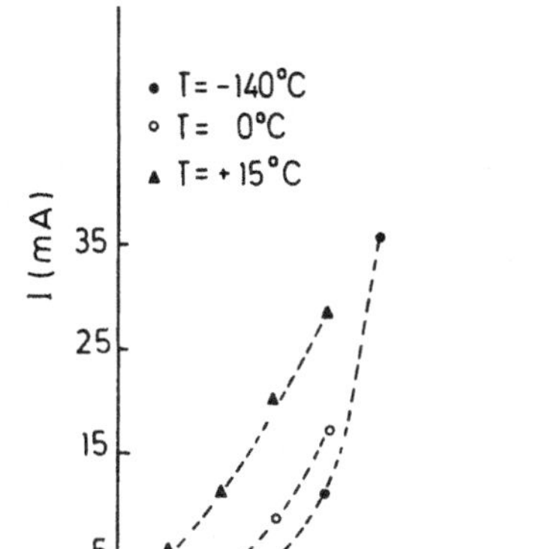

Fig. 12. Measured *I–V* characteristic of a grain boundary in p-type Silso for different temperatures.

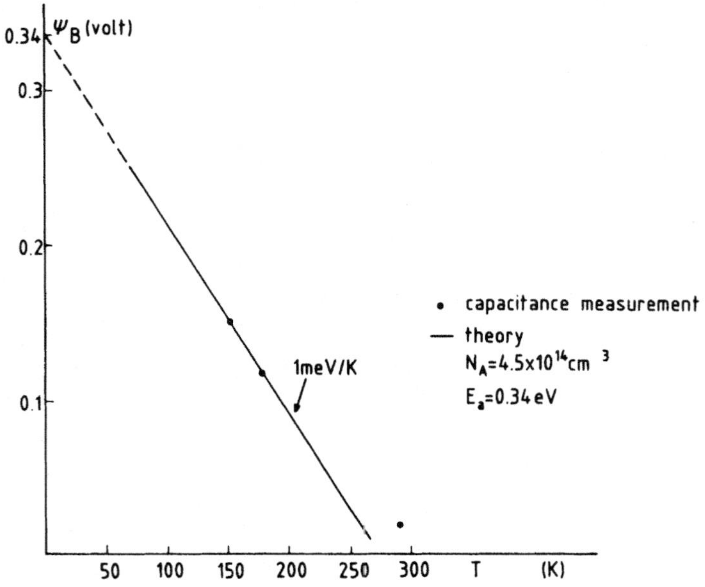

Fig. 13. Grain boundary barrier height as a function of temperature for a grain boundary in Silso from capacitance measurements, using eq. (2.1), (black dots) and from theory, $d\psi_B/dT = k \ln(N_V/N_A)$ (solid line).

boundary in a bicrystal (Mataré 1969). By crossing the space-charge layers on both sides of a grain boundary, a positive and a negative peak in the EBIC are observed, due to the opposite electric fields at the two sides of the grain boundary (fig. 14). By applying a sufficiently large lateral voltage, one space-charge layer disappears and only one current peak is observed. Changing the direction of the voltage changes the sign of the collected current. Mataré showed with this method that dislocations in germanium and silicon are also surrounded by a space-charge layer.

More recently, additional evidence for the double depletion layer grain boundary model in silicon was given by Seager and Castner (1978) and Pike and Seager (1979). They measured the conductance of a grain boundary in a silicon bicrystal as a function of temperature. Figure 15 shows similar results obtained on a Silso (Wacker) grain boundary. The activation energy, E_a, of this curve can be related to a barrier height. Derivation of the barrier heigth at room temperature from the measured activation energy is not straightforward. In the absence of barrier height fluctuations, the activation energy yields the barrier height at a temperature of 0 K. At room temperature, however, the barrier height will be smaller than E_a/q because the Fermi level is closer to midgap; a position where the neutral energy level seems to be located (Seager and Pike 1979, Seager et al. 1979, E.S. Yang et al. 1981). When the Fermi level is at this neutral energy level, the charge in the grain boundary is zero; when the Fermi level approaches the neutral energy level the charge in the grain boundary decreases. The measured variation of barrier height for a grain boundary in p-type Silso with temperature is presented in fig. 13. The data agree well with the calculated variation of the height, using the expression for the Fermi level shift: $dE_F/dT = d\psi_B/dT = k\ln(N_V/N_A)$ (Thomson and Card 1983). If the barrier has gross spatial nonuniformities, it will strongly influence the mea-

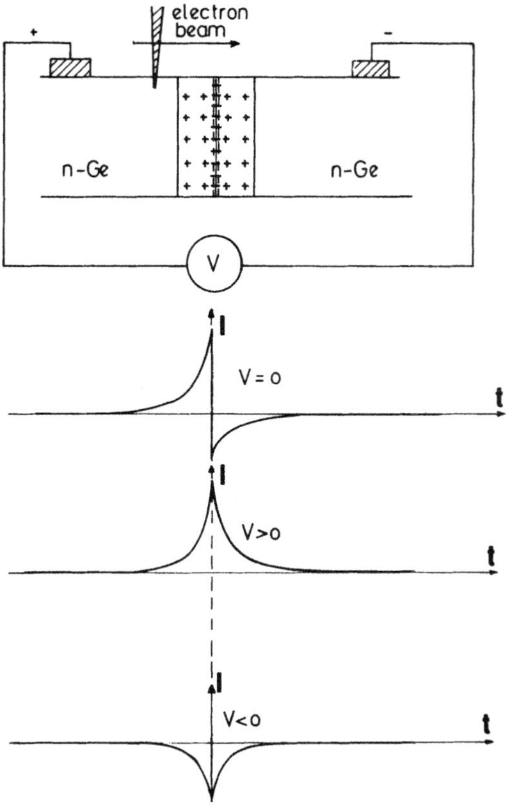

Fig. 14. Experiment of Mataré giving evidence for a space-charge layer around a grain boundary in a semiconductor.

sured activation energy, as shown by Thomson and Card (1983).

To account for the much smaller experimental currents when a voltage is applied over a grain boundary, a transmission factor for majority carriers has to be introduced. Measured transmission factors range from 0.06–0.2 (Seto 1975, Mandurah et al. 1979, Lu et al. 1980, 1981a).

Measurements on fine grained CVD material (Seto 1975, Baccarani et al. 1978, Mandurah et al. 1979) show that at high dopant concentrations the conductivity of polysilicon layers is enormously enhanced. This gives evidence for the decrease of the barrier height at higher doping levels, calculated in section 1. For this reason grain boundaries intersecting the emitter of a solar cell will hardly affect the series resistance. In fact, heavily doped polysilicon layers have been used to passivate emitters of high-efficiency solar cells. Although at higher doping levels the barrier height is very low, a much higher charge is accumulated in the grain boundary

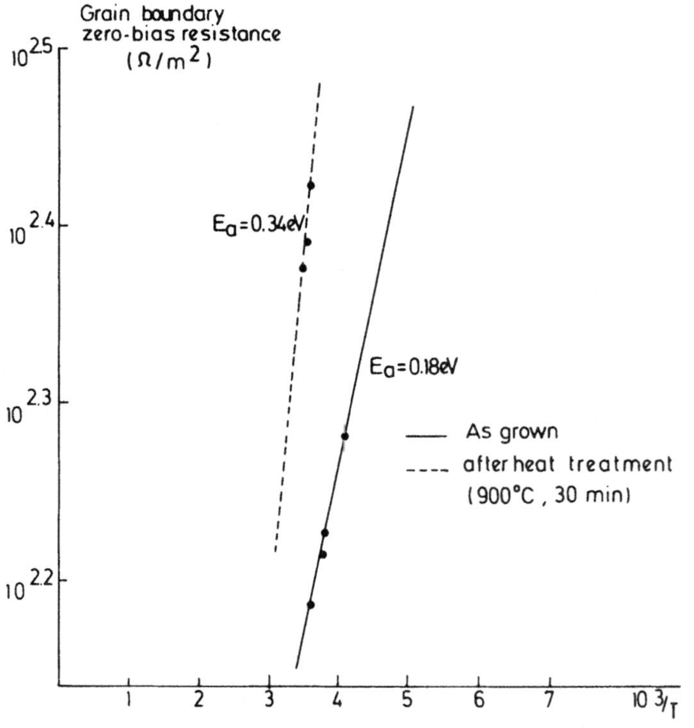

Fig. 15. Zero-bias grain boundary resistance as function of temperature (Silso silicon).

states. As a result, Coulombic scattering becomes important and will dramatically lower the carrier mobility in polycrystalline silicon, reducing the recombination current at semiconductor–metal contacts (Ning and Isaac 1980, Fossum and Shibib 1980).

2.3. Measurement of fundamental electronic parameters of grain boundaries

As calculated in section 1, the electrical properties of a grain boundary are entirely determined by the density of the grain boundary trapping states, their position in the energy gap and their capture cross-sections for holes and electrons. The reported measurements of trapping state densities are summarized in table 2 and drawn in fig. 16. They vary considerably in density and functional form. There are two reasons for this variation: variation of the grain boundary characteristics from sample to sample and measurement difficulties.

Figure 17 gives the EBIC response of an active Silso grain boundary, which shows a large variation in recombination activity along the grain boundary. Also the change in barrier height after a high-temperature anneal (see section 5.1) and the influence of hydrogen passivation (see section 5.2) support this observation. It is therefore not surprising that materials of different purity and different thermal history can differ significantly in electrical properties.

The other cause of the scattering in the data is measurement difficulties. One has to select a grain boundary of sufficient length, isolate it and make an electrical contact to it. If the grain size is small, only the average of a series of grain boundaries can be measured. In order to allow a determination of the grain boundary states, many methods require a large grain boundary barrier height (e.g., *I–V* deconvolution), which is not always the case. For the characterization of Si–SiO_2 interface states, many methods are well established. However, these methods cannot simply be applied to a grain boundary, because of the high grain boundary conductance and the orthogonal position of the grain boundary in the material. Therefore, several grain boundary characterization methods are still under development and, at this moment, it is still difficult to indicate the best method.

Information about the capture cross-sections is obtained from the study of transient phenomena, for example, DLTS (deep level transient spectroscopy). Reported values of capture cross-sections obtained by DLTS range from 10^{-14}–10^{-20} cm^{-2}.

The experimental observation of a low grain boundary barrier height at low doping levels, and the decrease of the grain boundary barrier height to zero under sufficient illumination (see section 2.5), strongly indicate that no fixed charge is present in the grain boundary. The absence of a fixed charge in grain boundaries, makes that many properties of a grain boundary in silicon differ from the properties of the $Si–SiO_2$ interface.

Table 2
The various GB trapping state distributions observed in polycrystalline silicon materials.

Distribution	Sample	Experimental technique	References
One dominant energy level	CVD poly-Si	Resistivity versus T for different doping levels combined with Hall mobility measurements	[1–4]
	Silso	Photoconductance decay as a function of T	[5]
U-shaped distribution	CVD poly-Si	MOS capacitance and conductance	[6]
		Resistivity as a function of doping level	[7]
	Bicrystals	Photocapacitance and photoconductance with subband light	[8]
¬-shaped distribution	Bicrystals	$I–V$ deconvolution	[9]
		$C–V$ recovery	[10]
Several important energy levels	Bicrystals	DLTS	[11, 12]
	Silso	DLTS	[13]
Continuous but increasing density from E_V to midgap	Silso	$I–V$ deconvolution	[14]
Constant throughout the energy gap	Silso	MOS capacitance and conductance	[15]

References:
[1] Seto (1975)
[2] Baccarani et al. (1978)
[3] Lu et al. (1981a,b)
[4] Kim et al. (1982)
[5] Poon et al. (1983)
[6] Hirae et al. (1980)
[7] De Graaff et al. (1982)
[8] Werner et al. (1981)
[9] Seager and Pike (1979)
[10] Seager et al. (1979)
[11] Cheng and Shyu (1981a,b)
[12] Shyu and Cheng (1982)
[13] Srivastava et al. (1982)
[14] Thomson and Card (1983)
[15] Rao et al. (1981)

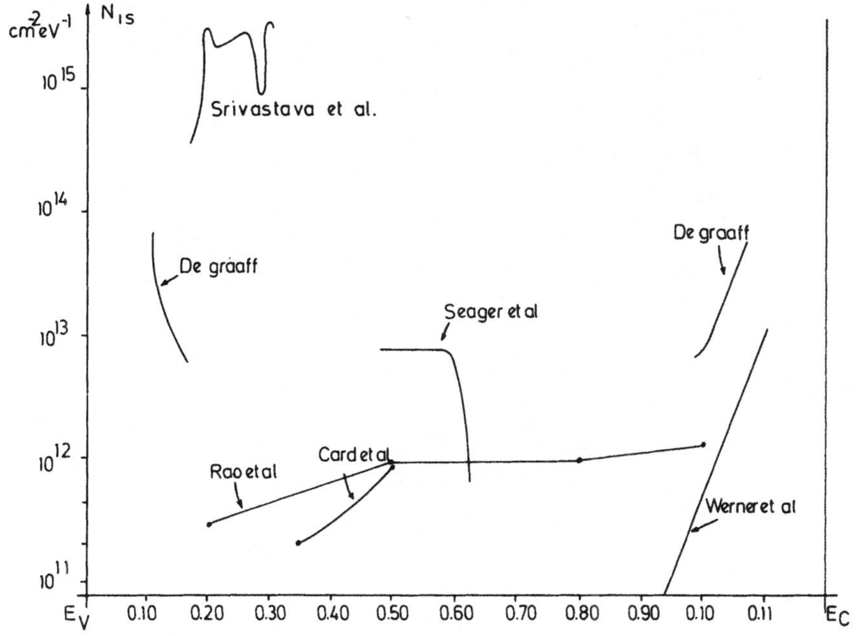

Fig. 16. Reported densities and distributions of interface states.

2.4. *Measurement of the effective grain boundary recombination velocity*

For the measurement of s_{eff} at a grain boundary mainly two direct methods are used: an electron or light beam induced current (EBIC or LBIC) method and a light beam induced conductivity method. Table 3 gives an overview of the reported values of s_{eff} in the dark or under weak illumination i.e. $s_{eff}(n_0(W))$. It is seen that on the cast silicon materials (Silso and HEM) the measured values of s_{eff} are much smaller than the theoretical maximum $\frac{1}{4}v_{th} \cong 2.5 \times 10^6$ cm/s. When the material is illuminated it is expected that s_{eff} will decrease even more, as will be discussed in section 2.5.

2.4.1. *EBIC and LBIC methods*

These methods to measure s_{eff} at a grain boundary were developed by Watanabe (1977), Von Roos (1979a,b), Zook (1980a) and Donolato (1983). For recombination at intragrain defects, Donolato (1978) developed a similar method. The experimental set-up is shown in fig. 18. A collecting junction (a diffused or a Schottky-barrier junction) is fabricated

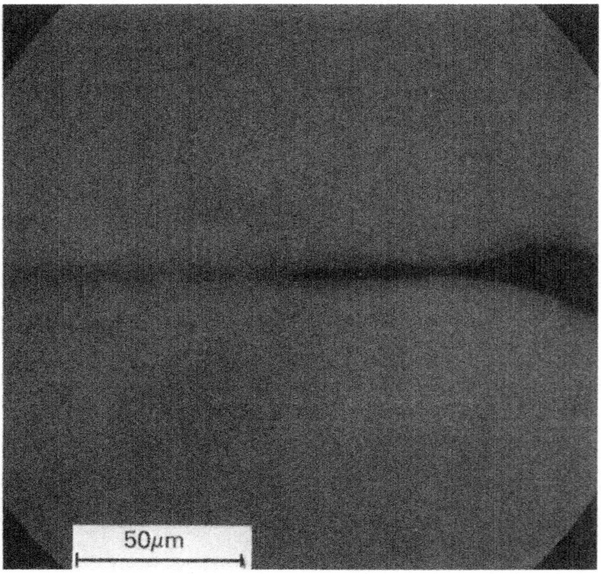

Fig. 17. EBIC response of a grain boundary in Silso, illustrating the large variation of the electrical properties over a single grain boundary.

on top of the sample. The grain boundary is perpendicular to this junction and a small source, creating electron–hole pairs is needed. This source can be either an electron beam or a light beam. The minority carriers are collected by the junction. The collected current is referred to as EBIC or LBIC (Electron or Laser Beam Induced Current). When the beam approaches the defect, more and more minority carriers recombine at the defect and the current collected by the junction decreases. When the beam is at the location of the grain boundary, the current collected by the junction is minimal. Figure 19 gives two EBIC curves for a grain boundary on Silso silicon. An accurate derivation of the grain boundary recombination velocity, s_{eff}, from these measurements is not difficult, if attention is paid to different points.

The first problem is that the shape of the curve is influenced by the excitation volume diameter, because the beam size is in practice non-negligible. As pointed out by Burk (1982), the surface recombination velocity cannot be determined correctly within a distance less than one excitation volume diameter from the grain boundary. A light beam of wavelength λ can be focussed to a minimal beam diameter of (M. Young 1977)

Table 3
Measured values for s_{eff}.

Group	s_{eff} (cm/s)	Method	Material
Russell et al. (1982)	10^3–10^4	EBIC	p-type Silso (Wacker)
Burk et al. (1983)	2–8 × 10^4	EBIC	p-type Silso (Wacker)
Marek (1982)	2 × 10^4	LBIC	p-type Silso (Wacker)
Cualid et al. (1982b)	10^4	LBIC	p-type Silso (Wacker)
	10^5	LBIC	n-type Silso (Wacker)
Faughnan et al. (1981)	4–8 × 10^3	LBIC	p-type Silso (Wacker)
	2 × 10^3	LBIC	HEM (Crystal Systems)
	5 × 10^3	LBIC	p-type UMG-Silicon (RCA)
Zook (1980b)	1 × 10^5	LBIC	SOC (Honeywell)
Seager (1982)	0.3–2 × 10^5	LBIC	SOC (Honeywell)
Donolato (1983)	1.6 × 10^4	EBIC	p-type Silso (Wacker)
Donolato and Bell (1982)	1.5–5.2 × 10^4	EBIC	p-type Silso (Wacker)
De Pauw (1984)	1 × 10^4	LBIC	p-type Silso (Wacker)
	0.6–3.5 × 10^4	EBIC	p-type Silso (Wacker)

$$d_\ell = \frac{2\lambda f}{\pi D_\ell}, \tag{2.2}$$

with f the focal distance of the objective lens and D_ℓ the Gaussian diameter of the laser beam before focussing. For $\lambda = 633$ nm spot diameters of 2–10 μm have been reported (Faughnan 1981). An electron beam can be focussed to very small diameters, as small as 10 nm. But here the generation volume in the semiconductor is limiting the resolution, due to the scattering of the highly energetic electrons. For an acceleration voltage of 25 keV the excitation volume diameter is about 10 μm in silicon (at 5 keV about 5 μm). Therefore, both methods are expected to have the same resolution.

Secondly, for the case of a grain boundary, the problem is not symmetrical. The excitation occurs now on one side of the grain boundary, so it is possible that carriers are transmitted through the grain boundary. Part of the excess minority carriers will be transmitted through the grain boundary, which results in a net diffusion current to the other grain. The boundary condition to be applied has been derived in section 1.4, and is given by

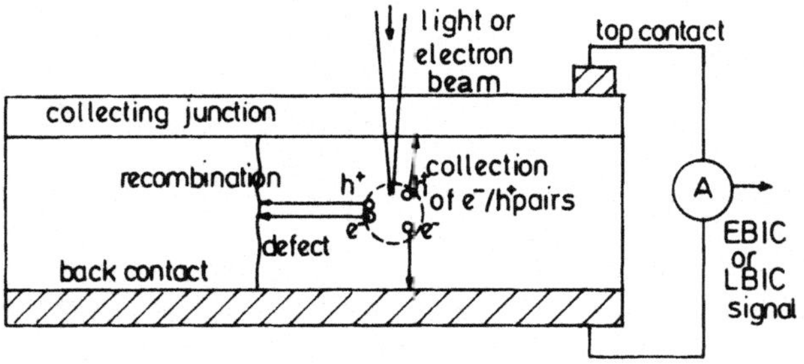

Fig. 18. Measurement set-up for EBIC or LBIC measurements.

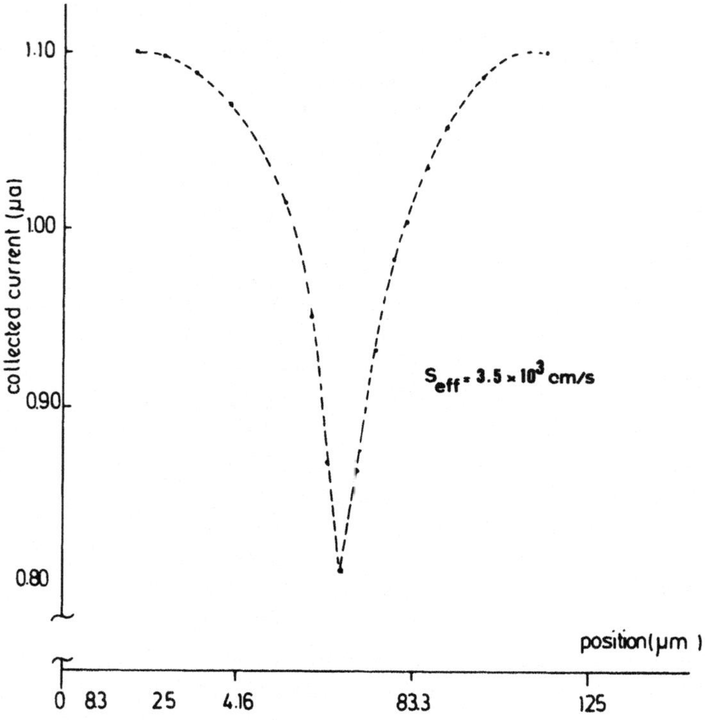

Fig. 19. Measured EBIC curve on a Silso grain boundary.

$$D\left.\frac{\mathrm{d}n}{\mathrm{d}x}\right|_{x=W_1} - D\left.\frac{\mathrm{d}n}{\mathrm{d}x}\right|_{x=-W_2} = s_{\mathrm{eff}\,1}(n(W_1) - n_0(W_1)) + s_{\mathrm{eff}\,2}(n(-W_2) - n_0(-W_2)), \tag{2.3}$$

assuming the x-coordinate to be zero at the location of the grain boundary, and s_{eff} defined as the effective recombination velocity on one side of the grain boundary.

Zook (1980a, 1983) gives a method to determine s_{eff} when the transmission factor t is zero. Zook derives the following formula to calculate the beam induced current, I_{BIC}, valid for excitation from a point source located at a depth z under the surface,

$$\frac{I_{\mathrm{BIC}}(z, L, s_{\mathrm{eff}}, x)}{I_{\mathrm{BIC}}^{\mathrm{MAX}}} = \mathrm{e}^{-z/L} - \frac{2}{\pi}\frac{s_{\mathrm{eff}}L}{D}\int_0^\infty \frac{(\sinh t)\exp[(\cosh t)x/L]\sin[(\sinh t)z/L]\,\mathrm{d}t}{(\cosh t)(Ls_{\mathrm{eff}}/D + \cosh t)}, \tag{2.4a}$$

where L is the intragrain diffusion length, the $x = 0$-plane is the grain boundary plane and the $z = 0$-plane represents the collecting junction.

For a light source of zero lateral width located at x, the generation density of the source is of the form $g(z) = \exp(-\alpha z)$. The LBIC expression is then given by integrating eq. (2.4a) over the absorption depth and the result is

$$\frac{I_{\mathrm{LBIC}}(\alpha, L, s_{\mathrm{eff}}, x)}{I_{\mathrm{LBIC}}^{\mathrm{MAX}}} = 1 - \frac{2}{\pi}(1 + \alpha L)\frac{s_{\mathrm{eff}}L}{D} \times \int_0^\infty \frac{(\sinh^2 t)\exp[(\cosh t)x/L]\,\mathrm{d}t}{(\cosh t)(Ls_{\mathrm{eff}}/D + \cosh t)((\alpha L)^2 + \sinh^2 t)}, \tag{2.4b}$$

where α is the absorption coefficient of the light of wavelength λ.

For an electron beam at x, also assuming a generation volume of zero lateral spread, the generation density as a function of depth is given by the polynomial of Everhart and Hoff (1971),

$$g(z) = \begin{cases} \dfrac{1}{R}\left[0.6 + 6.21\left(\dfrac{z}{R}\right) - 12.4\left(\dfrac{z}{R}\right)^2 + 5.69\left(\dfrac{z}{R}\right)^3\right], & 0 < \dfrac{z}{R} < 1.1, \\ 0, & \dfrac{z}{R} > 1.1, \end{cases} \tag{2.4c}$$

with R the primary electron range. By multiplying eq. (2.4c) with eq. (2.4a), and integrating over the absorption depth the expression for the EBIC signal is found. Fitting of eq. (2.4b) for the LBIC or the similar expression for the EBIC to the measured curve, gives a method to determine s_{eff} and L. However, in practice, the beam size is not negligibly small, and the shape of the beam induced curve is influenced by the lateral width of the beam. A method based on the integration of the beam induced current loss is expected to be less sensitive to the generation volume shape. The total loss in the collected current due to the grain boundary can be represented for both EBIC and LBIC methods by an effective width w, defined as

$$w = 2\int_0^\infty \frac{I_{BIC}^{MAX} - I_{BIC}(\alpha \text{ or } R, L, s_{eff}, x)}{I_{BIC}^{MAX}}\, dx \tag{2.5}$$

The effective width w of the boundary means that the decrease in J_{sc} due to the grain boundary recombination is the same as that due to an opaque strip of width w. The curves calculated with eq. (2.5) in the case of a laser beam have been plotted in fig. 20 for different values of the dimensionless quantities αL and w/L. In the case of an electron beam, similar curves could be calculated for the dimensionless quantities L/R and w/L. Figure 20 suggests that s_{eff} can also be determined as follows: w is obtained by integrating the grain boundary loss of the experimental curve according to

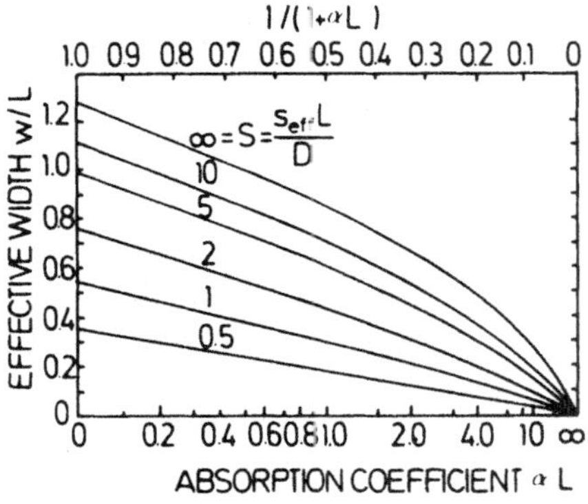

Fig. 20. Effective grain boundary width w/L versus αL, with $S = s_{eff}L/D$ as parameter. The effective grain boundary width w is obtained by integrating the grain boundary loss of the experimental curve according to eq. (2.5). L is obtained by an independent method. When the point (αL, w/L) is located in the figure, s_{eff} can be read-off immediately from the nearest-by curve (from Zook 1980).

eq. (2.5); L is obtained from another independent method, e.g., through the measurement of the quantum efficiency for long-wavelength light or simply from $I_{\mathrm{LBIC}}^{\mathrm{MAX}}$ if long-wavelength light is used for the scanning light beam. Now the point $(\alpha L, w/L)$ can be located in figure 20 and $S = s_{\mathrm{eff}}L/D$ can be read off immediately from the nearest-by curve. Important to note is that this analysis is independent of the lateral size of the generation volume. The LBIC method with the analysis of Zook has also been used by Seager (1982), Faughnan et al. (1981) and Oualid et al. (1982b) to measure s_{eff} in the dark.

The analysis of Zook has the drawback of requiring the determination of the intragrain diffusion length with an independent method. Donolato (1983) overcomes this problem by defining a new parameter, the variance σ^2 defined by

$$\sigma^2 = \frac{2}{w}\int_0^\infty x^2 \frac{I_{\mathrm{BIC}}^{\mathrm{MAX}} - I_{\mathrm{BIC}}(\alpha \text{ or } R, L, s_{\mathrm{eff}}, x)}{I_{\mathrm{BIC}}^{\mathrm{MAX}}}\, \mathrm{d}x \tag{2.6}$$

For this parameter to be useful, it should be independent of the lateral size of the exciting beam. Only in the case $\sigma \ll L$ this condition is fulfilled, and will the analysis given below be valid.

In the case of a laser beam, eq. (2.5) and (2.6) have the form

$$\begin{cases} w = Lf_1(\alpha L, Ls_{\mathrm{eff}}/D), \\ \sigma = Lf_2(\alpha L, Ls_{\mathrm{eff}}/D), \end{cases} \tag{2.7}$$

or by multiplying both sides with the known parameter α, we get

$$\begin{cases} \alpha w = Lf_1(\alpha L, Ls_{\mathrm{eff}}/D), \\ \alpha\sigma = Lf_2(\alpha L, Ls_{\mathrm{eff}}/D). \end{cases} \tag{2.8}$$

The analytical solution of L and Ls_{eff}/D from these equations is not feasible; hence, resort has to be made to a graphical method. In the plane $(\alpha\sigma, \alpha w)$, two families of curves can be obtained by keeping respectively αL and Ls_{eff}/D constant and regarding the other as a parameter. Figure 21a shows the set of curves generated in this manner. A couple of experimental values $(\alpha\sigma, \alpha w)$ defines a point in this figure; the corresponding curvelinear coordinates $(\alpha L, Ls_{\mathrm{eff}}/D)$ give the solution of eqs. (2.8). The curves of fig. 21 have been calculated for a grain boundary without depletion layers and a large transmission factor through the grain boundary.

An analogous figure for the case of electron excitation can be calculated, the electron range R playing the role of the reciprocal of the light absorption coefficient, α^{-1}. The results for the EBIC method are presented in fig. 21b. For a more accurate evaluation of s_{eff}, a much denser network is required. Both diagrams show that an accurate evaluation of high grain boundary recombination velocities is difficult, because of the reduced dis-

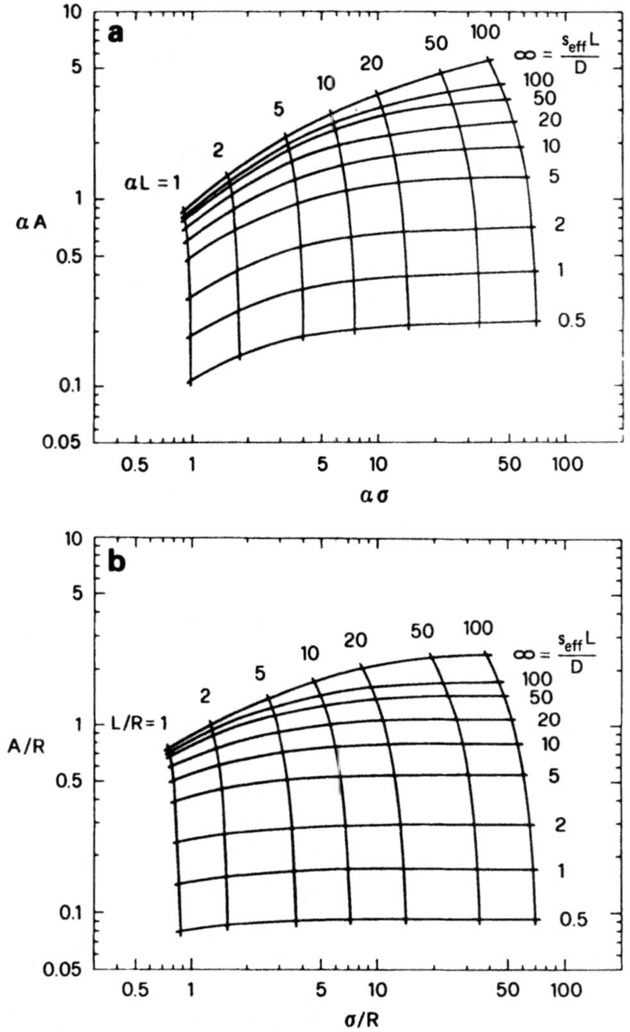

Fig. 21. Donolato's method to determine s_{eff} from LBIC (a) and EBIC (b) measurements (Donolato 1983).

persion of the curves of constant Ls_{eff}/D for large values of Ls_{eff}/D.

An overview of the measured values of s_{eff} at grain boundaries in defected silicon is given in table 3.

It is important that the analysis given for both EBIC and LBIC methods is only valid for low-level excitations, for which the GB height and s_{eff} do not change when approaching the grain boundary. Furthermore, the grain boundaries have to be perpendicular to the junction and separated by several diffusion lengths, so that they can be considered as independent.

When the collecting junction on top is a diffused junction, additional problems arise because of the preferential diffusion along the grain boundaries (see section 5.3). Unavoidable, the junction depth at a grain boundary is always larger. Until now this fact has not been recognized and taken into account. The error is significant if one uses a shallow generation volume, such as the 633 nm He–Ne laser beam or an electron beam. For this reason, only measurements with Schottky-barrier junctions or with deeply penetrating light are reliable.

In LBIC measurements, the absorption coefficient has to be accurately known. Oualid et al. (1982b) found that a stress field exists around a grain boundary, which influences the absorption coefficient. It is even possible that the grain boundary shows an increased quantum efficiency, probably resulting from an increased absorption coefficient or preferential grain boundary diffusion. An enhanced absorption coefficient at grain boundaries has been observed by Turner et al. (1982a), Storti et al. (1980) and S.M. Johnson et al. (1982).

2.4.2. *The photoconductivity method*

An other method to determine s_{eff} is a light beam induced photoconductive measurement. The experimental set-up is shown in fig. 22a. The measurement is usually performed with constant current. When the beam approaches the grain boundary, more minority carriers are trapped at the grain boundary, causing the grain boundary barrier height to decrease. The decrease in grain boundary barrier height is measured as a voltage decrease ΔV over the sample and, because of the constant current, this voltage decrease is proportional to the change in conductivity. A curve for a Silso grain boundary, is shown in fig. 22b. The logarithm of the change in conductance versus position yields a straight line and its slope gives the minority carrier diffusion length in the bulk of the grain (fig. 23). But the derivation of the grain boundary recombination velocity from these measurements is not straightforward. First, the barrier height and s_{eff} necessarily change during the measurement, which has to be accounted for in the

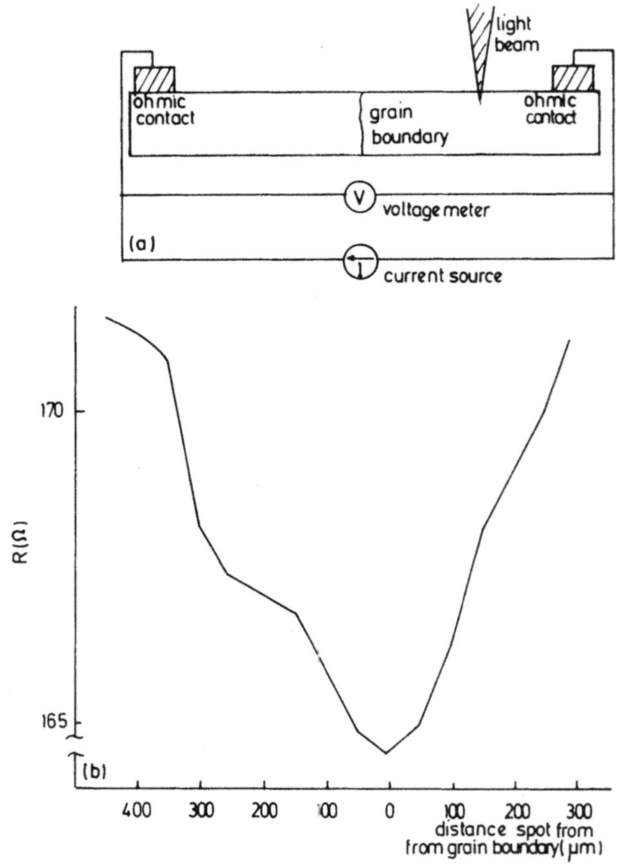

Fig. 22. Grain boundary resistance versus light beam position: (a) Measurement set-up, (b) Measured curve for a Silso grain boundary.

derivation. Furthermore, for each light spot location, the barrier height decrease is highly non-uniform, due to the small spot size. Nevertheless, this method has been used by Scholl (1981) and by Panayotatos et al. (1982), both using a different analysis of the data. Their results are also included in table 3.

2.5. *Dependence of the grain boundary characteristics on illumination*

This section reviews the data about the dependence of grain boundary characteristics on illumination. It is shown that ψ_B and s_{eff} decrease with increasing illumination level.

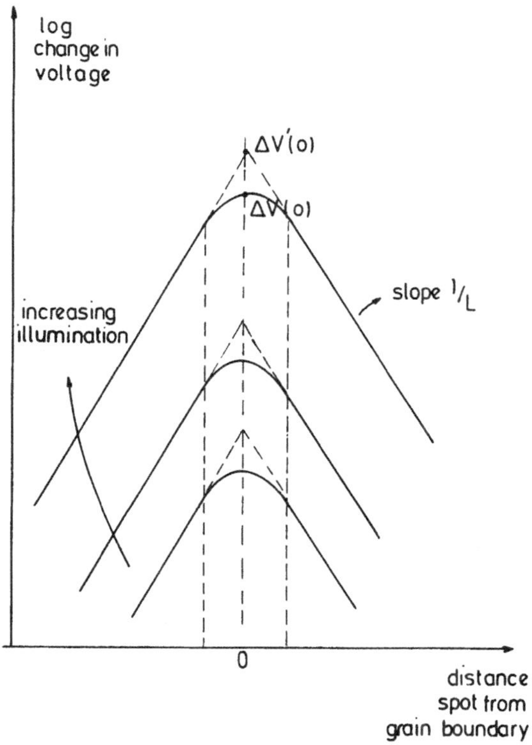

Fig. 23. Derivation of the intragrain diffusion length L from zero-bias resistance measurements like those of fig. 22b (Martinez et al. 1981).

2.5.1. Barrier height

The grain boundary barrier height and the effective grain boundary recombination velocity decrease with increasing illumination level (see fig. 6). As long as the grain-boundary remains in the minority carrier saturation regime, no decrease in recombination velocity and no change in the barrier height occur. This regime exists only for large barriers. Once out of this regime, a rapid decrease of GB barrier height is expected. The rate of decrease for small barriers is somewhat smaller.

Experimental evidence for the decrease of ψ_B with illumination is given by the measurements of the zero-bias grain boundary resistance versus illumination level. Figure 24 shows measurements for a grain boundary of p-type Silso polycrystalline silicon. For linearly increasing light intensity, the resistance drops exponentially. According to the thermionic emission

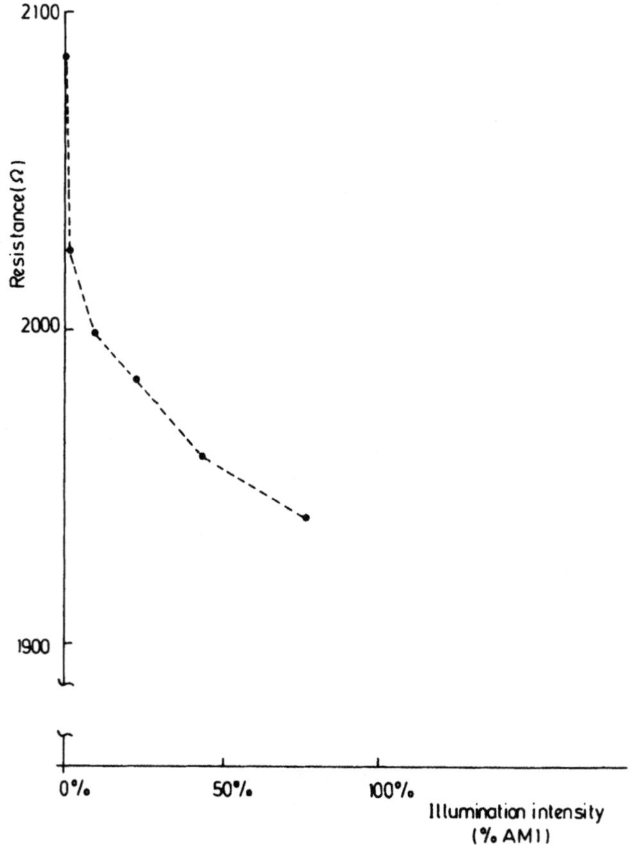

Fig. 24. Zero-bias grain boundary resistance as a function of white-light illumination for p-type Silso polycrystalline silicon.

theory and the double depletion layer model of Taylor et al. (1952) the grain boundary resistance is proportional to $\exp(q\psi_B/kT)$, indicating that ψ_B decreases approximately linearly with illumination. Derivations of grain boundary barrier height from such measurements have been made by Seager (1981) and Kazmerski (1982) using eq. (1.28). Seager's results are shown in fig. 25 and agree very well with the calculated values. Seager used monochromatic long-wavelength light to assure a spatially uniform decrease in ψ_B. Kazmerski (1982) used white-light illumination, while in his derivation of ψ_B he assumed a spatially uniform barrier height. This fact could

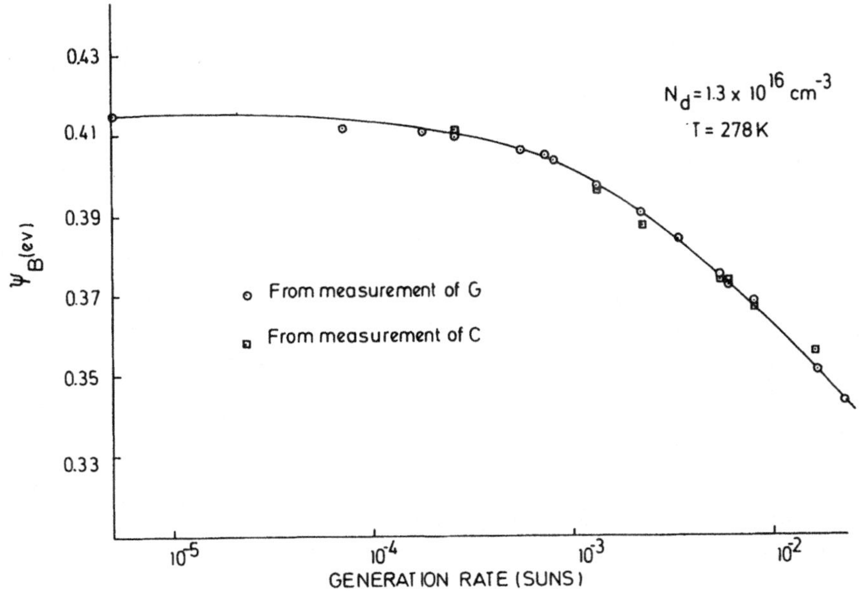

Fig. 25. Grain boundary barrier height ψ_B as a function of illumination intensity from capacitance and conductance measurements (from Seager 1981).

explain why his curves are generally more spread out. Seager also performed high-frequency capacitance measurements on the same sample. The barrier height can be derived from this measurement using eq. (2.1). The capacitance and the conductance results agree very well. A difficulty in doing the capacitance measurements is the high grain boundary conductance relative to the capacitance, which makes the method only useful for high bariers and at not too high illumination levels. These measurements agree with the calculated variation of ψ_B with the injection level, as shown in fig. 6.

Other evidence for the decrease of the barrier height with illumination is given by light beam induced conductivity measurements, with the experimental set-up of fig. 22a. The current is kept constant and the voltage ΔV over the sample is measured. A typical curve obtained in this way is shown in fig. 22b. When the focussed light beam approaches the grain boundary, the voltage ΔV decreases and attains a minimum when the spot is at the location of the grain boundary. The change in conductance indicates that ψ_B lowers with increasing minority carrier concentration.

2.5.2. *The effective grain boundary recombination velocity*

The earlier reported measurements of s_{eff} as a function of the injection level lead to contradictory conclusions. Panayotatos et al. (1982) found s_{eff} increasing with the injection level, by measuring the laser induced conductivity change of a grain boundary. The injection level is varied by changing the laser beam intensity. Using a chopped laser beam and lock-in amplification for the measurement of the grain boundary conductivity change, Scholl (1981) found a s_{eff} decreasing with increasing background illumination. It is very difficult to derive s_{eff} by this method because the measured phenomenon is the grain boundary barrier decrease, and s_{eff} is indirectly derived from this measurement. As the exciting beam is very small, the grain boundary height, ψ_B, does not decrease uniformly. To each different ψ_B corresponds, in principle, a different s_{eff}, which makes it not feasible to derive one s_{eff} corresponding to one injection level. Cheng and Shyu (1981a,b) fill the grain boundary states with a uniform light pulse, observe the conductivity decay after cutting off the light, and calculate s_{eff}. They find an increasing value with increasing illumination level. This result is not very conclusive in the light of the criticism on their derivation of s_{eff} (Cheng et al. 1982). Faughnan et al. (1981) use the LBIC method with varying intensity of the laser beam and found s_{eff} to be constant. However the intensity of their 1150 nm laser beam is not sufficient to saturate the grain boundary barrier height, which is a necessary condition in order to decrease s_{eff}. This is caused by the very low absorption by silicon for light of this wavelength.

An indirect observation of a s_{eff} decreasing with the injection level is given by the experimental observation of an increasing effective lifetime, τ_{eff}, of semicrystalline silicon with illumination. In section 3 the observed increase of τ_{eff} with illumination is explained by assuming a dependence of s_{eff} on illumination, as shown in fig. 5, for grain boundaries and for intragrain defects.

In order to directly measure the variation of s_{eff} with illumination the LBIC method is modified (De Pauw et al. 1984b). The laser beam is chopped and the induced current signal is obtained by lock-in amplification. This allows to add white-light illumination to the laser beam, in order to increase the injection level and saturate the grain boundary states. The laser intensity is kept constant and as low as possible, in order to eliminate effects of grain boundary barrier height change, when the laser beam approaches the grain boundary. To reduce noise, a bandpass filter with a quality factor of 100 is used at the input stage of the lock-in amplifier. The measurement set-up is shown in fig. 26. The solar cell is mounted on a

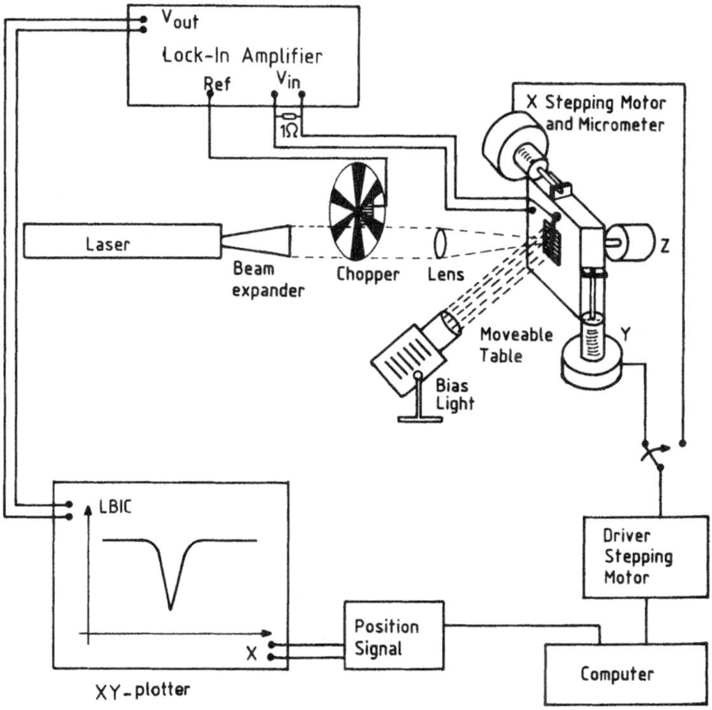

Fig. 26. Measurement set-up for the LBIC method, adapted for the measurement of s_{eff} under different injection levels.

table, driven by two stepping motors, one for the *X* and one for the *Y* direction. All stepping motor translations are computer controlled. A spatial resolution of 0.5 μm is obtained. The position signal is on the X-axis of the XY-plotter, and the output signal of the lock-in amplifier on the Y-axis. The position signal is automatically converted so that the total displacement during the scanning corresponds to the full X-range of the XY-plotter. A He–Ne laser with a wavelength of 633 nm and another He–Ne laser with a wavelength of 1150 nm are used. The intensity of both lasers is a few mW. The laser beam is focussed to a spot size of 20 μm for the 633 nm beam and approximately 100 μm for the 1150 nm beam. When the laser beam scans the cell surface, the grain boundary recombination activity is seen as a dip in the LBIC response. An eventual change in the recombination acitivity of a grain boundary with the injection level causes the depth of the LBIC-dip to change.

Using this measurement set-up the following experiments are done:

(i) Measurements with a 633 nm laser beam, with and without bias light, on solar cells on p-type Silso (Wacker) semicrystalline silicon substrates. The solar cells have been fabricated with an integral screen printing process (Cheek et al. 1984). The measurements reveal less than 6% variations of the LBIC signal due to grain boundaries. The LBIC scan under bias illumination is roughly the same as compared to the unilluminated case.

(ii) Measurements with a 633 nm laser beam on MIS solar cells on similar Silso substrates. The MIS junction has been made by covering the top surface with a thin 2 nm oxide layer and a thin 8 nm aluminum layer, so that the aluminum layer remains transparant for light. To decrease the series resistance, a thick aluminum finger pattern is evaporated through a nickel mask. At the back side aluminum is evaporated and sintered at 400°C to assure good ohmic contact. This temperature is the highest temperature the material has met. On these solar cells, a high grain boundary activity is observed. Indeed, the value of the photoresponse at the grain boundary is reduced to 10% of the value recorded in the intragrain region. From this observation, the very low grain boundary recombination activity under 633 nm light on the screen printed solar cells, is probably due to a preferential diffusion of phosphorus along the grain boundaries, resulting in a reduced recombination velocity. The depth of the preferential diffusion is at least the absorption depth of the 633 nm wavelength light in silicon, i.e. a few μm. Measurements with bias light on the MIS cell are difficult due to the still too high series resistance of the transparant metal layer.

(iii) Measurements with a 1150 nm laser beam on the screen printed and MIS solar cells on Silso material. As 1150 nm light is only slightly absorbed in silicon, a 1150 nm laser beam can be used to study the recombination activity of grain boundaries in the silicon bulk. These measurements on the MIS and screen printed solar cells on Silso material, reveal a strong and roughly comparable grain boundary recombination in the bulk. On the screen printed solar cells, the measurements of grain boundary recombination activity with and without bias light are compared. Figure 27a shows the curves obtained for one grain boundary. One observes the larger overall LBIC signal of the cell under bias light. If these two curves are shifted over each other (fig. 27b), it is clearly seen that the depth of the LBIC-dip is reduced when the cell is illuminated, indicating a decreasing s_{eff} with illumination. This measurement gives the first strong evidence that s_{eff} decreases with the injection level. The intensity of the

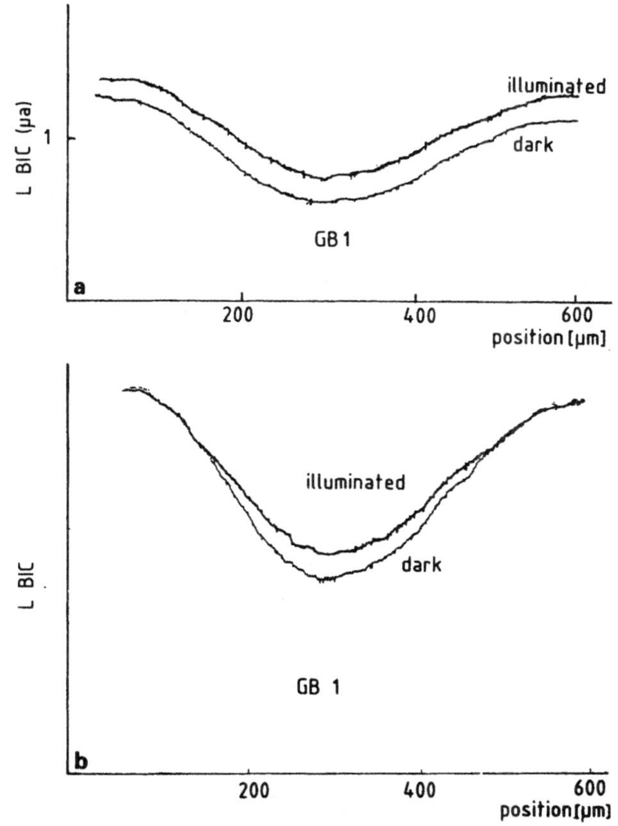

Fig. 27. LBIC-scan of a grain boundary on Silso silicon with and without bias illumination. (a) as measured, (b) the two curves shifted over each other.

white-light bias illumination is approximately 0.5 AM1. The value of s_{eff} for this grain boundary is estimated to be 1×10^4 cm/s without bias light and 5×10^3 cm/s under illumination of the bias lamps. On the Silso material, many grain boundaries have been observed to be in this case.

(iv) On screen printed solar cells on p-type HEM (Crystal Systems) semicrystalline silicon substrates similar observations as shown in fig. 27 can be obtained. These measurements indicate that the decrease of s_{eff} with the injection level is typical for most grain boundaries in silicon.

Remarkable is the increase of the overall LBIC response when bias light is applied. An increased absorption coefficient of the 1150 nm radiation due to a temperature increase under bias illumination has to be discarded, because the temperature of the solar cell is kept constant. Probably the increase of the overall LBIC response phenomenon is caused by saturation of the intragrain defects, which is responsible for the increase of L_{eff} with illumination (De Pauw et al. 1984a). In view of this, the increase of the overall LBIC response with light bias is another indication of the importance of intragrain defects.

3. Increase of the effective diffusion length with illumination

It is well known that the effective lifetime and diffusion length in defected silicon increase with the illumination level at low injection levels. A new explanation of this phenomenon is given by the recombination at intragrain defects and at grain boundaries. The calculations are performed under the simplified conditions of a uniform generation rate and a periodic distribution of local defect concentrations of one particular defect shape.

It will be shown below that the precise shape of the defects and the precise characteristics of the defect states are not a necessary element of the model.

3.1. Calculation of the dependence of L_{eff} on the injection level at low illumination levels

For the sake of simplicity, we consider the case of small spherical regions of high defect density. At the end of this section, we will show that other defect shapes yield qualitatively the same results. Also for simplicity, we assume that the defects are periodically distributed in the three-dimensional space of the crystal. A cross section of such a crystal is shown in fig. 28. Each defect is surrounded by a cube of side l. We approximate this silicon volume γ by a sphere of radius G such that

$$\gamma = \tfrac{4}{3}\pi G^3 = l^3 \tag{3.1}$$

We call this radius G the half interdefect distance.

Furthermore, we assume a uniform light generation rate g in each sphere γ. We assume also the defect model of section 1 to be valid and therefore that the $s_{eff}(n(W))$ versus $n(W)$ curve of figs. 4 and 5 is applicable. We now calculate the relation between $s_{eff}(n(W))$ and the generation rate g.

Because we consider a spherical defect, the minority carrier transport equation can be expressed in spherical coordinates and is given by

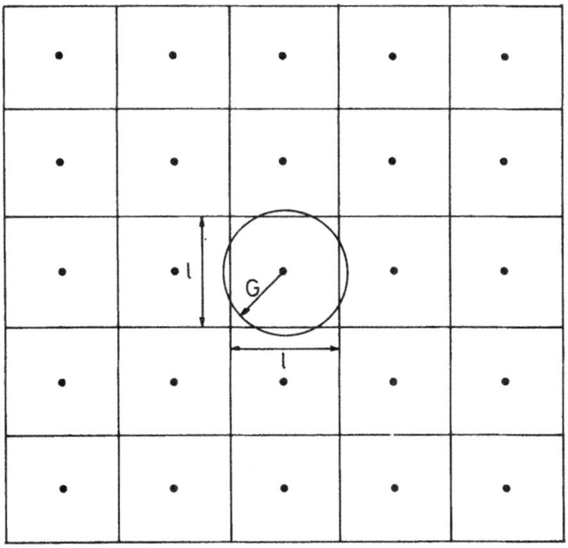

Fig. 28. Model of a defected silicon crystal. The defects are assumed to be periodically distributed throughout the crystal.

$$\frac{d^2n}{dr^2} + \frac{2}{r}\frac{dn}{dr} - \frac{n}{L^2} = -\frac{g}{D}, \tag{3.2}$$

with $L = (\tau_B D)^{1/2}$, the diffusion length in the silicon between the defects. This equation is the modified spherical Bessel equation with well-known solutions. We impose the following two boundary conditions:

$$\left.\frac{dn(r)}{dr}\right|_{r=G} = 0, \tag{3.3}$$

and

$$-D\left.\frac{dn(r)}{dr}\right|_{r=W} = -s_{eff}(n(W))\,[n(W) - n_0(W)], \tag{3.4}$$

where $s_{eff}(n(W))$ depends on $n(W)$. The first boundary condition results from symmetry considerations.

The solution of eq. (3.2) with boundary conditions (3.3) and (3.4) is the minority carrier profile in the sphere γ and can be written in the form

$$n(r) = g\tau_B \left\{1 - C\frac{W}{r}\left[\exp\left(\frac{r-G}{L}\right) + \exp\left(\frac{G-r}{L}\right)\left(\frac{G-L}{G+L}\right)\right]\right\}, \tag{3.5}$$

with

$$C = \left[\left(-\frac{D}{LS} + \frac{D}{WS} + 1\right)\exp\left(\frac{W-G}{L}\right) + \left(\frac{D}{LS} + \frac{D}{WS} + 1\right)\exp\left(\frac{G-W}{L}\right)\left(\frac{G-L}{G+L}\right)\right]^{-1}, \tag{3.6}$$

where we denote $s_{eff}(n(W))$ as S. The only unknown is the generation rate g. Putting $r = W$, and solving eq. (3.5) for g, we find g for a given set of $n(W)$ [and of $s_{eff}(n(W))$],

$$g = n(W)\left[\tau_B\left\{1 - C\left[\exp\left(\frac{W-G}{L}\right) + \exp\left(\frac{G-W}{L}\right)\left(\frac{G-L}{G+L}\right)\right]\right\}\right]^{-1} \tag{3.7}$$

We can now eliminate $n(W)$ from eq. (3.7) and the $s_{eff}(n(W))$ relation of fig. 5, and establish a relation between g and $s_{eff}(n(W))$.

The effective lifetime, τ_{eff}, is defined by

$$\frac{1}{\tau_{eff}} = \frac{1}{\tau_B} + \frac{|J_r| 4\pi W^2}{Q} = \frac{1}{\tau_B} + \frac{1}{\tau_d}, \tag{3.8}$$

where

$$Q = q4\pi \int_W^G [n(r) - n_0(r)] r^2 \, dr \tag{3.9}$$

is the total minority carrier density in the sphere γ and

$$J_r = q s_{eff}(n(W)) \, [n(W) - n_0(W)],$$

$n(r)$ is given by eq. (3.5) and $s_{eff}(n(W))$ is calculated with the method explained in section 1.2. τ_d is the minority carrier lifetime due to the defects. Note that J_r, Q, and in principle also τ_B, are functions of injection level.

If we write eq. (3.8) as

$$\frac{1}{\tau_{eff}} = \frac{1}{\tau_B} + C^* s_{eff}(n(W)) \, [1 - n_0(W)/n(W)], \tag{3.10}$$

with $C^* = 4\pi n(W)/Q$, we see that in materials where the lifetime is defect limited, i.e. $\tau_{eff} \cong \tau_d$, and under low injection, the effective lifetime is roughly proportional to $s_{eff}(n(W))$.

To calculate $L_{eff} = (\tau_{eff}D)^{1/2}$ we assume D to be independent of the injection level. Figures 29a,b show the result. We see that L_{eff} increases rapidly at low illumination intensities and, on a linear scale, seems to saturate at a generation rate of $3 \times 10^{19} s^{-1} cm^{-3}$. To produce the curves of fig. 29, we used the $s_{eff}(n(W))$ curve of fig. 5 together with $G = 10\,\mu m$, $r_c = 1\,\mu m$ and the bulk diffusion length $L = 70\,\mu m$; L was taken to be independent of illumination. The other parameters are listed in the first column of table 4. The curve of fig. 29a agrees well with the experimental results of fig. 7. Figure 29b shows that L_{eff} increases semi-logarithmically with illumination intensity, in agreement with the observations of Ho and Mathias (1981) and Shimokawa and Hayashi (1982). A difficulty is the assignment of a generation rate to the 1 sun (=AM1) intensity. Since the light absorption coefficient decreases exponentially with wavelength, there is a large gradient of the generation rate from the front side to the back side in a solar cell. However, the generated minority carriers diffuse quickly, so that the minority carrier concentration is leveled off. To compare accurately calculations with experiments, the L_{eff}-curve should be measured under high intensity monochromatic long-wavelength light alone. We approximated the average AM1 generation rate by calculating the average concentration of the photons absorbed in the bulk of a solar cell from the surface to a depth equal to the AM1 bulk diffusion length, as follows

$$g_{AM1} = \frac{1}{L_{AM1}} \int_0^{L_{AM1}} (1 - R(\lambda))\, N(\lambda)\, \{1 - \exp[-\alpha(\lambda) L_{AM1}]\}\, d\lambda. \qquad (3.11)$$

For $L_{eff} = 60\,\mu m$ an average AM1 generation rate of about $3.3 \times 10^{19} cm^{-2} s^{-1}$ is obtained.

Table 4
Modeling parameters with $N_A = 2.5 \times 10^{15} cm^{-3}$, $L = 70\,\mu m$, independent of injection level, $D = 25\, cm^2 s^{-1}$, and assuming a uniform distribution of the trapping states through the energy gap.

Parameter	Spherical defect with unequal capture cross-section	Cylindrical defect with unequal capture cross-sections	Spherical defect with equal capture cross-section
σ_C (cm^2)	1×10^{-14}	1×10^{-14}	1×10^{-21}
σ_N (cm^2)	1×10^{-16}	1×10^{-16}	1×10^{-21}
N_{is}^A (cm^{-2}eV^{-1})	1.5×10^{11}	1.5×10^{11}	5×10^{15}
N_{is}^D (cm^{-2}eV^{-1})	1.5×10^{11}	1.5×10^{11}	5×10^{15}
r_C (μm)	1	0.6	1
G (μm)	10	20	4.5

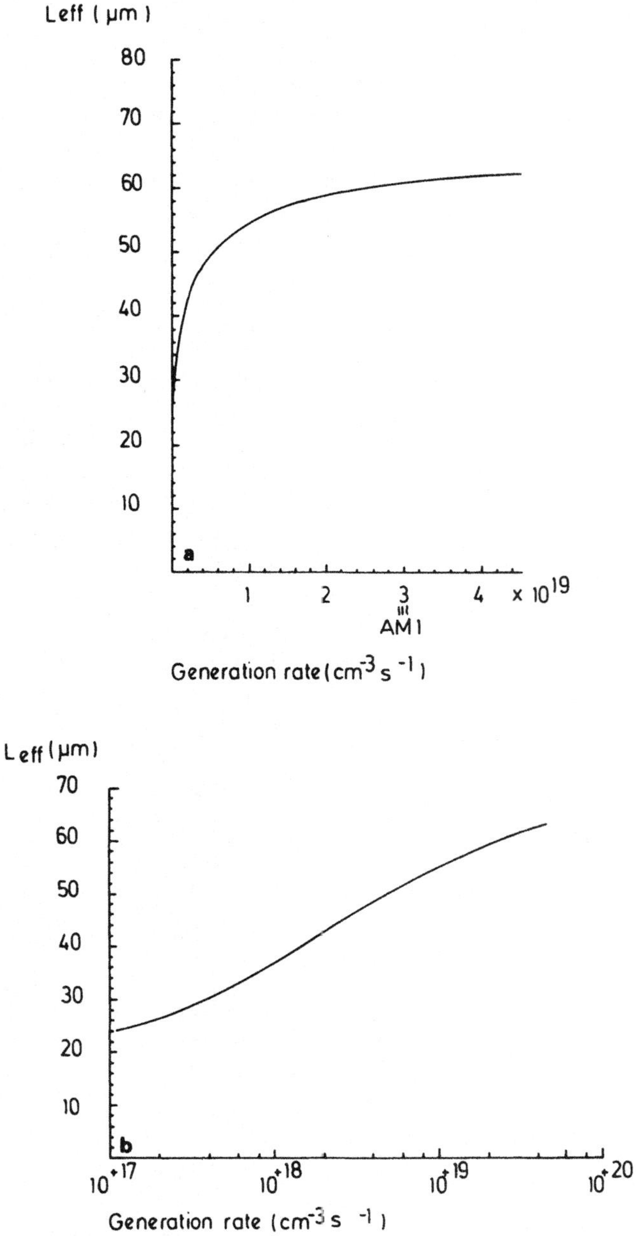

Fig. 29. Calculated effective diffusion length as a function of illumination, using the parameter values of table 4: (a) linear plot, (b) Semi-logarithmic plot.

The effect of modeling parameters on the L_{eff} versus generation rate curve has been studied. Figure 30a gives L_{eff} with the half interdefect distance G as parameter, and fig. 30b L_{eff} with the defect density on the defect surface N_{is} as parameter. These figures indicate that L_{eff} strongly depends on G and also on N_{is}. There is also a dependence on r_c, τ_B, the capture cross-sections and the shape of the defects. From this sensitivity analysis, the defected materials can be divided into two groups.

(i) Materials with a large interdefect distance ($G > \frac{1}{2}L$) or a low surface state density ($N_{is} < 10^{11}\,cm^{-2}eV^{-1}$). In this case the effective lifetime τ_{eff} is always determined by the bulk lifetime τ_B in the silicon between the defects. As indicated by the upper curve in fig. 30a this class of materials does not show an increase of L_{eff} with illumination. We suggest that the samples of figure 7 which do not show the L_{eff} increase belong to this group. Also the UMG-Si material (figure 8) is in this case because the bulk lifetime is much shorter than the lifetime determined by the defects.

(ii) Materials with a high defect density ($G < \frac{1}{2}L$) and a high interface state density ($N_{is} > 5 \times 10^{11}\,cm^{-2}eV^{-1}$). These materials show an increase of L_{eff} with illumination level. At very low injections τ_{eff} is determined by the recombination current J_r at the defects. When the illumination increases, the recombination at the defect decreases and τ_{eff} rises. Saturation finally occurs when τ_d becomes larger than τ_B. The p-type Wacker Silso and the HEM silicon of fig. 7 are examples of these materials. As most low-cost semicrystalline silicon materials have high defect densities (see section 2.1), these materials are likely to have an illumination dependent L_{eff}.

3.2. *Other defect shapes and other trapping state parameters*

The theory has also been applied to other defect shapes, such as line defects and planar defects. The rectangular filament, which contains the line defect, can be approximated by a cylinder. Grain and subgrain boundaries can be modeled one dimensionally.

The calculations for the case of a line defect are similar. The curves of fig. 29 also corresponds to the cylindrical defect model. The input data to generate these curves in the case of a line defect are: the $s_{eff}(n(W))$-curve of fig. 5, $G = 20$ μm, $r_c = 0.6$ μm and $L = 70$ μm (r_c and G are the radius of a cylinder in this case.) Here also, the bulk diffusion length L is taken independent of illumination level.

For each measured L_{eff} versus generation rate curve the same curve can be generated either by applying the spherical defect model or the cylindrical defect model. This indicates that the precise shape of the defects is not

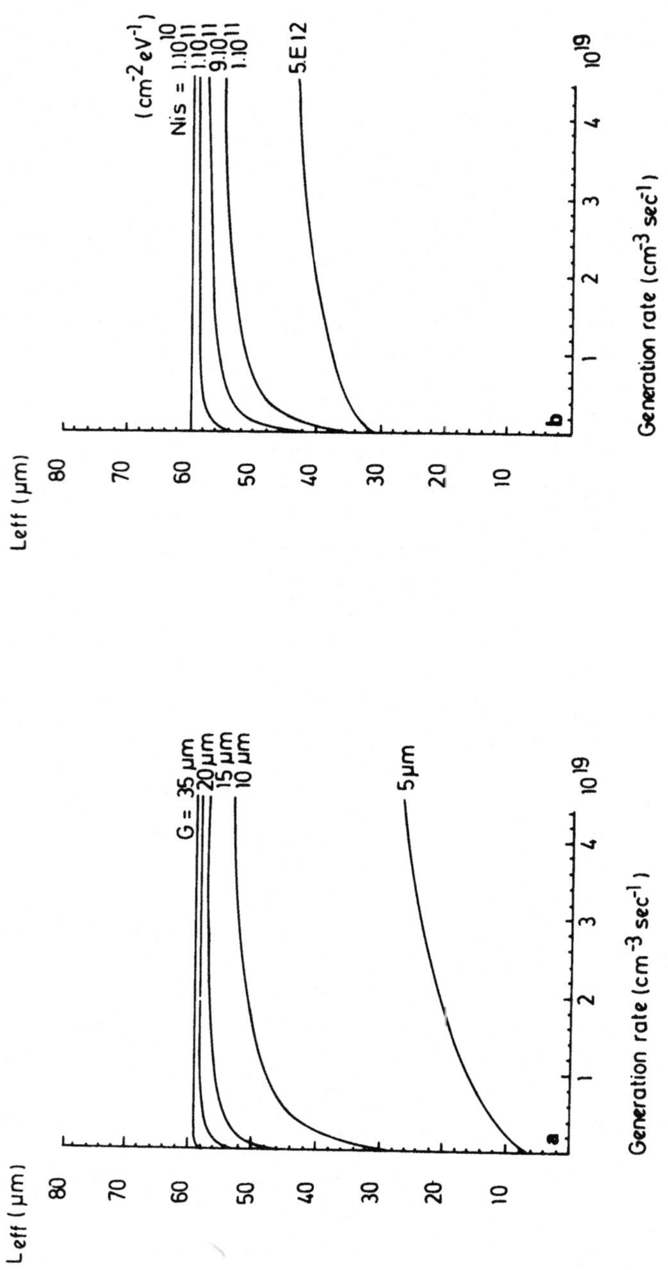

Fig. 30. Dependence of the L_{eff}-versus-generation-rate-curve on modeling parameters for a spherical defect. (a) The half interdefect distance G (b) The defect surface state density $N_{is} = N_{is}^{A} + N_{is}^{D}$.

important for the increase in L_{eff} to occur.

Another important parameter is the ratio of the carrier capture cross-sections σ_C/σ_N. If this ratio is unity, the increase of the effective diffusion length can also be obtained by an appropriate choice of the model parameters. In this calculation the values $\sigma_p = 1 \times 10^{-21}\,cm^2$, and a midgap trap density $N^A_{is} = N^D_{is} = 5 \times 10^{15}\,cm^{-2}eV^{-1}$ are used, as measured from a Silso grain-boundary by the DLTS technique (Srivastava et al. 1982), and further $\sigma_n = \sigma_p$ is assumed, which is equivalent to $\sigma_C = \sigma_N$. These data illustrate that the same L_{eff} curve can be obtained in several ways. Table 4 gives three sets of input data all generating the L_{eff} curve of fig. 29.

In conclusion, the only essential element of the model is a locally high density of trapping states.

4. Influence of grain boundaries and microdefects on solar cell characteristics

Until now, attention was only given to grain boundaries in the models of semicrystalline silicon. In this section, a model is presented which accounts for the influence of grain boundaries as well as intragrain defects on solar cell characteristics. In theory, a large increase in the first and the second exponential, and also in the leakage current is possible.

In this section a detailed analysis will be made of the characteristics of solar cells, fabricated on different semicrystalline silicon materials (Silso, HEM and UCP) with various fabrication processes. This analysis indicates that semicrystalline solar cells are very comparable. The most important degradation of the solar cell efficiency of these cells is the small value of the intragrain diffusion length. The low value of the effective diffusion length in these cells explains their low short-circuit current and their low value of the open-circuit voltage. Because of the necessarily large wafer thickness, a BSF cannot be made effective. Therefore, to improve the efficiency of these solar cells, the effective diffusion length of these substrates has to be improved.

An increased second exponential and shunt resistance are also found in semicrystalline solar cells. In addition, the screen printing process results in an increased series resistance. None of these degradation mechanisms is very important.

4.1. Introduction

The influence of defects on solar cell characteristics is simple: every defect is harmful and will degrade all solar cell parameters. The only reason why these defected materials have been developed is their potentially reduced cost. However, the kind and the amount of defects generated depends strongly on the silicon crystallization process.

The influence of grain boundaries on the solar cell efficiency depends on their orientation in the substrate. Let us first consider a solar cell containing some grain boundaries parallel to the junction (fig. 31a). Since the grain boundary barrier heights in the lowly doped base of the solar cell are high, these grain boundaries are not only an impediment for the majority carrier current flow, but also have a high effective recombination velocity. As a result, all minority carriers generated between the grain boundary and the back ohmic contact are collected by the grain boundary and will recombine, causing a reduction in the short-circuit current. Therefore, grain boundaries parallel to the junction are generally very harmful to solar cell efficiency. For this reason solar cells with only columnar grains are highly desirable. Since most potentially low-cost solar cell materials have this structure we will restrict ourselves for a more detailed analysis to the structure of fig. 31b.

For simplicity we assume that the superposition principle is valid. This means that the illuminated solar cell characteristic can be generated with the following components: a current source J_{sc}, a diode with dark current density J_{S1}, a series resistance and a shunt resistance. Then the solar cell can be represented by the equivalent electric scheme of fig. 32. All unknown parameters can be obtained by an independent measurement.

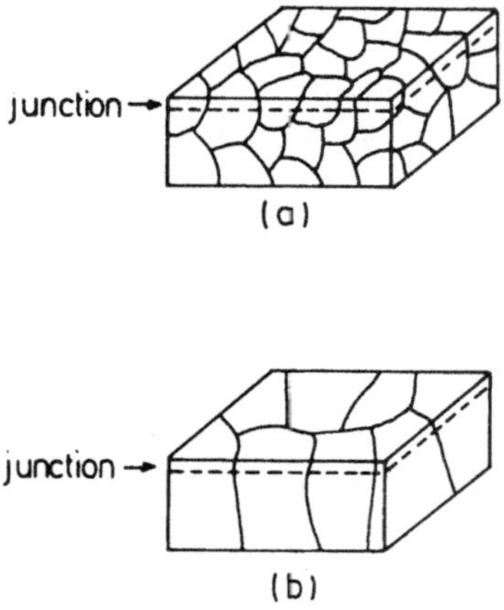

Fig. 31. Polycrystalline silicon with randomly oriented (a) and columnar grains (b).

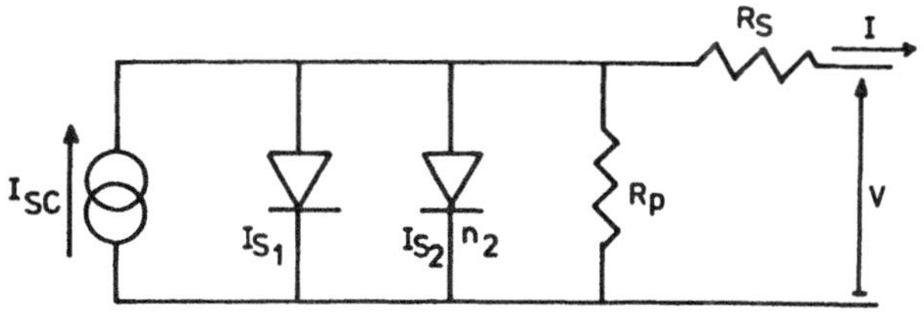

Fig. 32. Equivalent schematic diagram of a solar cell.

It is important to realize that the short-circuit current in a polycrystalline silicon solar cell is influenced by recombination in the neutral base of the cell, while on the other hand, the open-circuit voltage and fill-factor are strongly influenced by recombination in the neutral base and also in the space-charge region of the junction.

4.2. *Experimental results*

The results obtained on semicrystalline materials for three different processes are summarized in table 5. Between parentheses the relative reductions (in percent) as compared to single-crystal CZ silicon are indicated. These results indicate that the losses in V_{oc} and fill-factor are smaller than the losses in J_{sc}. Note that the values of table 5 are averages for one lot of material. It should be noted that materials characteristics may change from one lot to the other and that the material quality still steadily improves.

Hyland et al. (1982) compared different semicrystalline silicon materials in the same baseline process.

Hyland et al. (1982) also performed a more advanced process on the same materials. This included a shallower junction (0.2 μm), fine metal grid lines, multilayer AR coating, and a BSF (Back Surface Field) and BSR (Back Surface Reflector). AM1 efficiencies up to 13% were obtained both for the Silso and the UCP material.

Cheek et al. (1982) applied an integral screen printing process, which is a low-cost production process, on several defected materials. The obtained efficiencies are slightly lower. Also single crystalline but dislocated silicon is included for comparison. It is observed that dislocations have a pronounced deleterious effect on the solar cell efficiency of single crystalline silicon. The dependence of the efficiency on the solar cell area is also worth to mention.

The results of table 5 also indicate that the three cast materials, Silso, HEM and UCP, are very comparable and that AM1 efficiencies of 10% on semicrystalline materials are obtainable.

Solar cells made on defected silicon have a much smaller short-circuit current than the ones on single crystalline silicon. This is due to the recom-

Table 5
AM1 performance of solar cells fabricated on semicrystalline silicon.

	Area (cm^2)	FF (%)	V_{oc}(mV)	J_{sc} (mA/cm^2)	eff (%)
Baseline process[a]					
Silso		74.8	552	24.7	10.2
		(−2.5%)	(−6.3%)	(−14.8%)	(−21.5%)
HEM (Single)		73.0	579	27.7	11.7
(Poly)		74.7	561	26.0	10.9
		(−2.6%)	(−4.8%)	(−10.3%)	(−16.8%)
UCP		73.7	551	25.6	10.4
		(−3.9%)	(−6.5%)	(−11.7%)	(−20.6%)
CZ Control		76.7	589	29.0	13.1
Advanced process[a]					
Silso		77.7	564	29.9	13.1
		(+1.7%)	(−4.7%)	(−8.0%)	(−10.9%)
UCP		77.7	553	29.8	12.8
		(+1.7%)	(−6.6%)	(−9.1%)	(−12.9%)
CZ Control		76.4	592	32.5	14.7
Low-cost production process[b]					
Silso	3.2	70.7	545	26.1	10.1
	100	69.3	554	23.3	9.0
HEM	3.2	72.5	554	26.4	10.6
	100	67.4	574	23.6	9.1
MHO	3.2	75.7	561	27.8	11.8
$N_d^+ = 5 \times 10^4$[c]	86.3	70.1	551	25.4	9.9
$N_d = 5 \times 10^3$	86.3	73.8	569	26.6	11.2
CZ Control	78.5	72.5	595	29.2	12.7

[a] Hyland et al. (1982).
[b] Cheek et al. (1982).
[c] N_d^+ is the dislocation density per cm^2.

bination at grain boundaries and to the reduced intragrain diffusion length. For example, the same cell fabrication process results in AM1 short-circuit currents of 24.7 mA / cm^2 on Silso silicon (Wacker), 26.0 and 27.7 mA / cm^2 on HEM poly respectively single crystalline silicon (Crystal Systems), and 25.6 mA / cm^2 on UCP silicon (Semix) (Hyland et al. 1982). On CZ silicon the same process results in a short-circuit current of 29.0 mA / cm^2

4.3. Detailed analysis of the loss components in semicrystalline solar cells

A detailed analysis of the losses in semicrystalline solar cells is given in table 6. These results show that the second exponential is increased. The degradation of the fill-factor caused by the larger second exponential is more important than the series and shunt resistance losses. Also the V_{oc} is much degraded by the increased second exponential.

However, these three loss components cannot fully explain the lower fill-factor and V_{oc} of the semicrystalline solar cells. This suggests that also the first exponential, I_{S1}, is increased. The only material parameter, which is much different for semicrystalline and single crystalline silicon is the much lower effective diffusion length, L_{eff}, in semicrystalline silicon. Since this parameter has a large impact on I_{S1}, it is reasonable that the lower value of L_{eff} causes the increase of I_{S1}.

Table 6
Detailed analysis of the electrical characteristics of a 1.8 × 1.8 cm^2 Silso (Heliotronic) semicrystalline solar cell, solidified from highly pure silicon, and fabricated with the high-temperature screen printing process (Cheek 1984).

J_{S1} = 1.4×10^{-11} A/cm^2		
J_{S2} = 5.3×10^{-6} A/cm^2		
m = 2.34		
J_{sc} = 24.7 mA/cm^2		
R_s = 0.27 ohm		
R_p = 1×10^4 ohm		
FF = 64.8%	degradation by	R_s = 3.1 % (absolute)
		R_p = 0.0 % (absolute)
		J_{S2} = 10.9 % (absolute)
V_{oc} = 533 mV	degradation by	R_p = 0 mV
		J_{S2} = 27 mV
Efficiency = 8.6%	degradation by	R_s = 0.41% (absolute)
		R_p = 0.01% (absolute)
		J_{S2} = 1.94% (absolute)

The reductions in the fill-factor and the open-circuit voltage are less than the reduction in the short-circuit current. Since L_{eff} is the main parameter determining J_{sc}, we draw the conclusion that L_{eff} is the main parameter limiting the efficiency of a semicrystalline solar cell.

When a back surface field is included in the cell fabrication process, the difference between the single crystalline and the semicrystalline solar cell data increases. The low values of L_{eff} make it impossible to fabricate a semicrystalline solar cell with an effective BSF.

Therefore, more effort has to be put to increase the effective diffusion length in semicrystalline silicon. This can be obtained by an adjustment of the solidification process or by the application of gettering techniques, such as phosphorus and damage gettering.

4.4. *Splitting of the J_{sc}-loss*

It is important to determine whether a high grain boundary recombination or a small intragrain diffusion length is responsible for the low value of the short-circuit current, J_{sc}. These two components of the J_{sc}-loss can be obtained from the following measurements:

(i) the (AM1 or infrared) short-circuit current on a large area single crystalline solar cell, J_{sc_1},

(ii) the (AM1 or infrared) short-circuit current on a large area semicrystalline solar cell, J_{sc_2},

(iii) the (AM1 or infrared) short-circuit current of the intragrain material, J_{sc_3}.

The relative reduction due to the intragrain material is then J_{sc_3}/J_{sc_1}; the relative reduction due to the grain boundaries is then $(J_{sc_3}\text{-}J_{sc_2})/J_{sc_1}$. Measurements indicate that the relative reduction due to grain boundaries is small.

A direct determination of the J_{sc}-loss due to grain boundaries is also possible. Many of the semicrystalline solar cell models are based on the approximation that the effective lifetime in defected material can be written as

$$1/\tau_{eff} = 1/\tau_{bulk} + 1/\tau_{GB}, \tag{4.1}$$

where τ_{bulk} is the bulk lifetime, which may also be an effective lifetime when intragrain microdefects are present, and where τ_{GB} is a function of the dimensions of the grain and is given by $\tau_{GB} = d/4s_{eff}$ if the grain is columnar with side d (fig. 33). In this way the locally high defect concentrations are mathematically uniformly distributed over the bulk. The con-

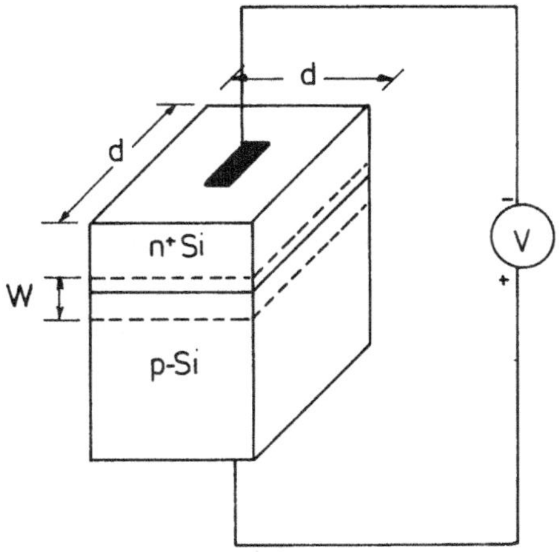

Fig. 33. Representative columnar grain of a n^+–p polycrystalline solar cell.

cept of using an effective lifetime is atttractive because of its simplicity. It is therefore frequently used (Makram-Ebeid 1979, Fossum and Lindholm 1980b, Ghosh et al. 1980).

However, the effective lifetime concept has to be carefully used. Jain et al. (1982, 1983) made a detailed study of the validity of the effective lifetime concept. They find that the effective lifetime is given by an infinite series. Under certain physical conditions, which can be easily met for J_{sc} calculations (but not for I_{sc} and V_{oc} calculations), the series is rapidly convergent and one value of the lifetime dominates the series. Using the effective lifetime approximation for calculating J_{sc}, a maximum error of 22% occurs when $s_{eff}d/2D$ is large, with s_{eff} the effective recombination velocity at the grain boundary depletion layer edge. Furthermore, the concept of one effective lifetime is in principle valid only if s_{eff} and τ_{bulk} are independent of the injection level. Since the use of an effective lifetime means that we have locally high defect concentrations and since locally high defect concentrations frequently give rise to an injection level dependent L_{eff}, an effective lifetime may only be used if it also is allowed to change with the injection level. Once determined the effective lifetime, the calculation of the short-circuit current and its reduction by grain boundary recombination is straightforward (see, for example, Hovel 1975). Since s_{eff} at the grain boundary is low, and because of the large grain size in semicrystalline silicon, also these calculations suggest small J_{sc}-losses due to grain boundaries.

Another method was suggested by Faughnan et al. (1981). As shown by the analysis of Zook, discussed in section 2.4, we can assign to each grain boundary an effective width w. The parameter w is obtained by integrating the loss in collected current according to eq. (2.5). With this definition, w has the following meaning: The decrease in J_{sc} at wavelength λ due to the recombination at the grain boundary is the same as that due to an opaque strip of width w. For intragrain microdefects, a similar definition is applicable. In this way the reduction of the short-circuit current by the defects can easily be calculated. For a square grain, with side d, the relative reduction in collection efficiency at wavelength λ is $[(d-w)/d]^2 \cong 1 - 2w/d$. This is indicated in fig. 34. This measurement is repeated for each wavelength. The reduction in the short-circuit current ΔJ_{sc} can then be calculated from

$$\Delta J_{sc} = \sum_{n=1}^{15} J_{sc}^{MAX}(1 - 2w/d)N(\lambda)\,R(\lambda), \tag{4.2}$$

where $n = 1, \ldots, 15$ corresponds to 15 filter wheel positions in the quantum efficiency measurement apparatus, $N(\lambda)$ is the spectral irradiance of the sun and $R(\lambda)$ is the cell's reflection at wavelength λ. To account for the

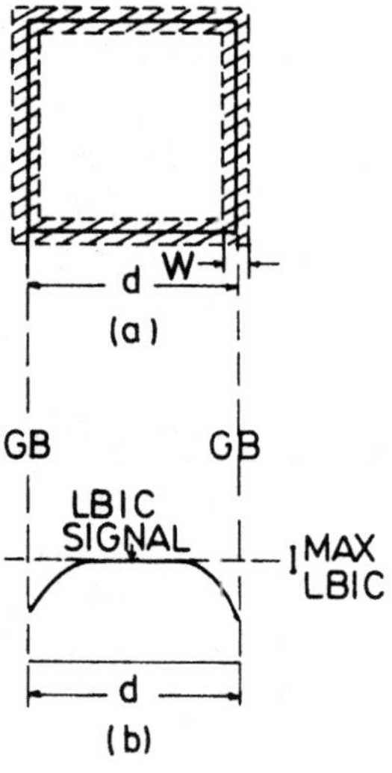

Fig. 34. Reduction of the short-circuit current due to recombination at grain boundaries: (a) Definition of the effective width w of a grain boundary for short-circuit current loss calculations, (b) LBIC response.

possible dependence of τ_{bulk} and s_{eff} on injection level, the measurement can be done with chopped monochromatic light and white-bias light. In this way one can calculate exactly what effect grain boundaries do have on J_{sc} under AM1 conditions. The calculation is accurate for columnar grains for $d < L$.

With a grain size of 1.0 mm and $S = s_{eff}L/D$ equal to 7.6 (a typical dark value found in LBIC measurements), grain boundaries reduce J_{sc} with 3.8% (Faughnan et al. 1981). Since the grain size is usually much larger and S is decreased under illumination, the J_{sc}-loss due to grain boundaries is even lower. However, the J_{sc}-difference between single crystalline and semicrystalline materials is on the average 8–15%. Therefore, the J_{sc}-loss by the intragrain (effective) diffusion length is larger than the grain boundary losses.

4.5. *Influence of defects on the open-circuit voltage and on the fill factor of a solar cell*

4.5.1. *Reduction of the short-circuit current*

The open-circuit voltage, V_{oc}, and fill-factor, FF, are determined by the diode dark current and the short-circuit current as follows (Green 1983)

$$V_{oc} = \frac{kT}{q} \ln\left(\frac{J_{sc}}{J_{S1}} + 1\right), \tag{4.3}$$

$$FF = \frac{qV_{oc}/kT + 1}{1.015\,qV_{oc}/kT + 5.7} \times 100. \tag{4.4}$$

Equation 4.4 is only valid for a negligible series resistance, leakage current, and second exponential.

Since the short-circuit current of a semicrystalline solar cell is lower than that of a single crystalline solar cell, this gives rise to a reduction in V_{oc}. However, with the short-circuit data of table 5, the V_{oc}-loss due to a lower short-circuit current can be calculated to be about 1 mV. Clearly this is insufficient to explain the much lower V_{oc} and FF values of semicrystalline solar cells as compared to single crystalline silicon. Let us therefore investigate the other degradation mechanisms.

4.5.2. *Increase of the second exponential*

We consider the columnar grain, shown in fig. 33, and calculate the dark current. Inside the grain we assume intragrain defects to be present. This calculation for a semi-crystalline solar cell is not an easy task because of the complicated nature of the defects and the poor knowledge of the defect structure.

The dark current can be split into two parts, as follows (Wolf et al. 1977)

$$I(V) = I_{S1}\exp(qV/kT) + I_{S2}\exp(qV/mkT), \tag{4.5}$$

with the ideality factor m close to 2.

Experimental values of I_{S2} are always much larger than I_{S1}. Because of the ideality factor is close to 2, the current of the second exponential increases less than the current of the first exponential. Therefore, at low voltages, the second exponential dominates the dark current, while at higher voltages, the first exponential dominates the dark current. In a practical large grain solar cell, at voltages around the maximum power point, the second exponential is still important, but around V_{oc} it may be dominated by the first exponential. In materials with high defect concentrations, the V_{oc} is also influenced by the second exponential. Typical dark and illuminated I–V characteristics for a solar cell with a low and a high defect density are shown in fig. 35.

The first exponential I_{S1} in eq. (4.1) is the diffusion current and can be split into the contributions $I_{S1}^{emitter}$ and I_{S1}^{bulk}, where each term is of the form $d^2qn_i^2D/(L_{eff}N)$, where D is the minority carrier diffusion constant, N is the doping level, L_{eff} is the effective diffusion length region and d^2 is the

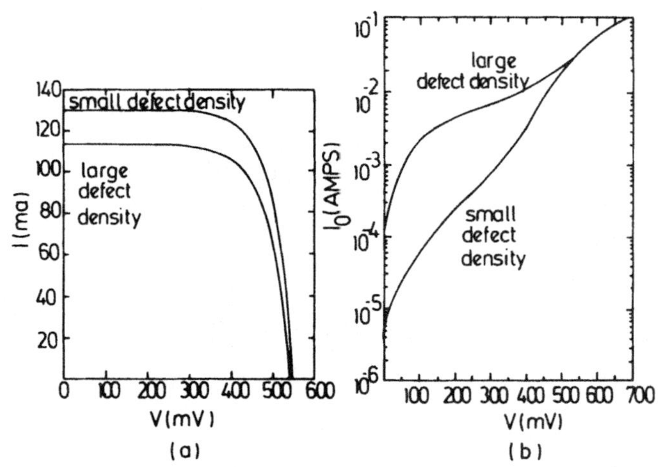

Fig. 35. Typical illuminated (a) and dark (b) I–V characteristics for a solar cell with a high and a small defect density.

intragrain surface area (fig. 33). For simplicity we assume the effective lifetime concept to be valid, although this is generally not the case for the calculation of the dark currents (Jain et al. 1982, 1983). The high dopant concentration in the emitter makes $I_{S1}^{emitter}$ small, while low dopant concentration and the low effective diffusion length in the bulk make I_{S1}^{bulk} much larger than $I_{S1}^{emitter}$. Therefore, the sum, I_{S1} is much larger for a semicrystalline solar cell, compared to the single crystalline solar cell.

The second component of the dark current, the second exponential I_{S2}, is caused by recombination in the junction space-charge region. Grain boundaries and intragrain defects intersecting the junction, will increase I_{S2}. The second exponential can be split into a grain contribution and a grain boundary contribution as follows

$$I_{S2} = I_{S2}^{grain} + I_{S2}^{GB}, \tag{4.6}$$

I_{S2}^{grain} is given by the Sah–Noyce–Shockley theory (Sah et al. 1957), while I_{S2}^{GB} is given by the theory of Henry et al. (1978) and revised by Fossum and Lindholm (1980b). Equation (4.6) can be written as

$$I_{S2} = (d^2)\,\frac{qWn_i}{2\tau_{bulk}} + \tfrac{1}{2}(4dW)qn_i s_{GB}, \tag{4.7}$$

where W is the junction space-charge region width, s_{GB} is the effective recombination velocity at one side of the grain boundary, and τ_{bulk} is the bulk lifetime, which can be an effective lifetime if intragrain defects are present. Equation (4.7) is based on many approximations, and can only be used as a first estimate. From eq. (4.7) it is easily seen that I_{S2} will increase when τ_{bulk} decreases, s_{GB} increases and d decreases. We can now write eq. (4.7) as

$$I_{S2} = d^2\,\frac{qWn_i}{2}\left[\frac{1}{\tau_{bulk}} + \frac{4s_{GB}}{d}\right]. \tag{4.8}$$

The expression between brackets in eq. (4.8) is very similar to the effective bulk lifetime for the calculation of the short-circuit current given by eq. (4.1). This similarity suggests that the influence of grain boundaries on the space-charge recombination can also be described by an effective lifetime,

$$I_{S2}^{GB} = d^2\,\frac{qWn_i}{2\tau_{eff}}. \tag{4.9}$$

We could also have split off in eq. (4.6) the intragrain defects intersecting the junction. However, as eqs. (4.8) and (4.9) show for the case of a grain boundary, the influence of the intragrain defects can be accounted for by an effective lifetime.

When the grain size $d < 4\tau_{bulk}s_{GB}$, eq. (4.8) shows that the recombination at the grain boundaries exceeds the intra-grain recombination. For example, if we take $s_{GB} = 10^4$ cm/s and $\tau_{bulk} = 1$ μs, which corresponds to $L = 50$ μm, I_{S2}^{GB} will only be larger than I_{S2}^{grain} for grain sizes d less than 400 μm.

As suggested by Fraas (1980), recombination in a diode containing grain boundaries can be much stronger than indicated by these simple formula. An overview of other mechanisms adding to an increased I_{S2} is given in the Appendix.

It is important to note that the second exponential is determined only by recombination at defects in the space-charge region, and not in the bulk. In a material where the open-circuit voltage is determined by the second exponential, already a shallow passivation is sufficient to strongly improve the FF and V_{oc}. J_{sc} is not influenced, since it is determined by the effective diffusion length in the neutral base (Fossum and Lindholm 1980a).

On very fine grain sized silicon, as obtained by evaporation, the defect density is much higher and the second exponential is reported to dominate the open-circuit voltage (Feldmann et al. 1982). For the other semicrystalline materials, not discussed here, it is still too early to give an analysis of their performance, because these materials are still under development and are not commercially available.

4.5.3. Degradation of the series and the shunt resistance

The series resistance highly affects the fill factor of a solar cell. The open-circuit voltage is never affected, because no current flows through the series resistor under open-circuit conditions. The short-circuit current is only affected in the case of a very high series resistance. A low shunt resistance mainly affects the fill factor and the open-circuit voltage of a solar cell.

The large values of the fill factors of semicrystalline solar cells, fabricated with a high-efficiency process, demonstrate that the defects do not increase the series resistance of a solar cell. Of the different types of defects, only two-dimensional defects, i.e. grain boundaries, present a barrier for the current. As calculated in section 1, the grain boundary barrier height is large when the doping level is in the range 10^{15}–10^{17} cm^{-3}. This situation occurs in the bulk. Therefore, grain boundaries parallel with the junction in the bulk will increase the series resistance. Since nearly all grains are columnar, this situation will rarely occur. Barriers in the emitter are important, since a relatively large current flows here laterally and the solar cell emitter is very shallow. The doping level in the emitter is high, and thus the grain boundary barrier height is very low, and only scattering at the high grain boundary charge may be important, resulting in a reduced mobility. This small resistance increase is not found to degrade solar cell efficiency, which is confirmed by the good fill factors observed in semicrystalline solar cells.

When using the screen printing technology, the series resistance of semicrystalline solar cells is much larger than on single-crystal ⟨111⟩ oriented silicon (Cheek 1984). The series resistance of solar cells made with the screen printing technology is known to depend on the crystal orientation of the silicon substrate. For example, solar cells made on ⟨100⟩ oriented silicon have a higher series resistance than solar cells on ⟨111⟩ oriented silicon. The larger series resistance of the solar cells on semicrystalline silicon is caused by the different crystal orientations present in this material.

In semicrystalline solar cells, the shunt resistance is found to be sufficiently high, so that it does not degrade the fill factor. Therefore, it can be concluded that the defects in semicrystalline silicon do not result in an appreciable leakage current across the p–n junction.

5. Reduction of the influences of defects on the solar cell characteristics

In this section we give an overview of the different techniques to reduce the deleterious effects of the crystal defects. Several techniques, or combinations of these are possible:

(i) A strong increase of the barrier height is found after a heat treatment. Therefore, to keep the grain boundary recombination at a low level, a *low cell-processing temperature* gives better results.

(ii) A large reduction of the grain boundary barrier height is observed after a hydrogen plasma passivation. Therefore, *hydrogen passivation* after solar cell fabrication is very attractive, if it can be done in an economic manner.

(iii) If grain boundaries are heavily doped their recombination activity is low. Therefore, *preferential diffusion* of grain boundaries yields a higher solar cell efficiency. In order to obtain the preferential doping of grain boundaries, fast diffusing species, such as aluminum and lithium, are interesting.

5.1. *Cold processing of solar cells*

5.1.1. *Influence of heat treatments on the electrical activity of defects in silicon*

Many experiments, done mainly on Silso material, indicate that the electrical activity of grain boundaries is strongly enhanced by a heating at 900–1000°C. In some cases, heating to 600°C is sufficient to activate the grain boundaries. This enhanced grain boundary activity after a heat treatment is seen in LBIC measurements (Temofonte et al. 1981) and is also observed as a decrease of L_{eff} (Legros et al. 1980, Ho et al. 1983) and J_{sc} (Rajeswaran et al. 1982). However, not all data are in agreement. For example, Turner et al. (1982b) find a decrease of the barrier height in Silso with thermal anneal and Sullivan and Ast (1981) find an increase of the intragrain EBIC response after a low-temperature heat treatment of EFG silicon ribbons.

The enhanced electrical activity can be correlated with an increase in barrier height of the grain boundaries after heat treatment. Figure 15 gives an example of such a measurement. It shows the zero-bias resistance versus

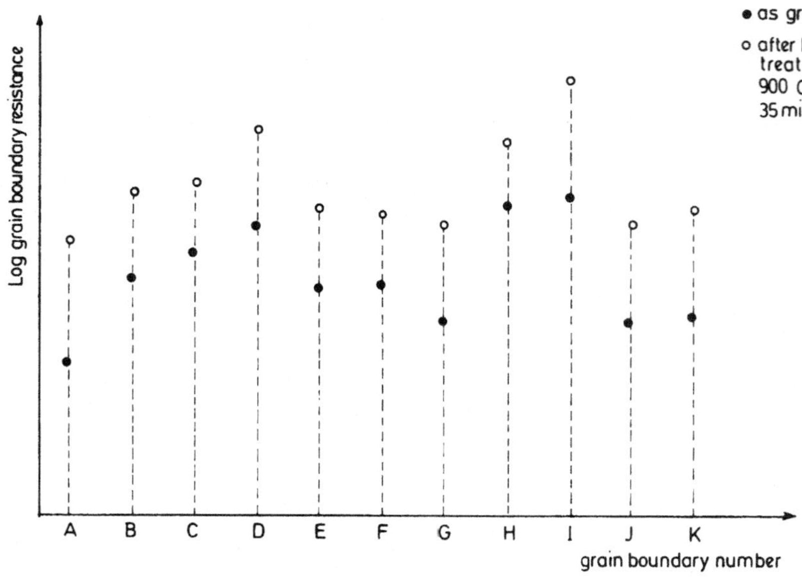

Fig. 36. Zero-bias resistance of a series of grain boundaries on the same wafer (Silso). The resistance values for the same grain boundary are given in the as-grown case (black dots) and after a heat treatment at 900°C for 30 min (open circles).

the inverse temperature for a Silso grain boundary before and after a heat treatment. An increase in the activation energy (i.e., the zero-temperature grain boundary barrier height) from 0.18 eV to 0.34 eV is found. An easier method to determine the influence of a heat treatment, is the measurement of the resistance between two points on a wafer separated by a grain boundary. Figure 36 gives the zero-bias resistance in as-grown condition and after heat treatment, for different grain boundaries on one wafer. The resistance increases here in all cases, indicating an increased barrier height.

What is the cause of this phenomenon? Redfield (1981) shows that after heat treatment and defect etching of a Silso (Wacker) slice many grain boundaries are surrounded by 'denuded bands'; these bands are free of defects, while the intragrain regions contain defects. Redfield suggests that in these bands the initially dissolved oxygen does not segregate at microdefects, like it does in the remaining part of the material, but segregates at the grain boundary. The segregation of oxygen at the grain boundary seems to enhance the barrier height and the electrical activity.

Helmreich et al. (1982) studied impurity precipitation in semicrystalline material of different oxygen and carbon content. On material of high oxygen and carbon content, they find a strong correlation between annealing time and the width of the denuded zones. An estimate of the diffusion coefficient based on these values shows that oxygen diffusion is causing the effect. Furthermore, only oxygen-rich samples show the formation of denuded zones. How the oxygen enters into the molten silicon is not clear at present.

Scanning microprobe analysis, using Auger electron and secondary ion mass spectroscopy (SIMS) (Kazmerski 1982), has detected significantly more oxygen at grain boundaries after heat treatment than at grain boundaries of the unheated material. Poon et al. (1983) measured the photoconductivity decay time of a grain boundary as a function of temperature. From these experiments they are able to derive the activation energy of grain boundary traps; they find a dominant trap of the donor type at 0.5 eV from the conduction band edge. This result is in remarkable agreement with the known energy level of oxygen in silicon, which is 0.51 eV below the conduction band edge (Chen and Milnes 1980).

Ginley (1981) shows that oxygen is able to modify the grain boundary barrier height of n-type silicon. To eliminate the effects of thermal anneal and to stabilize the grain boundary, the samples are first annealed at 600°C or 800°C. After oxygen plasma treatment (640°C, 1 h) a large increase of the barrier resistance was found.

Although these independent data give strong evidence for the important role of oxygen segregation at the grain boundaries during the heat treatment, this does not necessarily mean that oxygen is the cause of the grain boundary electrical activity. Also on FZ polysilicon, which is known to have a low oxygen content, a change of grain boundary characteristics has been observed (Young et al. 1982a). It is possible that the as-grown material is highly stressed and that the enhanced electrical activity is caused by stress relief at the grain boundaries.

5.1.2. Cold processing of solar cells

Since grain boundaries are more active after heat treatment, a cell fabrication process, which avoids high-temperature steps, may yield more efficient cells. Several low-temperature processes have been proposed: laser recrystallization, aluminum diffusion and MIS cell processing.

Young et al. (1982b) found that a very shallow (5 keV) ion implantation, followed by laser recrystallization yields higher solar cell efficiencies on semicrystalline material than a comparable furnace diffusion. During their experiments they found that the pulsed XeCl excimer laser gave the best results, due to its good uniformity, its very shallow melting depth, its high pulse repetition rate, and the low point-defect generation.

Turner et al. (1982a,b) proposed the gas immersion laser diffusion technique. In this method, a laser is used to decompose a dopant gas and to melt the silicon surface at the same time. This dopant enters the molten silicon and it is rapidly mixed in it. When the laser beam is cut off, the silicon solidifies and a junction is formed. This technique yields short-circuit currents about 10% higher than in the corresponding heat treated cells.

Wu and Yang (1981) and Tsaur et al. (1981) formed a junction in n-type silicon at a temperature of 400–500°C with aluminum. Such a structure is also possible with the screen printing technology (Cheek 1984). The only limitation is the use of n-type silicon, while the common used material for solar cells is p-type.

Also the MIS process is a candidate for the low-temperature processing of semicrystalline silicon (Van Halen 1981). On p-type Silso (Wacker) Schorner and Hezel (1981) obtained 12.3% AM1 efficiency using plasma enhanced CVD silicon nitride for creating the inversion layer.

On laser-recrystallized silicon, Turner et al. (1982a) found the surprising result that many grain boundaries have an increased quantum efficiency as compared to the bulk, especially at long wavelengths. This cannot be explained by an increased diffusion along the grain boundaries because the substrate remains completely cold during the pulsed laser recrystallization.

One explanation is that, during the slow cooling of the silicon after crystallization, impurities have diffused preferentially to the grain boundaries, leaving a small but better quality zone around the grain boundary. Another explanation is that the absorption coefficient is increased close to the grain boundaries. It is known that as-grown silicon is highly stressed. Even in as-grown CZ silicon the absorption coefficient is larger than after thermal anneal (Annual Book of ASTM Standards 1976). On Silso silicon in the vicinity of a grain boundary, the absorption coefficients for thermally stressed CZ silicon give a better fit to the spectral response curves than the absorption coefficients for thermally annealed CZ grown silicon (Oualid et al. 1982b). In the intragrain material, however, the absorption coefficient for thermally annealed CZ grown silicon gives the best fit. The presence of increased absorption can be proven by comparison of LBIC with EBIC measurements on a grain boundary showing an increased quantum efficiency.

Another interesting point is the observation of Young et al. (1982a) that an incoherent twin boundary can be converted into a coherent twin boundary by laser-induced melting. This suggests that the boundary tends to grow in the direction of the lowest energy, i.e., the coherent boundary.

5.2. Hydrogen passivation

Grain boundaries and intragrain defects limit the performance of semicrystalline silicon solar cells. To reduce their effect, the defect associated trapping states must be made electrically inactive. It is generally accepted that molecular hydrogen treatments do not influence the properties of the bulk defects in silicon. On the other hand, it is well established that the diffusion of atomic hydrogen into the silicon grain boundaries lowers the potential barriers.

Figure 37a shows the zero-bias resistance versus inverse temperature for two isolated Silso grain boundaries, in the as-grown condition. A zero-temperature barrier height of 313 mV can be derived for grain boundary A1, and 83 mV for grain boundary B1. Figure 37b shows the measurements after a hydrogen plasma passivation (400°C during 4 h). A much lower sample resistance is observed. Also the slope of the curve has changed. In the as-grown condition, the grain boundary resistance dominates the resistance of the sample. After passivation, the grain boundary resistance is so small, that only the silicon resistance is measured. Therefore, this measurement shows a strong reduction of the grain boundary barrier height after passivation.

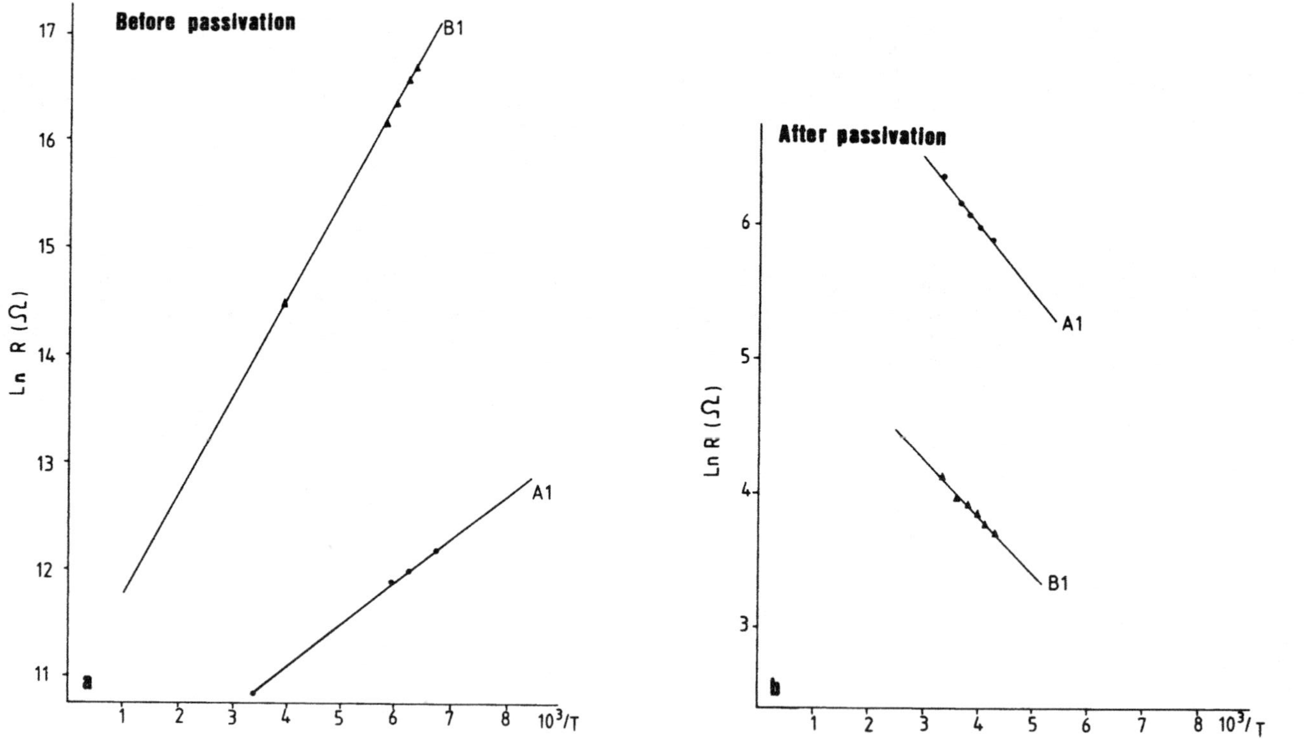

Fig. 37. Zero-bias grain boundary resistance as a function of temperature for a Silso grain boundary before (a) and after (b) hydrogen plasma passivation.

After hydrogen passivation, the EBIC signal at the grain boundary is increased (Seager et al. 1980), Donolato and Bell 1983). Also for the intragrain defects, an increased EBIC signal is found (Sullivan and Ast 1981, Schmidt et al. 1982). These observations indicate that the effective minority recombination velocity decreases after passivation. The hydrogen treatment is found to be more effective for material with intragrain defects; samples without intragrain defects have a much smaller improvement after passivation.

However, when the sample is heated above 400°C, the hydrogen escapes out of the sample, and the effect of the passivation is annihilated. For this reason the hydrogen passivation is always carried out when all high-temperature steps are finished. Since the hydrogen passivation is not effective on a solar cell with anti-reflective coating, the passivation has to be done before the deposition of the anti-reflective coating.

The apparatus normally used for hydrogenation consists of a quartz tube, surrounded by a Tesla coil. This RF glow discharge is very inefficient. To passivate most of the defects in a p–n junction requires hundreds of hours of hydrogen exposure, making it economically unattractive. Recently, Seager et al. (1982) used a Kaufman ion source and Young used a DC glow discharge apparatus. Both techniques reduce the time for passivation to only a few minutes due to the much higher hydrogen concentration in the near-surface region. Since this passivation technique is essentially a low energy (1.4 keV) implantation process, the accompanying damage and sputtering effects have to be controlled.

Table 7 summarizes the passivation results obtained on different solar cell materials. All AM1 efficiency data before and after hydrogen passivation are given for uncoated cells. To obtain the efficiency data with anti-reflective coating, these values have to be multiplied by 1.3 to 1.4 (the average reflection coefficient of silicon is about 30%).

Table 7 shows that the cells with the smallest AM1 efficiency show the largest improvement after hydrogenation. Schmidt et al. (1982) found that solar cells with initially lower photovoltaic parameters are more affected than better ones which are only slightly improved or remain unaffected. This results in a narrowing of the efficiency distribution after passivation. Schmidt et al. also found increasing relative improvements in J_{sc} and V_{oc} for increasing dopant density. As a result, the optimum doping level shifts to higher doping concentrations after passivation. Before passivation the best efficiencies are obtained on 2–5 Ωcm material and after passivation the optimum doping is found to be about 1 Ωcm. This result can be explained as follows: As was shown in section 4, J_{sc} and V_{oc} are limited by

Table 7
Hydrogen passivation results for cells without an anti-reflective coating.

Material	AM1 efficiency before passivation	AM1 efficiency after passivation	Relative change	Ref.
UMG-Si with epi-layer	6.5 %	7.3–8.0%	9 %	[1]
SOC (Honeywell)	2.7 %	6.6 %		[2]
Small grained polysilicon	2.33%	4.05%		[3]
EFG (Mobil Tyco)				[4]
poorer quality	7.0 %	8.4 %		
good quality	8.0–8.47%	9.53%		
Silso (Wacker)	7.83%*	8.5 %*	8.8%*	[5]
HEM (Crystal Systems)				[5]
with intragrain defects			9 %*	
without intragrain defects			3 %*	
Fast CZ pulled Silicon			14 %*	[5]
FZ silicon (reference)			0 %*	[5]

* These values are obtained under AM1.5 illumination.
References:
[1] Robinson and D'Aiello (1981)
[2] Ginley et al. (1981)
[3] Tsuo et al. (1982)
[4] Hanoka et al. (1983)
[5] Schmidt et al. (1982)

the bulk lifetime. When the curve bulk lifetime versus doping concentration shifts upward, always a shift to higher doping concentrations is found. In these materials the bulk lifetime is defect limited and is thus an effective lifetime τ_{eff}. From eq. (4.10) it is clear that in these materials τ_{eff} is proportional to $s_{eff}(n(W))$, which is roughly proportional to $\exp(-q\psi_B/kT)$. So the curves of fig. 3, which represent the dependence of ψ_B on the doping level for different N_{is}, present also qualitatively the dependence of $1/\tau_{eff}$ on the doping concentration in the lower doping region. At higher doping concentrations the lifetime of the crystal determines τ_{eff}. Before passivation the barrier heights are large. To get a reasonable effective lifetime, one has to take lower doping levels on the left side of the maximum. After passivation the maximum barrier height is reduced and shifts to lower doping levels. So we find now the optimum on the right side of the maximum, because here the dopant concentration still allows a reasonable lifetime.

The experiments of Schmidt et al. (1982) on low-purity silicon give higher efficiency improvements on some cells (up to 85%), while the improvement on other cells is rather poor.

As observed by Fossum and Lindholm (1980a), already a shallow passivation improves V_{oc} considerably, while J_{sc} remains unaltered. This can be explained by the fact that in many highly defected cells the open-circuit voltage is limited by recombination in the space-charge region. Only a deep passivation increases the minority carrier diffusion length in the bulk, which will result in an increased J_{sc}.

5.3. Preferential doping of grain boundaries

Smoluchowski (1952) and Queisser et al. (1961) already calculated that diffusion along a grain boundary occurs at a much higher rate than bulk diffusion. The lower activation energy for grain boundary diffusion explains this difference. Grain boundary diffusion coefficients have been measured in fine grained silicon for phosphorus and arsenic (Baumgart et al. 1982a,b), for aluminum (Hwang et al. 1980), and for boron (Horiuchi and Blanchard 1975). As an example, the diffusion coefficient for phosphorus in the bulk and along grain boundaries is plotted in fig. 38 as a function of temperature. As for single crystalline silicon, an Arrhenius behaviour is found, but the slopes for both curves are different. This means that at relatively low temperatures, the diffusion of phosphorus is three orders of magnitude larger along the grain boundaries than in single crystalline silicon. At high temperatures the difference between the two diffusion coefficients is much smaller. Furthermore, since grain boundary properties vary considerably, a large variation in the grain boundary diffusion coefficients is found. Generally, coherent twin boundaries do not show an enhanced diffusion (Faughnan et al. 1981), while very active grain boundaries show the largest diffusion coefficients (Legros et al. 1980). Using the EBIC method to detect grain boundary electrical activity (Cheng et al. 1982) a strong correlation between preferential grain boundary phosphorus diffusion and its electrical activity is found.

Preferential diffusion also occurs along intragrain defects. Lindmayer (1978) measured the junction capacitance per unit cell surface after a high-temperature diffusion and finds that the capacitance per unit area is higher for a junction made on semicrystalline silicon than for a cell made on good quality single-crystal silicon. The capacitance of a diode with grain bounda-

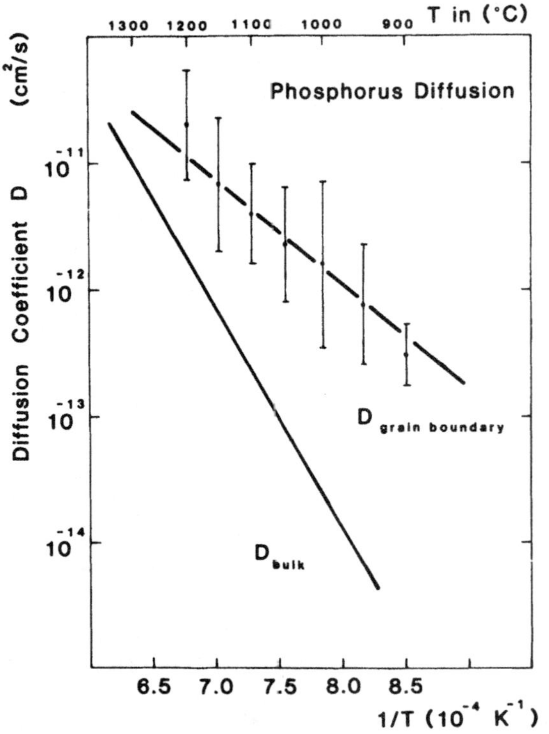

Fig. 38. Grain boundary and bulk diffusivity as a function of temperature for phosphorus (from Baumgart et al. 1982a,b).

ries and of a diode made on intragrain material was nearly identical, indicating a preferential diffusion along the intragrain defects.

One can take advantage out of this difference in grain boundary to bulk diffusion to passivate grain boundaries in a solar cell by producing, for example, a doped 10 μm long grain boundary with a junction depth in the bulk of only 0.2 μm. This can be done by a two-step diffusion as follows: The first diffusion is carried out at a lower temperature T_1 and for a long time. Since at this temperature grain boundary diffusion occurs orders of magnitude faster than bulk diffusion, the grain boundaries are preferentially doped. This is followed by a diffusion at a higher temperature T_2 which makes the front side junction. This procedure was first suggested by Distefano and Cuomo (1977). Two important benefits are expected from this preferential doping of the grain boundaries in a semiconductor:

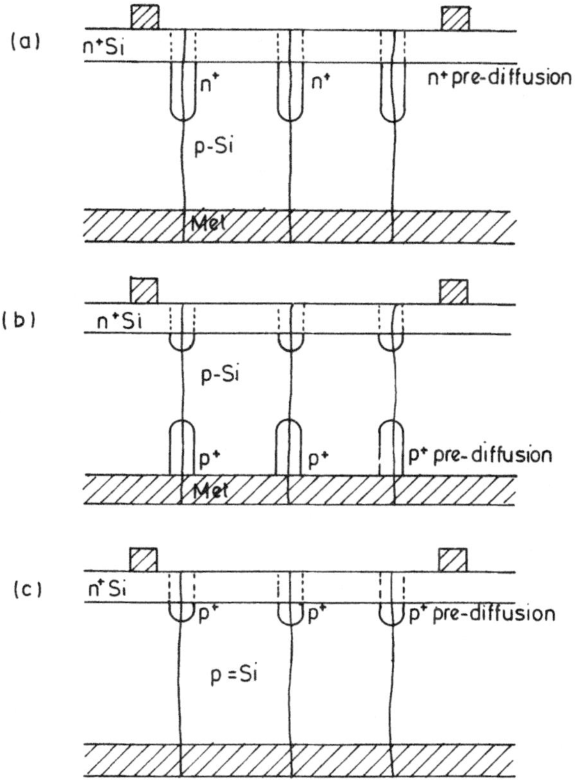

Fig. 39. Cell structures obtained by preferential diffusion of impurities along grain boundaries in p-type polycrystalline silicon: (a) the grain boundaries are collecting; (b) and (c) the grain boundaries act as a BSF.

(i) At higher doping levels ψ_B is reduced (see fig. 3). A lower barrier height means also a lower effective recombination velocity at the grain boundary depletion layer edge.

(ii) When a grain boundary is doped, it becomes part of the collecting junction, or acts like a back surface field (depending on the grain boundary doping type) (fig. 39).

Two types of polycrystalline solar cells with preferentially doped grain boundaries can be distinguished: a cell with grain boundaries converted into a junction (fig. 39a) and a cell with grain boundaries converted into a repelling field equivalent to a back surface field (BSF) at the rear side of a solar cell (fig. 39b,c).

Converting a grain boundary into a junction enhances the short-circuit current, but because of the larger junction area and space-charge recombination, the dark current increases which reduces V_{oc}. As calculated by Storti et al. (1980) and Elnahwy and Adeeb (1982), the net effect on the efficiency is always positive. The relative efficiency improvement increases with increasing penetration depth along the grain boundaries and decreasing grain size. However, for a grain size larger than 200 μm, assuming no intragrain defects, the obtainable gain is negligibly small. This may explain the very low improvements obtained on semicrystalline silicon (Silso, HEM and UCP). For grain sizes less than 10 μm, large efficiency improvements are possible (Elnahwy and Adeeb 1982). As a result, the passivation technique by preferentially doping the grain boundaries will only be cost effective on highly defected or fine grained material.

Another remark concerns the solar cell structure of fig. 39c. Since in present day semicrystalline cells the base diffusion length is too low to make an active BSF, the BSF of this structure will also not be very effective.

For deep phosphorus and boron penetration along grain boundaries, excessively long diffusion times are required, making the process economically unattractive. It is, however, not necessary to take for the preferential grain boundary diffusion the same dopant species as for the junction. Lithium, aluminum and copper offer interesting possibilities, since they diffuse preferentially and rapidly at much lower temperatures. This possibility is at the moment not fully exploited.

In conclusion, the passivation technique by preferentially doping the grain boundaries and intragrain defects will only be cost-effective on fine grained or highly defected material, using lithium, aluminum or copper as preferential dopant species. Since fine grained silicon is, in most cases, also a thin-film polycrystalline film, one must be careful to avoid that the preferential diffusion along the grain boundaries does not short-circuit the solar cell. Although data are scarce, we will now consider the specific properties of the different impurity species in more detail.

(i) *Phosphorus diffusion*. On p-type Silso a few two-step diffusion processes using phosphorus have been optimized. Ferraris et al. (1980) find that the best results with a first phosphorus diffusion step at 750°C for 9 hours, followed by a high temperature diffusion step. Cells of 12.7% efficiency have been obtained. Finetti et al. (1980) report efficiency improvements from 10.5–11.5%, by using a one-step phosphorus diffusion process (920°C, 5 min) to 11–12% by using a two-step diffusion process (750°C, 135 min; 920°C, 5 min).

(ii) *Boron diffusion.* Horiuchi and Blanchard (1975) reported the preferential diffusion of boron along grain boundaries. However, solar cell passivation results remain scarce. Neugroschel and Mazer (1982) reported difficulties for the preferential diffusion of boron along grain boundaries. Orr et al. (1980) diffused boron and phosphorus at low temperature (500°C, 12 hrs) in semicrystalline wafers. They found that a phosphorus pre-diffusion yields 9% smaller effective diffusion lengths, while a boron pre-diffusion yields 21% higher effective diffusion lengths than conventionally phosphorus diffused wafers.

(iii) *Aluminum diffusion.* Aluminum is a shallow acceptor and diffuses rapidly along silicon grain boundaries, making it suitable for diffusion at low temperatures (300–400°C). Furthermore, Al has an eutectic point with silicon at 577°C. The liquid Al–Si alloy, which is formed above this temperature, has interesting properties.

Gay (1978), Arienzo et al. (1980), Orr and Arienzo (1982), Krishna (1982) and Sundaresan et al. (1982) report that the Al–Si alloy seems to act as a sink for metallic impurities (copper, gold and iron) present in the silicon substrate. This gives rise to an increased bulk lifetime and a better long-wavelength quantum efficiency. Such diffusion is probably enhanced by the presence of dislocation pipes and grain boundaries which can act as high diffusivity paths for atomic migration. When Al-alloying is applied at the backside of a p-type wafer, the solar cell structure of fig. 39b is obtained. In this way, this method may allow to produce an effective BSF in defected silicon substrates.

Aluminum can also be diffused from the front side of a p-type wafer before junction formation. In this way the structure of fig. 39c is obtained. To avoid junction shorting, the Al diffusion has to be done at temperatures lower than the Al–Si eutectic temperature. EBIC measurements of Sundaresan et al. (1982) on this type of solar cell reveal a reduced s_{eff} at the grain boundaries close to the junction.

(iv) *Lithium diffusion.* Lithium is a shallow, fast diffusing, interstitial donor in silicon (Crouch and Gilmer 1969). Several lithium doping techniques are known already for a long time and were developed for space solar cells to increase their radiation hardness (Wysocki et al. 1966). Li-doped cells can recover spontaneously at room temperature from radiation damage. Li doping can be achieved in several ways: deposition from molten $LiNO_3$ or LiCl/KCl baths at a few hundred °C, lithium metal evaporation or ion implantation. As found by Miller and Orr (1980), Li doping lowers the grain boundary barrier height and increases the effective lifetime of fine grained n-type polycrystalline material. LBIC measurements on large grain n-type poly-Si (Young et al. 1981) reveal that after lithium diffusion, recombination at the grain boundaries is drastically reduced. By using low-temperature Li diffusion together with boron implantation and pulsed laser annealing, Young et al. (1982a) were able to make a solar cell with 2.66% efficiency on fine grained (10 μm) silicon. With phosphorus doping, cell efficiencies of 1.76% were obtained. When hydrogen passivation was carried out after lithium passivation, efficiencies of 4.05% were reported (Young et al. 1982b). These results indicate that Li is an interesting passivation species for n-type silicon. Also on p-type material interesting results were reported. Kai et al. (1982) used low-temperature lithium diffusion combined with low-temperature MIS processing on Silso silicon. Efficiencies from 5.05% without lithium to 10.9% with lithium diffusion were measured.

However, Miller and Orr (1980) reported stability problems. They find that the resistivity of lithium-doped polycrystalline silicon increases monotonically over a period of several weeks.

(v) *Copper diffusion*. A few years ago a passivation process by copper diffusion has been proposed (Koliwad et al. 1980). The copper is introduced into the silicon by thermal diffusion or by melt doping before crystal growth. The use of copper is surprising because in single crystalline silicon it is a harmful element (Davis et al. 1980). However, in fine grained silicon copper seems to improve considerably all solar cell parameters. For small grain sizes (0.1 mm) a relative increase of solar cell efficiency of 14.9% after copper diffusion is found.

Zehaf et al. (1983) found that the best results are obtained when copper is diffused at very low temperature (400°C); in order that the atomic diffusion occur via the crystal defects and that the copper concentration remain low. Zehaf et al. found that copper passivates drastically the intragrain defects, especially the dislocations.

Maybe copper segregates at the crystal defects, thereby saturating the dangling bonds and reducing the minority carrier recombination. In the case of large grain size solar cells, on the other hand, copper will precipitate also in the intragrain regions, causing the observed efficiency degradation.

5.4. *The quasi grain boundary free cell structure*

A grain boundary intersecting the junction space-charge region, enhances the space-charge recombination and therefore decreases the solar cell efficiency. Gonzales and Neugroschel (1981, 1982) proposed a new structure which avoids these intersections. A diffusion mask is formed, covering the grain boundaries with a thick oxide layer (fig. 40). The grain boundaries are first preferentially etched with a $HF-HNO_3-H_2O$ mixture. On the surface an oxide layer is deposited, followed by a layer of positive photoresist. The entire wafer is illuminated and the illuminated photoresist is removed. The photoresist above the grain boundaries remains, due to the larger thickness. The oxide layer not covered by the photoresist is then removed, while it remains above the grain boundaries. In the subsequent diffusion step, the junction is only formed in the intragrain material. Important is that the oxide layer remains when the metal contacts are deposited. On small grained silicon one has to cover the cell with a transparant and conducting ITO-layer or a polysilicon layer to interconnect the isolated areas.

Gonzales and Neugroschel also found that this method gives no improvement for intragrain defects. As a result, the method is expected to improve

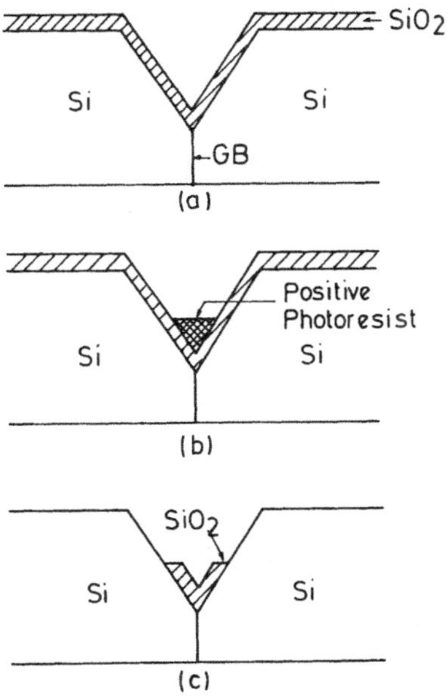

Fig. 40. Basic processing steps in the fabrication of a quasi-grain-boundary-free cell structure. (a) After preferential grain boundary etching and deposition of an oxide layer. (b) after the photoresist step. (c) after removal of the oxide, not covered by the photoresist, and after removal of the photoresist. The grain boundaries are covered by an oxide layer and the diffusion mask is ready.

the efficiency of fine grained silicon, whereas on coarse grained silicon (e.g., Silso, HEM, Semix) nearly no improvement is expected due to the large grain size.

Appendix. Additional mechanisms contributing to an increased second exponential

In a diode containing grain boundaries and intragrain defects, I_{S2} can be stronger than indicated by eq. (4.7). Three phenomena may result in an increased dark current: band-gap narrowing, reduced built-in potential, and conduction along the defects. Since these recombinations occur in the junction space-charge region, these processes are also characterized by a $2kT$ dependence.

(i) *Bandgap narrowing.* Most defect models assume distorted bonds. Distorted bonds in the defect core shift the conduction band downwards and the valence band upwards, which results in a band-gap narrowing (see, for example, Marklund 1978). This gap narrowing results in a larger value of the intrinsic carrier concentration n_i at the defect, and therefore in larger dark currents.

(ii) *Reduced built-in potential,* V_{bi}. Defects are surrounded by a depletion layer in p-type as well as in n-type material. This is shown in fig. A1, which gives a two-dimensional diagram of the valence band, for a grain boundary or dislocation intersecting the junction. As is seen, the depletion

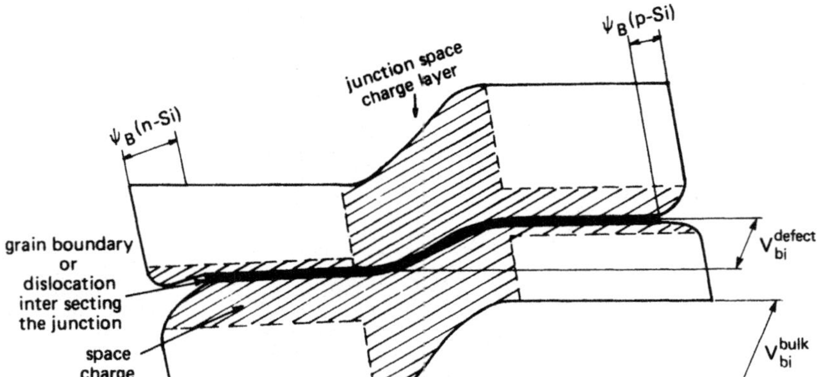

Fig. A1. Two-dimensional valence band diagram of a grain boundary, or dislocation, intersecting a junction. We clearly see that the built-in potential at the defect, V_{bi}^{defect}, is reduced in comparison with the built-in potential of the bulk, V_{bi}^{bulk}.

layer around grain boundaries and dislocations results in a reduced built-in potential along the defect.

(iii) *Conduction along the defects.* Studies of 'hopping conductivity' of polycrystalline silicon at low temperatures (Taniguchi et al. 1980) present evidence that silicon grain boundaries have a significant sheet conductivity, estimated to be between 10^{-5} and 10^{-11} mhos/square. Morrison (1982) shows theoretically that these shunt currents can attain significant importance in silicon solar cells. Once a majority carrier is captured at the core of a defect (which is a thermally activated process), it can arrive in the junction region by hopping from defect state to defect state along the grain boundary. The probability of capturing a majority carrier is high, because of the relatively low barrier around the defect. Three different space-charge recombination processes are indicated in fig. A2 (Fraas 1980).

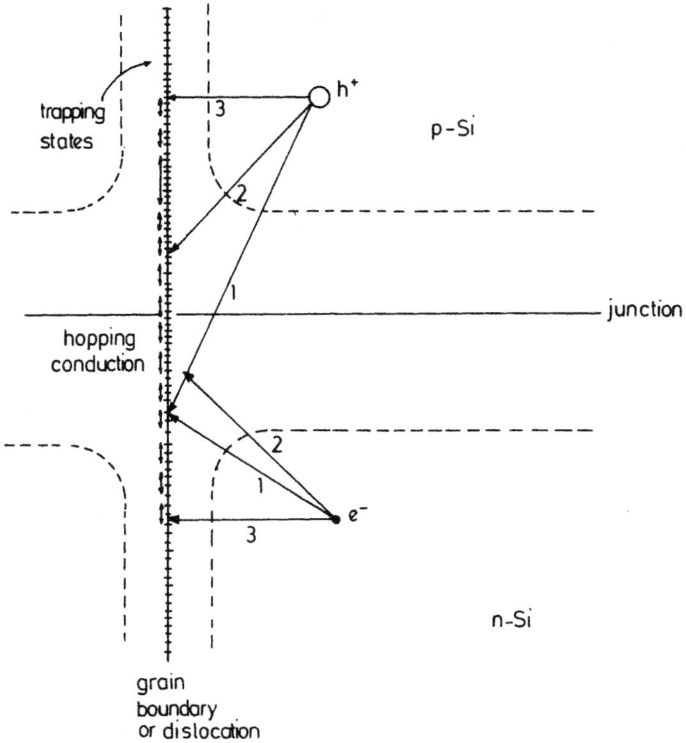

Fig. A2. Enhanced recombination in the space-charge region, due to the intersection of the junction with a grain boundary or dislocation.

The defect core itself can be considered as a diode in parallel with the diode in the bulk grain. The *I–V* characteristics of a polycrystalline p–n junction are then modeled in terms of parallel bulk and defect diodes with different areas, intrinsic carrier concentrations, mobilities, and built-in potentials. It is clear that, although the area of the bulk diode is usually large in comparison to the area of the defect diode, this factor can be compensated by the larger value of the intrinsic carrier concentration and the smaller value of the built-in potential at the defect.

References

Arienzo, M., A. Muleo and W.A. Orr, 1980. Proc. 3rd EC Photovoltaic Spec. Conf., October 1980, Cannes (Reidel, Dordrecht) p. 650.

Baccarani, G., B. Ricco and G. Spadini, 1978, J. Appl. Phys. **49**, 5565.

Baumgart, H., H.J. Leamy, L.E. Trimble, C.J. Doherty and G.K. Celler, 1982a, Proc. Materials Res. Soc. Ann. Meeting, Boston November 1981, eds H.J. Leamy, G.E. Pike and C.H. Seager (North-Holland, Amsterdam) p. 311.

Baumgart, H., H.J. Leamy, G.K. Celler and L.E. Trimble, 1982b, J. Phys. (France) **43**, C1-363.

Burk, D.E., 1982, IEEE Trans. Electron Devices **ED-29**, 1887.

Burk, D.E., S. Kanner, J.E. Muyshondt, D.S. Shaulis and P.E. Russell, 1983, J. Appl. Phys. **54**, 171.

Card, H.C., and W. Hwang, 1980, IEEE Trans. Electron Devices **ED-27**, 700.

Cheek, G., 1984, Ph.D. Thesis, ESAT Laboratory, University of Leuven, Belgium.

Cheek, G., R. Janssens, M. Leempoels, R. Mertens and R. Van Overstraeten, 1982, Proc. 4th EC Photovoltaic Solar Energy Conf., May 1982, Stresa (Reidel, Dordrecht) p. 926.

Cheek, G., R. Mertens, R. Van Overstraeten and L. Frisson, 1984, IEEE Trans. Electron Devices **ED-31**, 602.

Chen, J.W., and A.G. Milnes, 1980, Ann. Rev. Mater. Sci. **10**, 157.

Cheng, L.J., and C.M. Shyu, 1981a, Semiconductor Silicon – 1981, eds H.R. Huff, R.J. Kriegler and Y. Taklishi (Electrochemical Society, Pennington, NJ) p. 390.

Cheng, L.J., and C.M. Shyu, 1981b, Proc. Materials Res. Soc. Ann. Meeting, Boston, November 1981, eds H.J. Leamy, G.E. Pike and C.H. Seager (North-Holland, Amsterdam) p. 105.

Cheng, L.J., C.M. Shyu, K.M. Stika, T. Daud and G.T. Crotty, 1982, Proc. 16th IEEE Photovoltaic Spec. Conf., September 1982, San Diego (IEEE, New York) p. 554.

Ciszek, T.F., 1982, Proc. 16th IEEE Photovoltaic Spec. Conf., September 1982, San Diego (IEEE, New York) p. 316.

Crouch, R.K., and T.E. Gilmer, 1969, J. Phys. & Chem. Solids **30**, 2037.

Crowell, C.R., and S.M. Sze, 1966, Solid State Electron. **9**, 1035.

Culik, J.S., P. Alexander, K.A. Dumas and J.W. Wohlgemuth, 1983, Proc. 5th EC Photovoltaic Solar Energy Conf., October 1983, Athens (Reidel, Dordrecht).

Cunningham, E., H. Strunk and D.G. Ast, 1982, J. Electrochem. Soc. **129**, 1128.

Daud, T., K.A. Dumas, G.H. Schwuttke, P. Smetana and K.M. Kim, 1982, Proc. 16th IEEE Photovoltaic Spec. Conf., September 1982, San Diego (IEEE, New York) p. 63.

Davis Jr, J.R., A. Rohatgi, R.H. Hopkinss, P.D. Blais, P. Rai-Choudhury, J.R. McCormick

and H.C. Mollenkopf, 1980, IEEE Trans. Electron Devices **ED-27**, 677.
De Graaff, H.C., M. Huybers and J.G. De Groot, 1982, Solid State Electron. **25**, 67.
De Pauw, P., 1984, Ph.D. Thesis, ESAT Laboratory, University of Leuven, Belgium.
De Pauw, P., R. Mertens, R. Van Overstraeten and S.C. Jain, 1984a, Solid State Electron. **27**, 573.
De Pauw, P., R. Mertens and R. Van Overstraeten, 1984b, IEEE Electron Devices Lett. **EDL-5**, 234.
Dietl, J., D. Helmreich and E. Sirtl, 1981, Crystals: Growth, Properties and Applications 5 (Springer, Berlin) p. 43.
Distefano, T.H., and J.J. Cuomo, 1977, Appl. Phys. Lett. **30**, 351.
Donolato, C., 1978, Optik **52**, 19.
Donolato, C., 1983, J. Appl. Phys. **54**, 1314.
Donolato, C., and R.O. Bell, 1983, Rev. Sci. Instrum. **54**, 1005.
Elnahwy, S., and N. Adeeb, 1982, Solid State Electron. **25**, 1111.
Everhart, T.E., and P.H. Hoff, 1971, J. Appl. Phys. **42**, 5837.
Fabre, E., and M. Mautref, 1975, Acta Electron. **18**, 331.
Fabre, E., M. Mautref and A. Mircea, 1975, J. Appl. Phys. Lett. **27**, 237.
Faughnan, B.W., W. Phillips and D. Redfield, 1981, SERI Rep. Contract No XS9-8276-1.
Feldman, C., F.G. Satkiewicz and N.A. Blum, 1982, Thin Solid Films **90**, 461.
Ferraris, F., F.C. Matacotta, S. Pidatella and L. Sardi, 1980, Proc. 3rd EC Photovoltaic Solar Energy Conf., October 1980, Cannes (Reidel, Dordrecht) p. 625.
Finetti, M., P. Ostoja, S. Solmi and Q. Zini, 1980, Sol. Cells **2**, 101.
Fossum, J.G., and F.A. Lindholm, 1980a, IEEE Electron Device Lett. **EDL-1**, 267.
Fossum, J.G., and F.A. Lindholm, 1980b, IEEE Trans Electron Devices **ED-27**, 692.
Fossum, J.G., and M.A. Shibib, 1980, IEEE Electron Device Meeting Tech. Digest (IEEE, New York) p. 280.
Fraas, L.M., 1980, Proc. 14th IEEE Photovoltaic Spec. Conf., January 1980, San Diego (IEEE, New York) p. 1316.
Frisson, L., P. Lauwers, P. Bulteel, L. De Smet, R. Mertens, R. Govaerts and R. Van Overstraeten, 1978, Proc. 13th IEEE Photovoltaic Spec. Conf., June 1978, Washington DC (IEEE, New York) p. 530.
Gay, C.F., 1978, Proc. 13th IEEE Photovoltaic Spec. Conf., June 1978, Washington DC (IEEE, New York) p. 444.
Ghosh, A.K., C. Fishman and T. Feng, 1980, J. Appl. Phys. **51**, 446.
Ginley, D.S., 1981, Appl. Phys. Lett. **39**, 624.
Ginley, D.S., D. Haaland and C.H. Seager, 1981, Electrochemical Society Meeting, October 11–16, Denver CO (Electrochemical Society, Princeton, NJ).
Gonzales, F., and A. Neugroschel, 1981, IEEE Electron Device Lett. **EDL-2**, 141.
Gonzales, F., and A. Neugroschel, 1982, IEEE Electron Device Lett. **EDL-3**, 134.
Green, M.A., 1983, Sol. Cells **7**, 337.
Grove, A.S., and Fitzgerald, 1966, Solid State Electron. **9**, 183.
Gummel, H.K., and H.C. Poon, 1970, Bell Syst. Tech. J. **49**, 827.
Hanoka, J.I., 1979/80, Sol. Cells. **1**, 123.
Hanoka, J.I., C.H. Seager, D.J. Sharp and J.K.G. Panitz, 1983, Appl. Phys. Lett. **42**, 618.
Hari Rao, C.V., R.O. Bell, M.C. Cretella, J.C. Ho, F.V. Wald and K.V. Ravi, 1978, Proc. 13th Photovoltaic Spec. Conf., June 1978, Washington DC (IEEE, New York) p. 1213.
Helmreich, D., H. Seiter, D. Huber and R. Wahlich, 1982, Proc. 16th IEEE Photovoltaic Spec. Conf., September 1982, San Diego (IEEE, New York) p. 405.

Henry, C.H., R.A. Logan and F.R. Merrit, 1978, J. Appl. Phys. **49**, 3530.
Hirae, S., M. Hirose and Y. Osaka, 1980, J. Appl. Phys. **51**, 1043.
Ho, C.T., and J.D. Mathias, 1981, Solid State Electron. **24**, 115.
Ho, C.T., R.O. Bell and F.V. Wald, 1977, Appl. Phys. Lett. **31**, 463.
Ho, C.T., G. Moeller and J.D. Mathias, 1983, Solid State Electron. **26**, 247.
Horiuchi, S., and R. Blanchard, 1975, Solid State Electron. **18**, 529.
Hovel, H.J., 1975, Solar Cells, Semiconductors and Semimetals, Vol. 11 (Academic Press, New York) p. 15.
Huber, D., R. Wahlich and D. Helmreich, 1980, Proc. 14th Photovoltaic Spec. Conf., January 1980, San Diego (IEEE, New York) p. 316.
Hwang, J.C.M., P.S. Ho, J.E. Lewis and D.R. Campell, 1980, J. Appl. Phys. **51**, 1576.
Hyland, S., P. Iles, D. Leung, G. Schwuttke and J.A.A. Engelbrecht, 1982, Proc. 16th IEEE Phtovoltaic Spec. Conf., September 1982, San Diego (IEEE, New York) p. 68.
Indenbom, V.L., 1979, Krist. & Tech. **14**, 493.
Inoue, N., C.W. Wilmsen and K.A. Jones, 1981, Sol. Cells **3**, 35.
Jain, S.C., R. Janssens, G. Cheek, P. De Pauw, R. Mertens and R. Van Overstraeten, 1982, Proc. 4th EC Photovoltaic Solar Energy Conf., May 1982, Stresa (Reidel, Dordrecht) p. 432.
Jain, S.C., R. Janssens, G. Cheek, P. De Pauw, R. Mertens and R. Van Overstraeten, 1983, Sol. Cells **9**, 345.
Johnson, S.M., and J.S. Culik, 1982, Proc. 16th IEEE Photovoltaic Spec. Conf. (IEEE, New York) p. 548.
Johnson, S.M., R.W. Armstrong, R.G. Rosemeier, G.M. Storti, H.C. Lin and W.F. Regnault, 1982, Proc. Materials Res. Soc. Ann. Meeting, Boston, November 1981, eds H.J. Leamy, G.E. Pike and C.H. Seager (North-Holland, Amsterdam) p. 179.
Kai, F., G. Rajeswaran, M.A. Jackson, V.J. Rao and W.A. Anderson, 1982, Proc. 16th IEEE Photovoltaic Spec. Conf., September 1982, San Diego (IEEE, New York) p. 543.
Kazmerski, L.L., 1982, J. Vac. Sci. & Technol. **20**, 423.
Kim, D., A. Khondker, S. Ahmed and R. Shah, 1984, IEEE Trans. Electron Devices **ED-31**, 480.
Kim, D.M., A.N. Khondker, R.R. Shah and D.L. Crosthwait, 1982, IEEE Electron Device Lett. **EDL-3**, 141.
Koliwad, K.M., T.Daud and J.K. Liu, 1980, Proc. 3rd EC Photovoltaic Solar Energy Conf., October 1980, Cannes (Reidel, Dordrecht) p. 710.
Krishna, S., 1982, IEEE Trans. Electron Devices **ED-29**, 430.
Legros, B., G.R. David, M.C. Boissy, J. Lebailly, T.P. Ngo and E. Fabre, 1980, Proc. 3rd EC Photovoltaic Solar Energy Conf., October 1980, Cannes (Reidel, Dordrecht) p. 593.
Lindholm, F.A., and J.G. Fossum, 1981, Proc. 5th Potovoltaic Spec. Conf., May 1981, Kissimee, FL (IEEE, New York) p. 422.
Lindmayer, J., 1978, Proc. 13th IEEE Photovoltaic Spec. Conf., June 1978, Washington, DC (IEEE, New York) p. 1096.
Lu, N.C.-C., L. Gerzberg and J.D. Meindl, 1980, IEEE Electron Device Lett. **EDL-1**, 38.
Lu, N.C.-C., L. Gerzberg, C.-Y. Lu and J.D. Meindl, 1981a, IEEE Trans. Electron Devices **ED-28**, 818.
Lu, N.C.-C., L. Gerzberg, C.-Y. Lu and J.D. Meindl, 1981b, IEEE Electron Device Lett. **EDL-2**, 95.
Makram-Ebeid, S., 1979, Proc. 2nd EC Photovoltaic Solar Energy Conf. Berlin 1979 (Reidel, Dordrecht) p. 792.

Mandurah, M.M., K.C. Saraswat and T.I. Kamins, 1979, J. Electrochem. Soc. **126**, 1019.

Marek, J., 1982, Proc. 16th IEEE Photovoltaic Spec. Conf., September 1982, San Diego (IEEE, New York) p. 627.

Marklund, S., 1978, Phys. Status Solidi (b) **85**, 673.

Martinez, J., A. Criado and J. Piqueras, 1981, J. Appl. Phys. **52**, 1301.

Mataré, H.F., 1971, Defect Electronics in Semiconductors (Wiley-Interscience, New York).

Mataré, H.F., and C.W. Laakso, 1969, J. Appl. Phys. **40**, 476.

McKelvey, J.P., 1966, Solid State and Semiconductor Physics (Harper & Row, New York NY) p. 146 and p. 348.

Miller, G.L., and W.A. Orr, 1980, Appl. Phys. Lett. **37**, 100.

Morrison, S.R., 1982, Proc. Materials Res. Soc. Ann. Meeting, Boston, November 1981, eds H.J. Leamy, G.E. Pike and C.H. Seager (North-Holland, Amsterdam) p. 193.

Neugroschel, A., and J.A. Mazer, 1982, IEEE Trans. Electron Devices **ED-29**, 225.

Ning, T.H., and R.D. Isaac, 1980, IEEE Trans. Electron Devices **ED-27**, 2501.

Orr, W.A., and M. Arienzo, 1982, IEEE Trans. Electron Devices **ED-9**, 1151.

Orr, W.A., G. Mabilia and V. Paraggio, 1980, Proc. 3rd EC Photovoltaic Solar Energy Conf., October 1980, Cannes (Reidel, Dordrecht) p. 646.

Oualid, J., M. Zehaf, H. Amzil, J.P. Crest, G. Mathian, J. Dugas, J. Gervais, F. Minari, B. Pichaud, S. Martinuzzi and J. Fally, 1982a, Proc. 4th EC Photovoltaic Solar Energy Conf., May 1982, Stresa (Reidel, Dordrecht) p. 421.

Oualid, J., M. Bonfils, J.P. Crest, J. Mathian, J.H. Amzil, J. Dugas, M. Zehaf and S. Martinuzzi, 1982b, Rev. Phys. Appl. (France) **17**, 119.

Panayotatos, P., and H.C. Card, 1980, IEEE Electron Device Lett. **EDL-1**, 263.

Panayotatos, P., E.S. Yang and W. Hwang, 1982, Solid State Electron. **25**, 417.

Pike, G.E., and C.H. Seager, 1979, J. Appl. Phys. **50**, 3414.

Pogany, A., 1980, Proc. 14th Photovoltaic Spec. Conf., January 1980, San Diego (IEEE, New York) p. 410.

Poon, E., E.S. Yang, H.L. Evans, W. Hwang and R.M. Osgood Jr, 1983, Appl. Phys. Lett. **42**, 285.

Queisser, W.J., K. Hubner and W. Schockley, 1961, Phys. Rev. **123**, 1245.

Rajeswaran, G., M. Thayer, V.J. Rao and W.A. Anderson, 1982, Proc. Materials Res. Soc. Ann. Meeting, Boston, November 1981, eds H.J. Leamy, G.E. Pike and C.H. Seager (North-Holland, Amsterdam) p. 235.

Rao, V.J., W.A. Anderson, G. Rajeswaran and M. Thayer, 1981, Phys. Status Solidi (a) **67**, 709.

Redfield, D., 1981, Proc. 15th IEEE Photovoltaic Spec. Conf., May 1981, Kissimee, FL (IEEE, New York) p. 1179.

Robinson, P.H., and R.V. D'Aiello, 1981, Appl. Phys. Lett. **39**, 63.

Russell, P.E., C.R. Herrington, D.E. Burk and P.H. Holloway, 1982, Proc. Mater. Res. Soc. Ann. Meeting, Boston, November 1981, eds H.J. Leamy, G.E. Pike and C.H. Seager (North-Holland, Amsterdam) p. 185.

Sah, C.T., R.N. Noyce and W. Shockley, 1957, Proc. IRE **45**, 1228.

Schmidt, W., K.-D. Rasch and K. Roy, 1982, Proc. 16th IEEE Photovoltaic Spec. Conf., September 1982, San Diego (IEEE,New York) p. 537.

Scholl, F.W., 1981, J. Appl. Phys. **52**, 3439.

Schorner, R., and R. Hezel, 1981, IEEE Trans. Electron Devices **ED-28**, 1466.

Seager, C.H., 1981, J. Appl. Phys. **52**, 3960.

Seager, C.H., 1982, J. Appl Phys. **53**, 5968.

Seager, C.H., and T.G. Castner, 1978, J. Appl. Phys. **49**, 3879.
Seager, C.H., and D.S. Ginley, 1981, J. Appl. Phys. **52**, 1050.
Seager, C.H., and G.E. Pike, 1979, Appl. Phys. Lett. **35**, 709; Erratum: Appl. Phys. Lett. **37**, 251 (1980).
Seager, C.H., G.E. Pike and D.S. Ginley, 1979, Phys. Rev. Lett. **43**, 532. Errata: Phys. Rev. Lett. **44**, 1365 (1980).
Seager, C.H., D.S. Ginley and J.D. Zook, 1980, Appl. Phys. Lett. **36**, 831.
Seager, C.H., D.J. Sharp, J.K.G. Panitz and R.V. D'Aiello, 1982, J. Vac. Sci & Technol. **119**, 948.
Secco d'Aragona, F., 1972, J. Electrochem. Soc. **119**, 948.
Seto, J.Y., 1975, J. Appl. Phys. **46**, 5247.
Shimokawa, R., and Y. Hayashi, 1982, Jpn. J. Appl. Phys. **21**, L558.
Shimokawa, R., and Y. Hayashi, 1983, IEEE Trans. Electron Devices **ED-30**, 1770.
Shockley, W., and T. Read Jr, 1952, Phys. Rev. **87**, 835.
Shyu, C.M., and L.J. Cheng, 1982, Proc. Materials Res. Soc. Ann. Meeting, Boston 1981, eds H.J. Leamy, G.E. Pike and C.H. Seager (North-Holland, Amsterdam) p. 131.
Shyu, C.M., and L.J. Cheng, 1982, Proc. Materials Res. Soc. Ann. Meeting, Boston, November 1981, eds H.J. Leamy, G.E. Pike and C.H. Seager (North-Holland, Amsterdam) p. 131.
Smoluchowski, R., 1952, Phys. Rev. **87**, 482.
Sopori, B.L., R.J. Ellis, R.N. Legge, K.R. Sarma, I.A. Lesk and C. Ackerman, 1982, Proc. 16th IEEE Photovoltaic Spec. Conf., September 1982, San Diego (IEEE, New York) p. 431.
Srivastava, P.C., J.C. Bourgoin, F. Rabajo and J.M. Arrayo, 1982, J. Appl. Phys. **53**, 8633.
Storti, G.M., S.M. Johnson, H.C. Lin and C.D. Wang, 1980, Proc. 14th IEEE Photovoltaic Spec. Conf., Jan. 1980, San Diego (IEEE, New York) p. 191.
Sullivan, T.D., and D.G. Ast, 1981, Proc. Materials Res. Soc. Ann. Meeting, November 1980, Boston, Defects in semiconductors, eds J. Narayan and T.Y. Tan (North-Holland, Amsterdam) p. 303.
Sundaresan, R., D.E. Burk and J.G. Fossum, 1982, Proc. 16th IEEE Photovoltaic Spec. Conf., September 1982, San Diego (IEEE, New York) p. 421.
Taniguchi, M., M. Hirose, Y. Osaka, S. Hasegawa and T. Shimuzu, 1980, Jpn. J. Appl. Phys. **19**, 665.
Taylor, W.E., N.H. Odell and H.Y. Fan, 1952, Phys. Rev. **88**, 867.
Temofonte, T.A., J.R. Szedon and W.J. Biter, 1981, Proc. 15th IEEE Photovoltaic Spec. Conf., May 1981, Kissimee, FL (IEEE, New York) p. 1304.
Texier, C., M. Mautref, S. Makram-Ebeid and C. Belouet, 1980, Proc. 3rd EC Photovotaic Solar Energy Conf., October 1980, Cannes (Reidel, Dordrecht) p. 563.
Thomson, D.J., and H.C. Card, 1983, J. Appl. Phys. **54**, 1976.
Tsaur, B.-Y., G.W. Turner and J.C.C. Fan, 1981, Appl. Phys. Lett. **39**, 749.
Tsuo, Y.S., R.W. Hardy and J.B. Milstein, 1982, Proc. 16th IEEE Photovoltaic Spec. Conf., September 1982, San Diego (IEEE, New York) p. 616.
Turner, G.B., D. Tarrant, D. Aldrich, R. Pressley and R. Press, 1982a, Proc. Materials Res. Soc. Ann. Meeting, Boston, November 1981. eds H.J. Leamy, G.E. Pike and C.H. Seager (North-Holland, Amsterdam) p. 241.
Turner, G.B., D. Tarrant, D. Aldrich, R. Pressley and R. Press, 1982b, Proc. 4th EC Photovoltaic Solar Energy Conf., May 1982, Stresa (Reidel, Dordrecht) p. 1039.
Van Halen, P., 1981, Ph.D. Thesis, ESAT Laboratory, University of Leuven, Belgium.

Van Overstraeten, R., and Van De Wiele, 1969, Introduction to Quantum Physics of Electronics, Part II (Wouters, Leuven) p. 8–40.
Von Roos, O., 1979a, Appl. Phys. Lett. **35**, 408.
Von Roos, O., 1979b, Solid State Electron. **22**, 773.
Watanabe, M., G. Actor and H.C. Gatos, 1977, IEEE Trans. Electron Devices **ED-24**, 1172.
Werner, J., W. Jantsch, K.H. Frochner and H.J. Queisser, 1982, Proc. Materials Res. Soc. Ann. Meeting, Boston, November 1981, eds H.J. Leamy, G.E. Pike and C.H. Seager (North-Holland, Amsterdam) p. 99.
Wohlgemuth, J.H., J.S. Culik and P. Alexander, 1983, Electrochemical Society, Spring Meeting, 1983 San Francisco (Electrochemical Society, Pennington, NJ) Abstract No. 219.
Wu, C.-M.W., and E.S. Yang, 1981, Appl. Phys. Lett. **38**, 813.
Wu, C.-M.W., and E.S. Yang, 1982a, Appl. Phys. Lett. **40**, 49.
Wu, C.-M.W., and E.S. Yang, 1982b, IEEE Trans. Electron Devices **ED-29**, 1598.
Wysocki, J.J., P. Rappaport, E. Davidson, R. Hand and J.J. Loferski, 1966, Appl. Phys. Lett. **9**, 44.
Yang, E.S., E.K. Poon, C.M. Wu, W. Hwang and H.C. Card, 1981, IEEE Trans. Electron Devices **ED-28**, 1131.
Yang, K., G.H. Schwuttke and T.F. Ciszek, 1980, J. Cryst. Growth **50**, 301.
Young, R.T., M.C. Lu, R.D. Westbrook and G.E. Jellison Jr, 1981, Appl. Phys. Lett. **38**, 628.
Young, R.T., J. Narayan and Y.K. Chang, 1982a, Proc. Materials Res. Soc. Ann. Meeting, Boston, November 1981, eds H.J. Leamy, G.E. Pike and C.H. Seager (North-Holland, Amsterdam) p. 111.
Young, R.T., G.A. Van Der Leeden, R.F. Wood and R.D. Westbrook, 1982b, Proc. 16th IEEE Photovoltaic Spec. Conf., Sept. 1982, San Diego (IEEE, New York) p. 427.
Zehaf, M., G. Mathian and S. Martinuzzi, 1983, Proc. 5th EC Photovoltaic Solar Energy Conf., October 1983, Athens, Greece (Reidel, Dordrecht) p. 950.
Zook, J.D., 1980a, Appl. Phys. Lett. **37**, 224.
Zook, J.D., 1980b, Proc. 3rd EC Photovoltaic Solar Energy Conf., October 1980, Cannes (Reidel, Dordrecht) p. 569.
Zook, J.D., 1983, Appl. Phys. Lett. **42**, 602.

CHAPTER 2

THE WACKER INGOT CASTING PROCESS

D. HELMREICH

Heliotronic GmbH, D-8263 Burghausen, Fed. Rep. Germany

Silicon Processing for Photovoltaics II, edited by C.P. Khattak and K.V. Ravi

Contents

1. Introduction

To meet the requirements of a future 'solar' silicon industry, the chances of unconventional ingot growth methods have to be correlated with their large-scale potential in terms of practicability, productivity, production costs and material quality. In the mid-seventies it was recognized that monocrystalline silicon may be replaced by melt-grown polycrystalline silicon. In the meantime it has been demonstrated that such material with an average grain size beyond a certain threshold value can be applied successfully for solar cells, with promising trends in terms of efficiency.

There are various approaches to the preparation of coarse-grained polycrystalline (multicrystalline) silicon sheets:

- silicon layers peeling off foreign or appropriate silicon substrates,
- self-supporting silicon slices prepared from ingots by cutting,
- self-supporting silicon sheet prepared directly by ribbon or foil growth.

Prior to the industrial application of electronic-grade silicon in the second half of the fifties, only those crystal growth methods based on directional solidification could be utilized which needed a melt container. Subsequent development of electronic-grade silicon concentrated only on CZ and FZ pulling because these two processes met best the requirements for a high-quality standard of semiconductor silicon monocrystals. Three aspects play a major role:

(1) microscopic and macroscopic crystal perfection,
(2) chemical purity,
(3) homogeneity and extent of doping.

With the beginning of terrestrial photovoltaic programs (see, e.g., Photovoltaic Program, Program Summary, January 1978 (DOE/ET–0019/1), Division of Solar Energy, US Department of Energy, Washington, DC) another aspect superseded the former points: the material from which the terrestrial solar cells are prepared has to be cheaper by about two orders of magnitude (Wolf 1975). Therefore, the Bridgman–Stockbarger technique and a growth method described by Stöber (which had already been used in the thirties with considerable success to grow large single crystals of halides for optical applications) had been reinvestigated in a simplified manner.

Different methods based on directional solidification are now in use as cost-effective alternatives to conventional Czochralski pulling. The techniques are: HEM (heat exchanger method; Khattak and Schmid 1978), UCP (ubiquitous crystallization process; Lindmayer 1976), Polyx (Fally and Guenel 1980), DS (directional solidification; Ciszek et al. 1979, Fujisaki et al. 1984), GAS (gas-assisted solidification; Hwang et al. 1984) and they are described in detail in the relevant literature [see, e.g., key references (section 8)]. A compilation of the afore-mentioned crystallization methods is given in fig. 1.

In contrast to all these procedures emerging from the Bridgman–Stockbarger technique, the Wacker ingot casting process (WICP) uses two different containers (and areas) for melting and crystallization.

WICP allows the preparation of silicon ingots from the melt to be carried out in a simpler, more flexible and more cost-effective way.

The topic of this chapter is the preparation of silicon ingots from the melt through ingot casting by aid of WICP. After a short outline of the history of this particular ingot casting process (section 2) its fundamentals will be described in more detail (section 3). The technical aspects of the individual processing steps – melting, casting and crystallization – will be

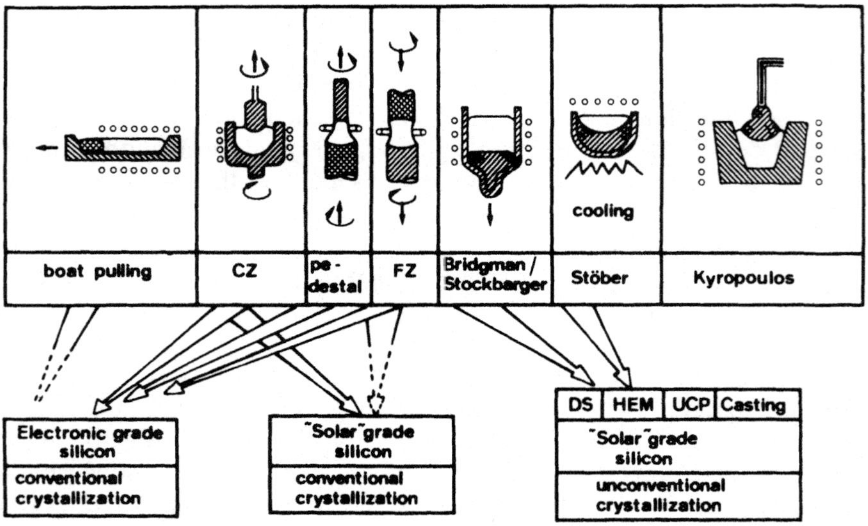

Fig. 1. Past and present crystal growth techniques applied to silicon; some of these have been abandoned, while others are still used in production of electronic-grade silicon. Conventional crystallization techniques were rather directly applied to solar-grade silicon; unconventional crystallization techniques had to be modified (Helmreich 1982a).

presented in section 4, together with a description of the WICP equipment in different stages of development. An efficient ingot technology has to be complemented by qualified cutting and wafering steps to produce low-cost silicon wafers at minimum material loss. Therefore, section 5 will be dedicated to cutting/wafering techniques which are in development at Wacker. In sections 6 and 7 characteristic data on Wacker's multicrystalline silicon (SILSO® or similar lab specimens) will be given. In the last two sections some examples of the behaviour during heat treatment (section 8) and device processing (section 9) are described.

2. Historical background of silicon casting

There are several difficulties in silicon casting which are due to the unique characteristics of silicon in the solid as well as in the molten form. One main problem is the high reactivity of the silicon melt and the resulting contamination of the silicon melt by impurities from the container. Furthermore, the solidifying melt mainly adheres to the container walls, and the ingot and/or container will crack during the cooling process due to the difference in thermal expansion coefficients. Protective layers of compacted Si_3N_4, SiO_2, Si, and graphite powder, respectively, or of a low-melting slag can be of help here. The volume increase of about 10% during solidification complicates the fabrication of flawless ingots.

Various methods of silicon ingot casting have been proposed in the past to overcome these obstacles by specific solutions. Already thirty years ago the best choice was the lining of the container with Si_3N_4 powder (see fig. 2). A second method is the pouring of silicon melt in an adequately preheated mold so that a reaction between mold wall and molten silicon is minimized and that on the other hand there is no severe chill crystalliza-

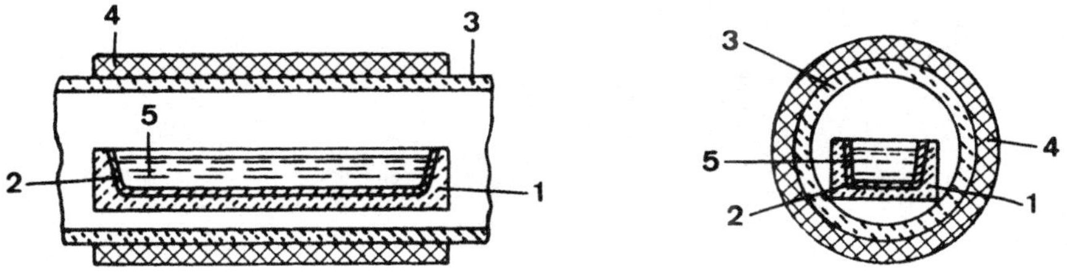

Fig. 2. Early set-up for silicon ingot casting. The melt is crystallized by directional solidification. (1) graphite crucible, (2) Si_3N_4-liner, (3) graphite heater, (4) thermal insulation, (5) silicon melt (Bradshaw and Sargeant 1955).

tion. This walking on a ridge is realized in the Wacker ingot casting process.

Activities at Wacker date back to the late sixties. At that time, polycrystalline silicon of sufficient transparency for infrared optics were fabricated by casting in form of prismas and domes (fig. 3). Some years later, first solar cells were prepared from cast silicon with good success. In 1974, the first wafers from $5\times5\times5\,cm^3$ ingots were supplied to AEG-Telefunken (Authier 1974, 1978) which had at that time already a lot of experience in manufacturing high-efficiency cells for space application. It could be shown that, based on this casting technique, a starting material can be obtained which leads to efficient solar cells (Pschunder and Bebermeier 1974, Fischer and Pschunder 1976, Fischer 1978).

Since fall 1977 the further development of the casting process has been taken care of by Heliotronic GmbH, a subsidiary of Wacker-Chemitronic in the framework of a government supported research program. In the meantime melt charges have been increased from about 300 g to more than 100 kg, and ingot dimensions have been enlarged to $43\times43\times24\,cm^3$. 20 kg ingots with $22\times22\,cm^2$ cross-section have been fabricated on a pilot-production basis at Wacker-Chemitronic for a number of years now.

Fig. 3. Photograph of an infrared-dome produced in 1974 by casting: in crude form (left) and finished by grinding (right). (Courtesy of B.A. Authier, Heliotronic GmbH).

3. Fundamentals of the Wacker ingot casting process

The preparation of silicon ingots by casting is a technique which offers the chance to perform the particular processing steps separately and individually. Especially melting with subsequent crystallization which, in the case of directional solidification (DS) techniques [HEM (Khattak and Schmid 1978), Polyx (Fally and Guenel 1980), GAS (Hwang et al. 1984), UCP (Lindmayer 1976), DS (Ciszek et al. 1979, Fujisaki et al. 1984)], are carried out in one single container, can be decoupled in the WICP and carried out in different containers and at different places as well. Along these lines a series of problems caused by the simultaneous application of the receptacle for both melting and crystallization can be avoided:

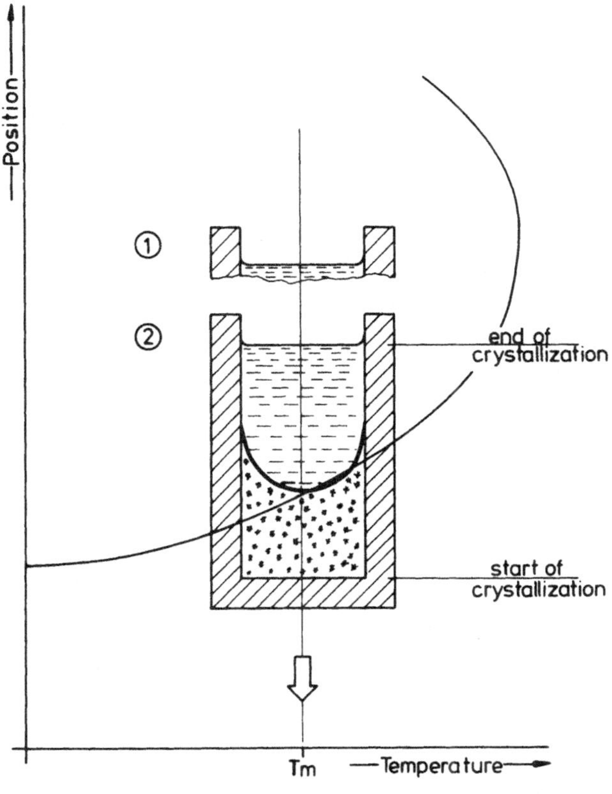

Fig. 4. Schematic graph of ingot growth by directional solidification (Bridgman–Stockbarger method). The crucible is moved in a fixed temperature field (ref. E1).

(1) due to the considerable time of residence of the melt in the crucible, especially during crystallization, contamination of the silicon melt by impurities from the container is quite severe.

(2) In the absence of particular precautions the solidifying melt adheres to the container walls.

The temperature curves given in the schematic drawings (figs. 4–7) represent the vertical temperature gradient in the middle axis of the crucible. If a (relatively) cold crucible wall is used (as is the case in figs. 4, 5 and 7) the solid–liquid interface (SLI) is concave with respect to the already crystallized portion of the silicon melt. In the case of HEM (fig. 6) a hot crucible is used, and the SLI is convex.

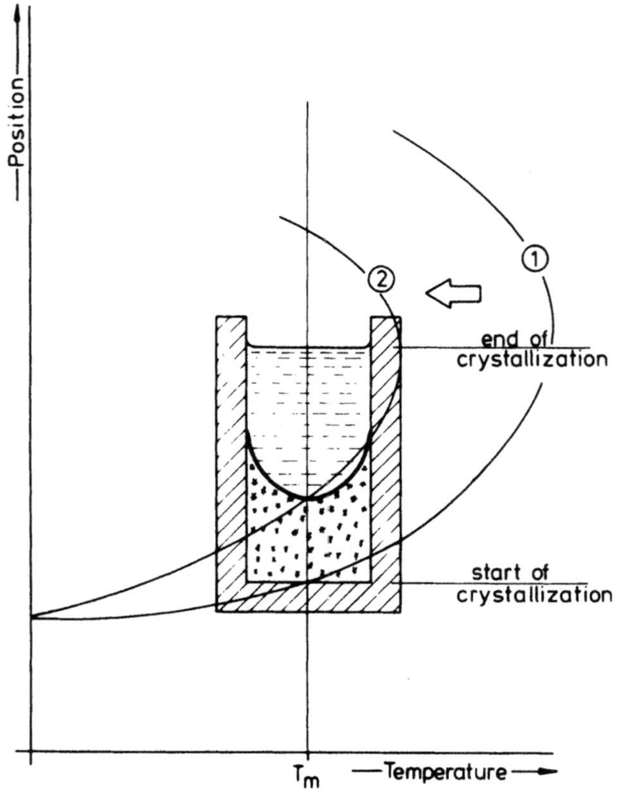

Fig. 5. Schematic graph of ingot growth by directional solidification. The temperature profile is 'moved' while the crucible is held fixed (ref. E1).

In general the form of the SLI can be modified by adequate manipulation of the horizontal temperature gradient. The form of the SLI is a parameter which influences critically the stress/strain situation in the solidified ingot as well as the concentration of crystal defects (especially dislocations). In UCP the dislocation density within individual crystal grains can be reduced dramatically if the SLI is transformed to a nearly planar one (Regnault and Yoo 1984). The characteristic feature of WICP is that crystallization takes place in a mold with a liner of crystalline silicon. This liner is formed from the Si-melt in a first step in the crystallization process through adequate preconditioning of the mold and the melt to be cast. The temperature of the mold is preset at a value which guarantees that it will not reach, or even exceed, the melting temperature of silicon if the silicon melt gets in contact with the mold. In this way the silicon melt crystallizes without contamination from the mold, which gives some freedom in the

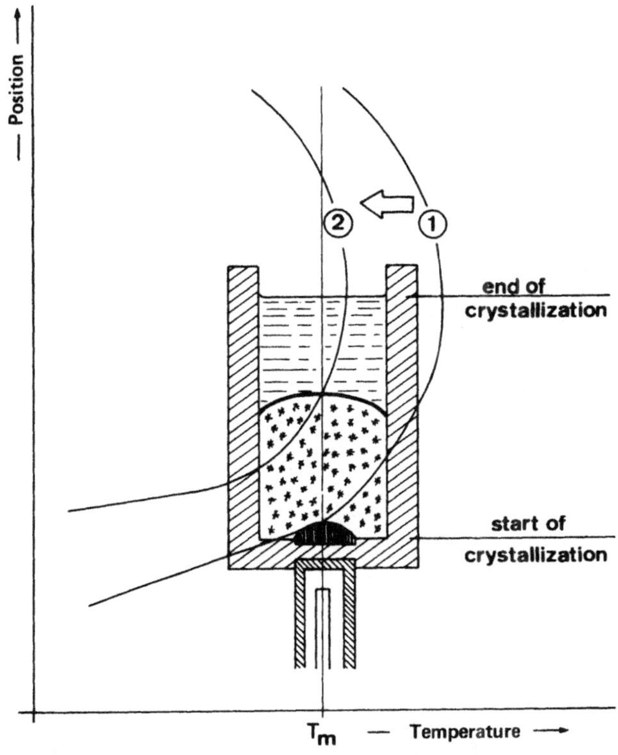

Fig. 6. Schematic graph of ingot growth by heat exchanger method (HEM) (ref. E1).

choice of the mold material. In addition a main portion of fast diffusing impurities are gettered by the chill layer which is formed by this mode of processing. As a result of low linear crystallization velocity there is a very efficient macroscopic segregational effect. Those parts of the ingot showing increased impurity levels, i.e. bottom, sides and top are cut off and discarded.

The principle set-up of the equipment for the WICP is shown in fig. 8. The crucial point is that this set-up allows the individual adjustment of temperature level and gradient in the empty mold as well as in the mold filled with crystallizing melt.

The different sizes and geometries of ingots, shown in figs. 9–11, demonstrate the flexibility of the WICP.

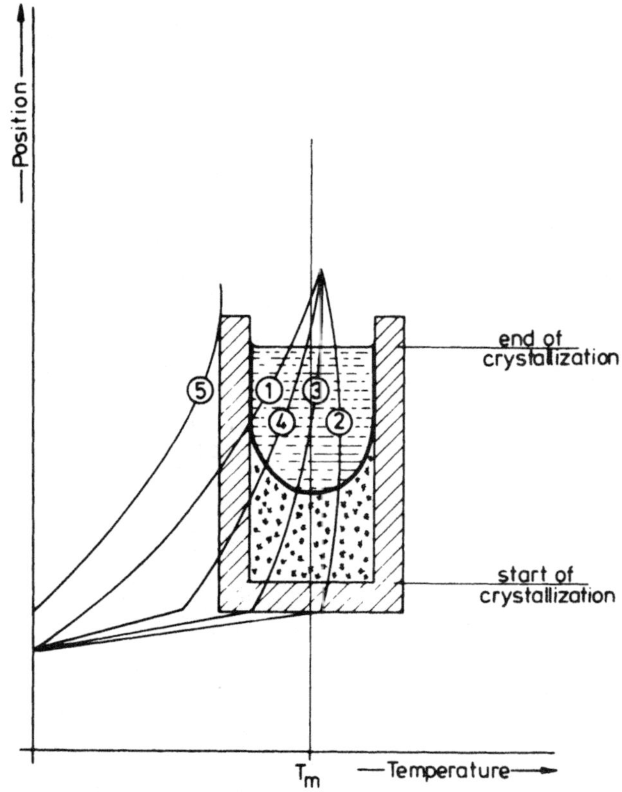

Fig. 7. Schematic graph of Wacker ingot casting process. The different stages of temperature profiles are given by curves 1–5 (ref. E1).

In order to guarantee higher throughput, however, a large cross-section should be preferred over the increase of linear growth velocity, because a growth velocity exceeding a certain threshold value is of influence on material quality through decrease of grain size, and changes in nature and concentration of crystal defects. In the meantime ingot dimensions of $43 \times 43\,cm^2$ cross-section and a height of 25 cm have been proven advantageous. This ingot size still allows further processing through conventional and/or unconventional cutting/wafering techniques (see section 5) in an easy way.

The possibility of decoupling of processing steps in the WICP offers the

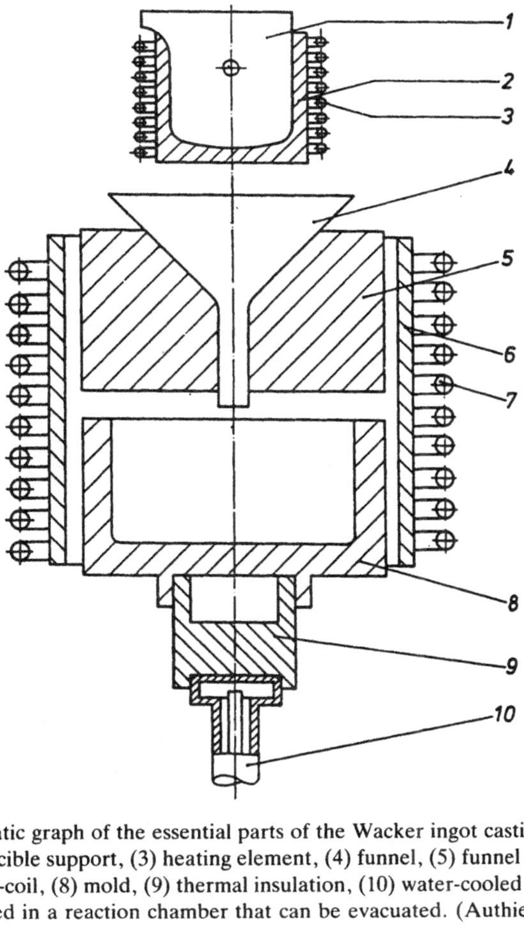

Fig. 8. Schematic graph of the essential parts of the Wacker ingot casting process. (1) melt crucible, (2) crucible support, (3) heating element, (4) funnel, (5) funnel support, (6) heating element, (7) RF-coil, (8) mold, (9) thermal insulation, (10) water-cooled rotating shaft. They are installed in a reaction chamber that can be evacuated. (Authier 1975, 1978).

Fig. 9. Some geometric forms of ingots: circular and quadratic cross-section (Helmreich 1980).

Fig. 10. Some geometric forms of ingots: single and double casting (Helmreich 1980).

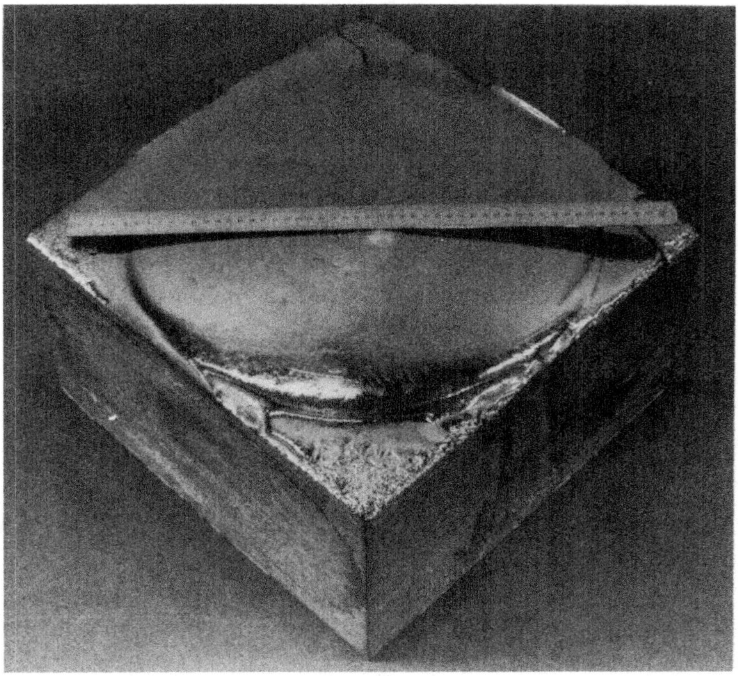

Fig. 11. A 100 kg ingot with $43 \times 43\,cm^2$ cross-section.

chance to produce silicon wafers for terrestrial solar cells in a cost-effective manner:

(1) WICP is an energy saving process. Whereas up to 60 kWh per kg silicon are consumed in advanced CZ methods or in techniques of directional solidification, values as low as 8 kWh/kg have already been realized for SILSO® in an R&D equipment with 100 kg melt charges. In the case of casting it is not necessary to keep a great mass of silicon in the molten state during an extended period of crystallization.

(2) The particular parts of the equipment can be individually adapted to the respective subtechnologies. In this way the installed equipment can be used more favorably (see also Helmreich 1982b).

Figure 12 shows a very schematic tentative graph of an equipment in which the production of multicrystalline silicon ingots can be carried out semicontinuously. The additional components are omitted: recharging chamber, crystallization chambers, cooling chambers as well as loading/un-

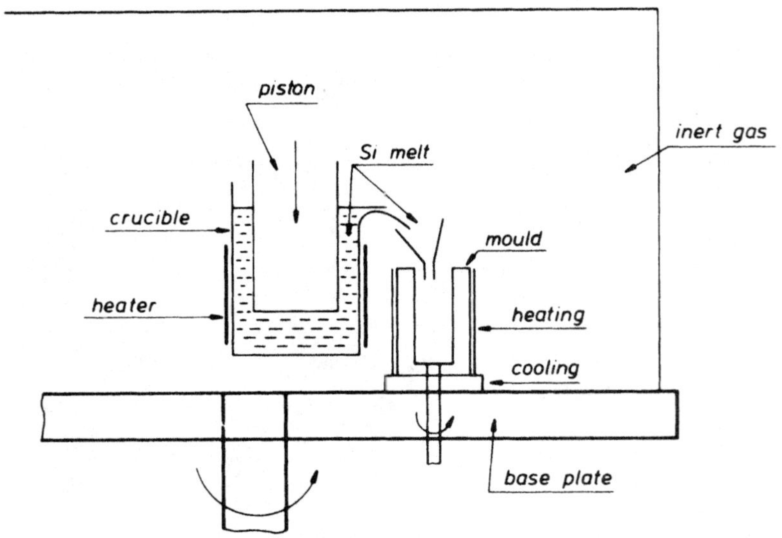

Fig. 12. Schematic graph of an equipment for semicontinuous production of multigrain silicon ingots (ref. E1).

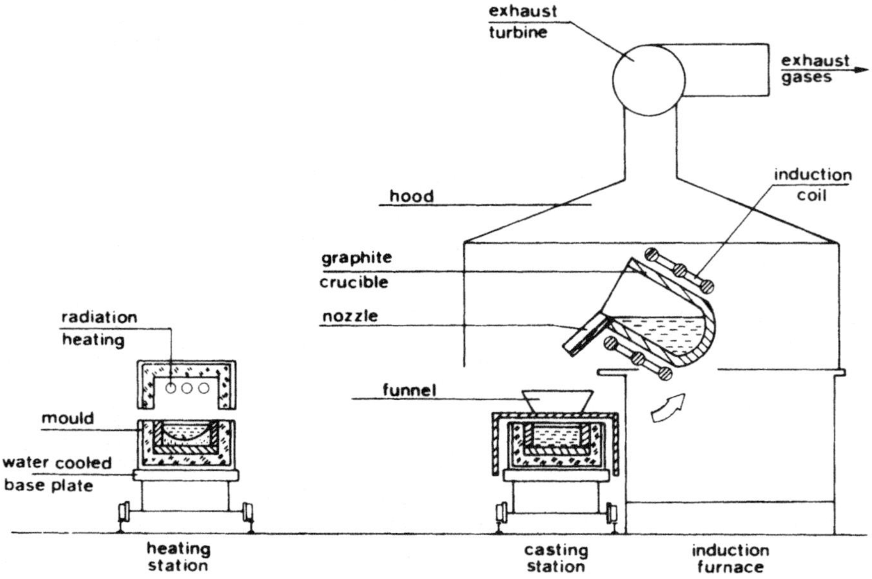

Fig. 13. Schematic graph of an open-hearth casting equipment (Helmreich and Suk 1982).

loading chambers for withdrawing the molds with solidified ingots and introducing empty molds (Helmreich 1984).

In addition to the afore-mentioned techniques, which are practized within a recipient in vacuum or inert gas atmosphere, further process and equipment austerity studies have been undertaken. Our work in open-hearth casting (Helmreich and Suk 1982) is an attempt to explore the limits of upscaling and the limitations of melting and casting in a truncated equipment (fig. 13).

4. Technical aspects of the Wacker ingot casting process

There are two design modifications which have been successfully applied in the WICP. In both cases the procedure consists of pouring silicon melt into a preheated mold and subsequent directional solidification.

In the set-up given in fig. 14 the silicon melt is prepared via radiant heating with a graphite susceptor which receives its energy from an RF-coil (several hundred kHz). The set-up is contained in a quartz-recipient thus allowing the adjustment of specifically defined conditions during preconditioning, melting, casting and crystallization.

The principal construction of a set-up implementing induction heating (2–4 kHz) is shown in fig. 15. In this case the RF-energy is coupled directly into the graphite crucible which acts as a susceptor for the actual melt crucible.

Due to the good electrical conductivity of graphite, the susceptor absorbs a large part of the RF-energy. At low frequencies, however, the

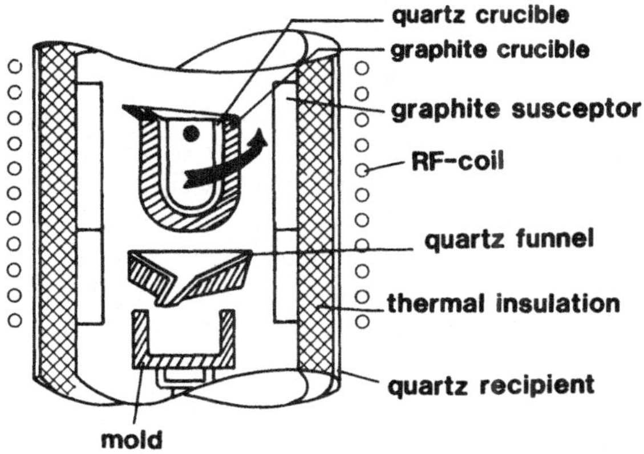

Fig. 14. Schematic graph of a melt crucible set-up for small melt charges.

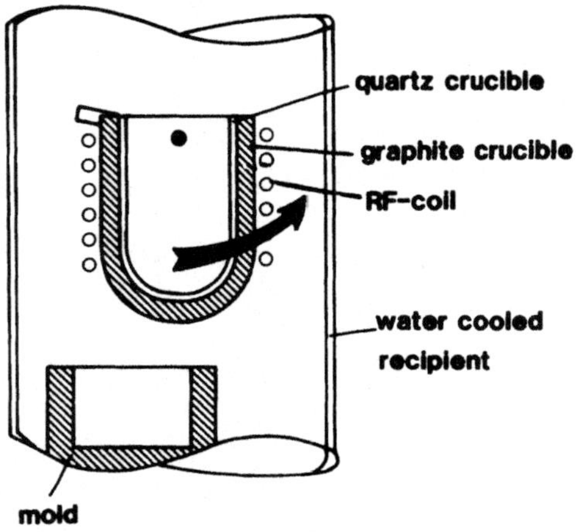

Fig. 15. Schematic graph of a melt crucible set-up for big melt charges.

RF-field strongly penetrates the graphite wall. In this way it directly heats up the melt and creates a strong stirring effect. The thickness of the graphite crucible is chosen such as to avoid the motion of the melt bath due to RF-stirring. This set-up is applied to the handling of bigger charges (100 kg and more) and the preparation of bigger ingots (50 kg, $33 \times 33\,cm^2$ cross-section or 100 kg, $43 \times 43\,cm^2$ cross-section).

4.1. *Melting*

The basic principles of the melting procedure refer to the technique used in Czochralski-pulling. Therefore, the design of the equipment and the processing are quite similar to those of the latter technique (see ref. E3).

Most of the melt energy which has to be put into the silicon to be molten is lost through the free surface of the melt. In a crucible where the height of the melt, h, equals the diameter of the crucible d ($h/d \approx 1$) the energy loss through the surface is given by curve $\dot{Q}_s$ in fig. 16. The bigger the melt charge the smaller is the free surface in relation to the melt volume and the smaller is the energy loss. From the ratio of energy loss to melt charge, $\dot{Q}_s/w$, a threshold power for keeping silicon in the molten state can be inferred. For a 100 kg melt charge this threshold power is about 22 kWh/kg. The theoretical amount of energy for melting silicon is about 0.75 kWh/kg solid silicon. With big melt charges of 100 kg energies as low as 3 kWh/kg have already been obtained in the case of an efficient thermal insulation, especially atop the melt crucible.

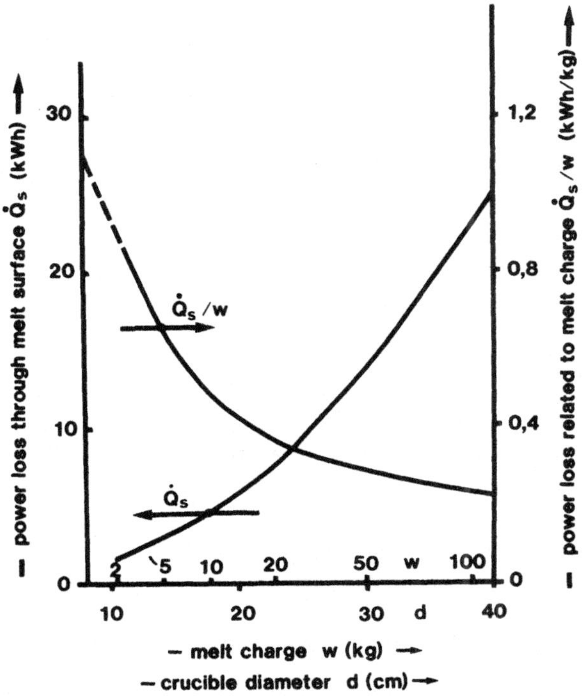

Fig. 16. Energy loss through radiation from melt surface versus crucible diameter and size of melt charge, respectively.

Up to now crucibles of different quartz qualities have been proved best with respect to melt pollution, flexibility of geometry, dimension and costs. Other alternatives like (silicon-resistant) graphite, silicon nitride or silicon carbide are restricted in their use because it is not yet possible to produce crucibles from these materials that have big enough dimensions. In the case of large quartz crucibles a good choice is the snow ball quality. With crucible diameters of up to 40 cm the temperature of the crucible wall can be held still below the softening temperature of quartz. Under normal melting conditions the temperature of crucible wall is about 1450°C in the case of a 100 kg melt charge.

The starting material is a chunky silicon of semiconductor-grade quality (CVD-poly or clippings of Czochralski and float-zone silicon, respectively). It is properly arranged in the crucible to avoid cracking of the crucible during the melting phase and to obtain a high degree of filling.

Silicon powder and small granules have already been molten successfully. However, due to the low degree of filling, recharging of the starting material is necessary here.

In all cases preparation of the melt takes place in a controlled atmosphere of inert gas at reduced pressure. With the help of regulated inert gas flows the condensation of SiO in the space above the melt and the back streaming of CO are avoided, thus preventing excessive contamination of the silicon melt with oxygen and carbon.

4.2. Casting

Most important for the quality of the ingot is the mold that has to be adjusted in terms of material and construction to the nature of the molten material and to the individual casting process.

Materials used in contact with molten silicon must exhibit a high degree of inertness against the silicon melt. Although Si_3N_4, SiC, quartz or high melting ceramics like mullite are suited as mold material, graphite has proved to be the most qualified one for the following reasons:

(1) Its heat-conducting properties are very favorable.

(2) The small coefficient of thermal expansion is very close to that of silicon, thus preventing the cracking of the ingot during its cooling-down period.

(3) Graphite parts are very insensitive to temperature shocks, and maintain their shape under the conditions of pouring and crystallization.

(4) One of the prominent properties of graphite is its good workability which allows a more complex construction of the mold and an easier repairing.

(5) A protective reducing atmosphere is established automatically during the casting and solidification process. Thus, gas bolsters of this kind between mold and ingot help to avoid the formation of stress situations in mold and ingot.

The methods of directional solidification generally use one-way molds. Multiple-use molds can be obtained by lining them with powders of Si_3N_4, Si and graphite (to name just a few materials) or by protecting them with a low melting slag layer.

In the WICP the molds are built with graphite elements. The construction has a well-defined geometry, inside and outside, to ensure the appropriate dimension of the ingot as well as a homogeneous radial temperature distribution. The outer surface of the side walls and, sometimes, the bottom are covered with a heat insulation consisting of carbon felt, graphite foil or carbon foam, respectively. With the side-wall insulation crystal growth from the periphery can be controlled with respect to orientation and size of the crystallites (see also section 4.3). The shape of the solid/liquid interface is mainly influenced by the side-wall insulation. The bottom insulation sets the heat transfer coefficient which determines the linear growth rate and thus the overall output in kg/min. The preconditioning of the mold is done in a preheating station. There the temperature level and the temperature distribution within the mold are adjusted correctly. The silicon melt is poured through a graphite-supported quartz funnel and fed into the appropriately preheated mold. Rotating the mold ensures uniform distribution of the heat which helps to develop a homogeneous chill-layer of solid silicon before the silicon melt can react with the mold walls. In this way it is possible to use the mold repeatedly. The silicon melt can also be poured directly into the mold without the use of a funnel but a thermally stabilized funnel helps to balance inhomogeneities in the silicon melt. The subsequent solidification process is carried out in a controlled vertical temperature gradient.

4.3. Crystallization

The crystallization rate depends predominantly on the vertical heat transfer through the already crystallized silicon mass, the bottom of the mold, and, in the case of thermal insulation, through the latter one into the water-cooled shaft (as it is used in the normal case). The different heat transfer situations can be described by one heat transfer coefficient k_a which can vary between about 50–100 W/m^{2}K. In fig. 17 three different calculated examples are given with the twin-curves a, b, and c. A 50 kg ingot with 33×33 cm^2 cross-section has a height of about 20 cm. With an assumed heat transfer coefficient $k_a = 100$ W/m^{2}K a crystallization time of about 3 h is to be expected (curve c). In the case of a naked melt surface the heat loss from the free surface is high enough that a solid crust is formed. In this case, the last region to solidify would be in the interior. As silicon expands during freezing, melt occlusion leads to a highly stressed zone or to the cracking of the ingot. It is, therefore, necessary to keep the melt surface open until the end of the solidification process is reached. This is done with a radiation heater (graphite) atop the mold set-up. When the crystallization process has come to an end, a temperature difference of about 600°C between bottom and top of the ingot is expected which results in a mean vertical temperature gradient of 3°C/mm.

The other curves refer to a 10 cm high ingot with a heat transfer coeffi-

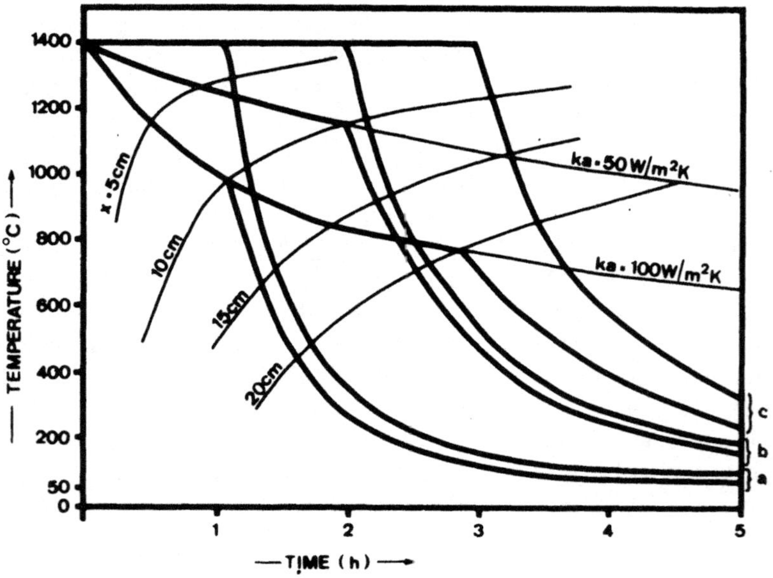

Fig. 17. Temperature versus time diagrams for different ingot heights and heat transfer coefficients.

cient of 100 (curve a) and 50 W/m^{2}K (curve b), respectively. Decreasing the heat transfer coefficient results in a decrease of temperature gradient and an increase of solidification time. As fig. 18 points out, the medium local growth rate is at a lower level with smaller heat transfer coefficients, but its variation along the ingot is smaller, too. In the experimental set-up, shown the inset of the fig. 19, the temperature curves of a solidifying silicon melt (5 kg melt charge) in an appropriately insulated graphite mold with 162×162 mm^2 inner cross-section have been measured. The temperature curves demonstrate that solification of the ingot at a height of about 80 mm is finished after one hour. The local growth rates achieved in this experiment are given in fig. 20. The values given in this figure are calculated from the temperature curves in fig. 19.

Mean local growth rates have also been determined in a 20 kg ingot with 22×22 cm^2 cross-section by inserting a probe from atop to determine the position of the solid/liquid interface. The calculated values of the linear growth rates are given in fig. 21, ranging from 1 to 1.85 mm/min. The SLIs at different time intervals are drawn as trajectories to the main orientation of the grain boundaries.

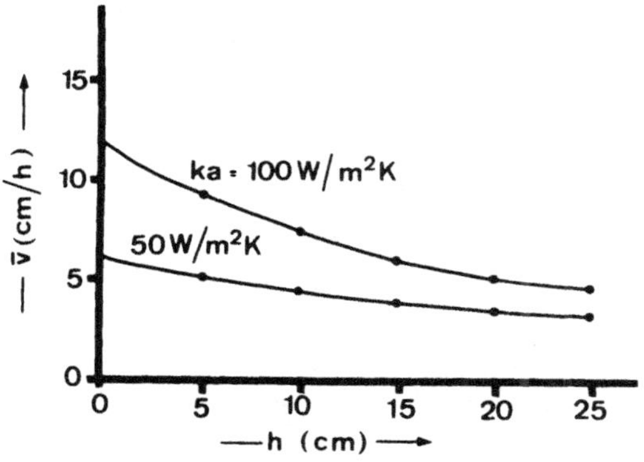

Fig. 18. Medium local growth rates versus ingot height.

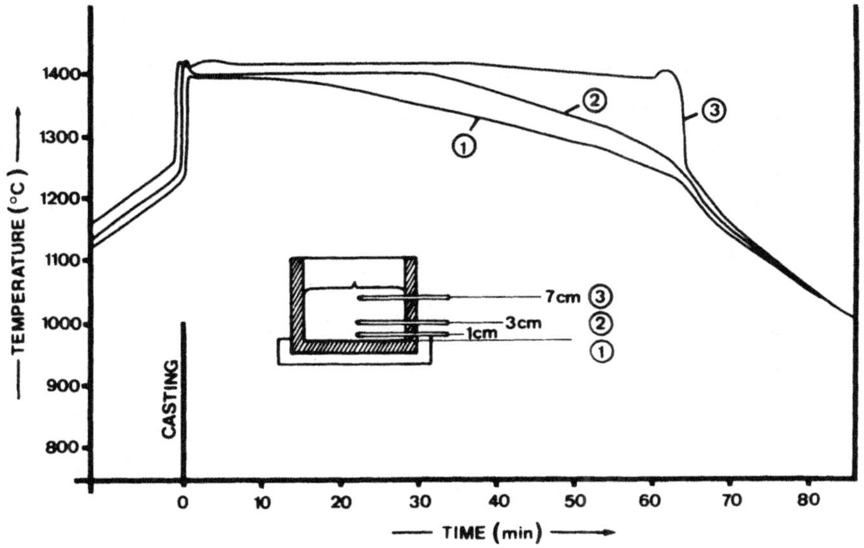

Fig. 19. Experimental temperature curves of a crystallizing ingot (5 kg) in a graphite mold (cross-section $162 \times 162\,mm^2$). Inset shows experimental set-up and location of thermocouples (Helmreich 1980).

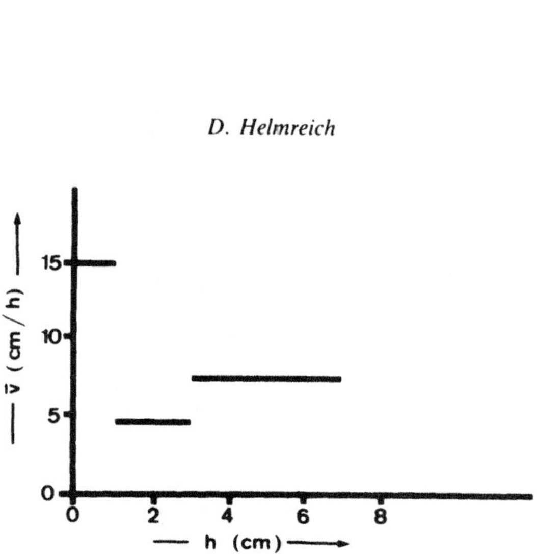

Fig. 20. Measured local growth rates versus ingot height. The values are taken from fig. 19.

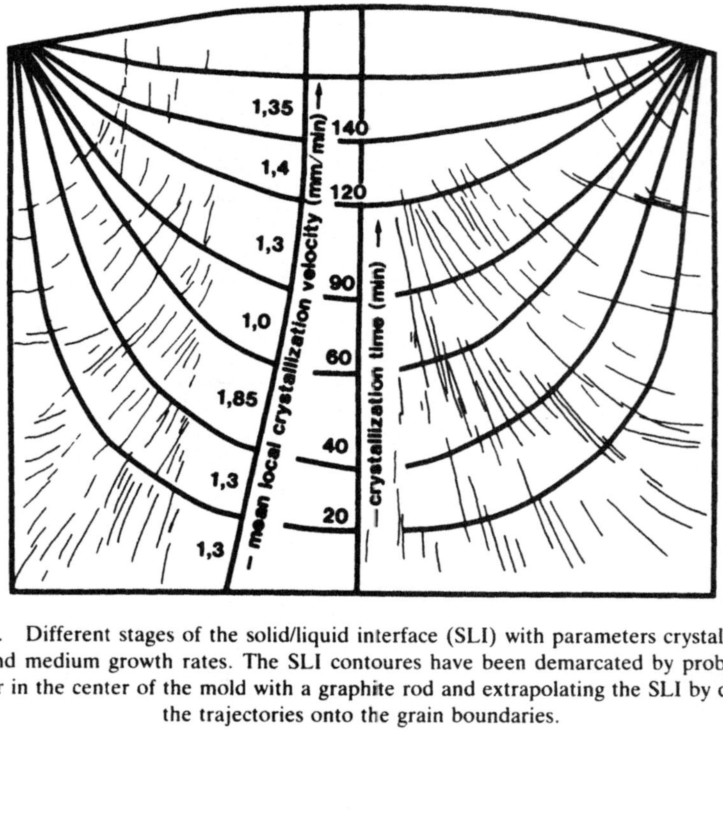

Fig. 21. Different stages of the solid/liquid interface (SLI) with parameters crystallization time and medium growth rates. The SLI contoures have been demarcated by probing the contour in the center of the mold with a graphite rod and extrapolating the SLI by drawing the trajectories onto the grain boundaries.

As it turned out, linear growth rates should not exceed a threshold of about 1 mm/min. It is not worthwhile to increase the throughput by increasing the growth rate, because a high growth rate spoils the material with respect to crystallinity and thermally induced stresses. From 100 kg charges 100 kg ingots of $43 \times 43\,cm^2$ cross-section have been obtained with a crystallization velocity of about 0.5 kg/min (Helmreich 1985). Due to the scaffold time in the yet non-continuously operating R&D equipment throughput is reduced to about 50% of that value. In a large-scale production, growth and cooling rates must not affect the production rate: crystallization and cooling can be decoupled from the melting and casting steps.

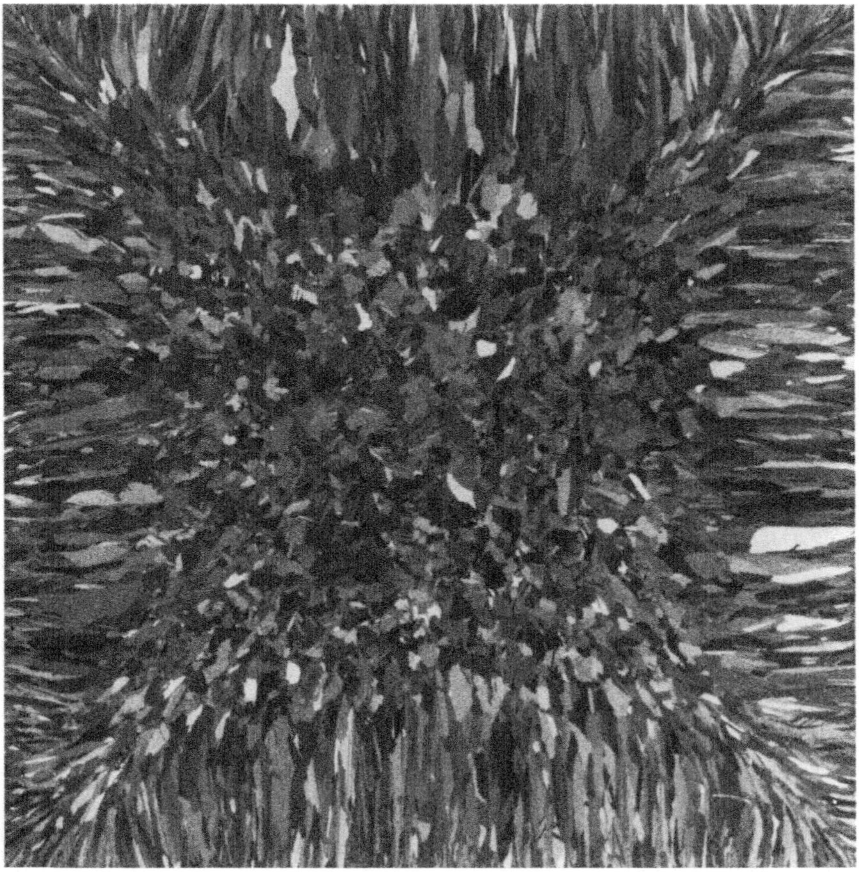

Fig. 22. Horizontally cut wafer ($10 \times 10\,cm^2$) from an ingot cast in a mold with little thermal insulation and low preheating temperature.

A further increase of throughput through further increase of ingot cross-sections seems not rational; at the present stage of development a 100 kg ingot with $43\times43\,cm^2$ cross-section is the utmost dimension that can be handled easily.

In general a columnar structure is found in pure metals with a short freezing range as it is realized in the case of silicon. In order to suppress the nucleation centers in the melt, the crystallization front has to move fast enough. The limiting factor of crystallization rate is the heat extraction which, in turn, is governed by the thermal properties of the mold.

The influence of vertical and lateral growth rates can be seen in the horizontal cuts of ingots with $10\times10\,cm^2$ cross-section presented in figs. 22 and 23. The fine crystalline structure of the wafer given in fig. 22 is the

Fig. 23. Horizontally cut wafer ($10\times10\,cm^2$) from an ingot cast in a mold with considerable thermal insulation and high preheating temperature.

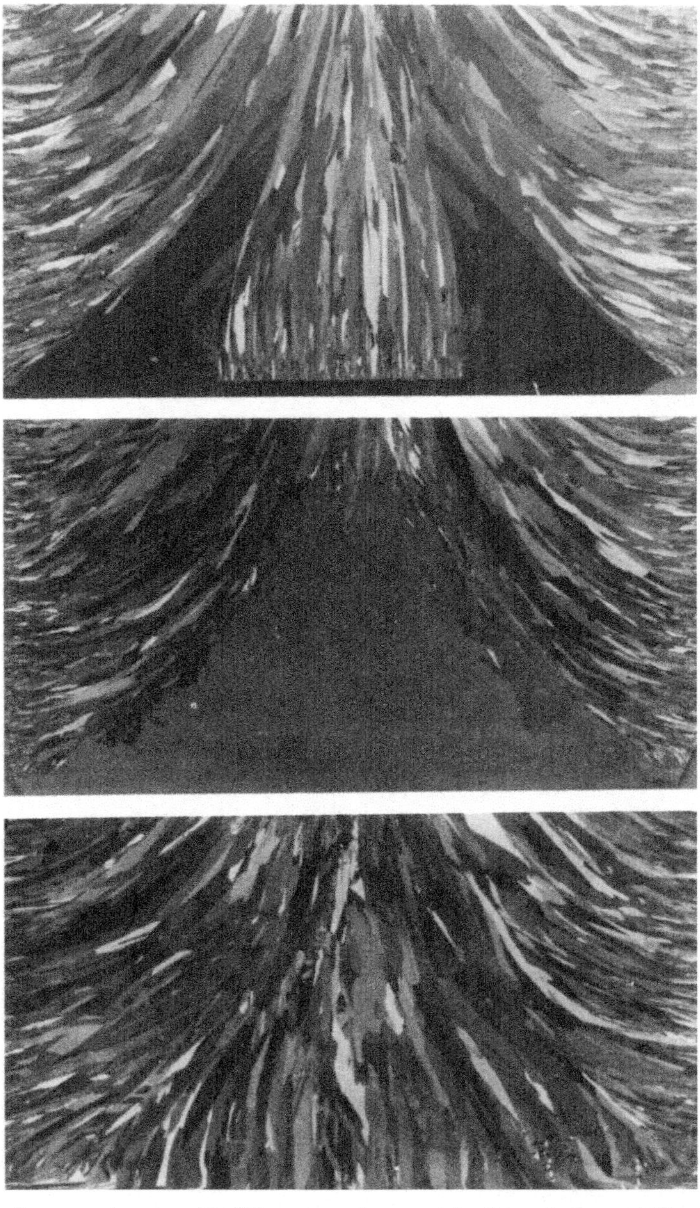

Fig. 24. Seeding experiments with different seeding crystals. Top: single crystalline seed with insufficient melt back. Middle: single crystalline seed. Bottom: seed with columnar structure (Helmreich 1980).

result of a raised growth rate due to a lower preheating temperature and a larger heat transfer coefficient ($>100\,W/m^2K$). Contrary to that the coarse crystalline structure of the wafer given in fig. 23 results from a higher preheating temperature and a smaller heat transfer coefficient ($<50\,W/m^2K$).

The crystallites that were nucleated at the bottom and on the walls of the mold usually are randomly oriented. They grow in competition with each other, and preferred orientations develop as a result of the suppression of those crystals that are least favorably oriented. The rate of lateral growth seems to be at least as important as the rate of vertical growth. The elimination of unfavorably oriented grains is gradual and the increase of grain size as well as the sharpening of the orientation develop simultaneously.

If the thermal gradient does not change throughout the solidification process, the columnar structure proceeds. It is most desirable to avoid the

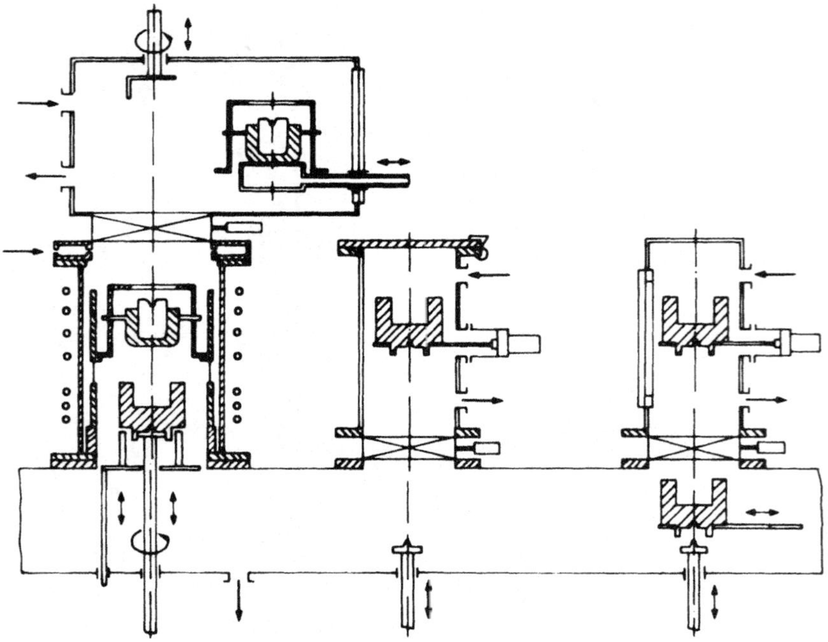

Fig. 25. Schematic graph of an equipment for semicontinuous production of silicon ingots with an assembly like that given in fig. 14. On top of the melting/casting station (left) is a recharging chamber. Cooling chamber (middle) and loading/unloading chamber (right) are connected to the recipient (Zauhar et al. 1979, Helmreich 1980).

transition stage and to reach stable columnar growth quite from the beginning of solidification. Therefore, seeding experiments have been carried out with a variety of seed crystals (Helmreich 1980).

A few examples are given in fig. 24. They show that sufficient back-melting is necessary in order to obtain effective seeding. Monocrystalline growth from a corresponding seed can be preserved only over a limited distance. In the given temperature field a stable columnar growth is favored which can be predetermined by using a seed of already columnar structure.

4.4. *Equipment for the casting process*

The WICP as it is principally sketched in fig. 8 is presently carried out on a larger scale in a pilot-production line. The schematic drawing of the equipment for the semicontinuous production of silicon ingots is shown in fig. 25 (see Zauhar et al. 1979, Helmreich 1980) supplemented by a photographic view (fig. 26). In addition to the central unit where melting, casting, and solidification take place some further components are included:

Fig. 26. Photograph of the equipment given schematically in fig. 25 (Helmreich 1980).

a recharging chamber where the empty crucible can be removed and replaced by a filled one is added on top of the central unit. Connected with the recipient are the cooling chamber in the center and to the right the loading/unloading chamber for withdrawing the mold with the solidified ingot and introducing an empty mold. This equipment is operated semicontinuously and the ingots produced in it from semiconductor-grade silicon have a weight of 8 kg.

A schematic sketch of the ingot casting equipment where larger melt charges (up to 120 kg) can be handled and bigger ingots can be fabricated (50 kg, $33\times33\,cm^2$ and 100 kg, $43\times43\,cm^2$) is given in fig. 27 (Helmreich 1984, 1985). The equipment is contained in a water-cooled recipient (11).

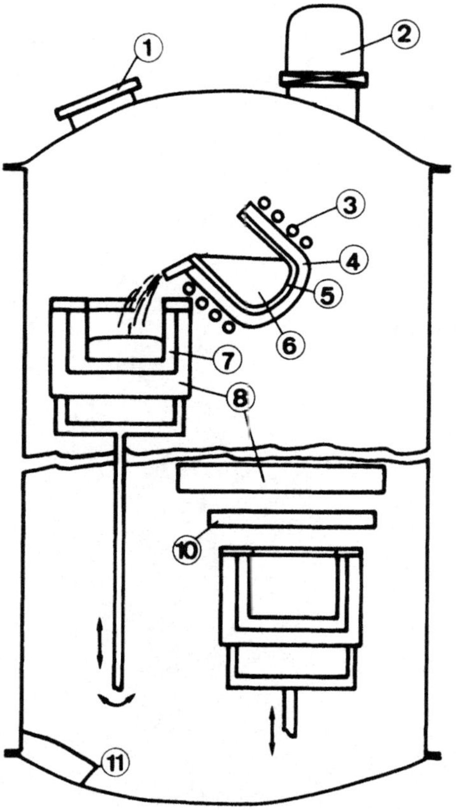

Fig. 27. Schematic graph of an equipment for semicontinuous production of silicon ingots with an assembly like that given in fig. 15 (Helmreich 1984, 1985).

Fig. 28. Photograph of the equipment given schematically in fig. 27 (Helmreich 1984).

In the upper part the melting and casting take place, while in the lower part the preconditioning of the mold and the crystallization are carried out. The starting material is melted in a quartz crucible (5) with RF-heating (2–4 kHz). The RF-energy is coupled directly from the induction coil (3) into a graphite 'crucible' (4) which acts as a susceptor for the quartz crucible. The recharging equipment (2) allows increase of the degree of filling as well as a repeated recharging of the empty crucible.

After casting of molten silicon which can be observed through a view port (1) the mold (7) with its thermal insulation is moved to the crystallization station where a vertical temperature gradient is built up by aid of a radiation heater (10). There, the crystallization is carried out in a controlled manner so that a crystalline shell is generated which caulks the mold from the silicon melt. Crystallization of the remaining silicon melt is then finished in this crystalline container. A photograph of this three storey equipment is given in fig. 28. Its individual parts can be traced with the help of the schematic in fig. 27.

5. Cutting/wafering

An efficient ingot technology has to be complemented by qualified cutting and wafering techniques which produce low-cost silicon wafers at minimum material loss. Material utilization is of major importance with respect to productivity. Depending on ingot size and cutting/wafering technique, conversion ratios (m^2 silicon sheet per kg ingot) are achieved which range from 0.35–0.6 m^2/kg for the as-grown ingot. The latter value applies to bigger ingots where the surface/volume ratio of the ingot is very low (Gessert and Helmreich 1985). Contouring and portioning of large-sized ingots is favorably carried out with a band-saw technique which allows the machining of bigger dimensions at low values of material loss as compared to the conventional OD (outer diameter) sawing. A 100 kg-ingot being cut into 16 columns each of $10\times10\,cm^2$ cross-section is shown in fig. 29. Further development with respect to equipment, working tool and procedure has still to be carried out to adapt this technology, which is well established in the stone and wood industry, to the extreme sensitive and brittle material silicon.

The technical state of the art in silicon slicing is represented by four different methods:

(1) ID (internal diameter sawing);
(2) MBS (multiblade slurry) sawing (Werner 1982, Ross 1980, Fleming et al. 1979, Werner et al. 1982);

Fig. 29. A 100 kg ingot cut into columns with cross sections of $100 \times 100\,mm^2$ ready for wafering (Ermer et al. 1984).

(3) MWS (multi wire slurry) sawing (Anderson 1980, Pryor 1980, Lauvray et al. 1980);
(4) FAST (fixed abrasive slicing technique (Schmid et al. 1982).

A comprehensive review including a qualitative assessment of the operational feasibility of the alternative wafering methods mentioned has been published recently (Werner 1982). Because of the high technological and operational reliability, the conventional ID technique still is the prevailing slicing method for solar silicon wafer production (Fiegl and Bonora 1980, Daud et al. 1981). Besides ID sawing, MBS sawing is a slicing technique which has a definite chance to play a major role in the future. As compared with other slicing techniques, it provides the highest production rates and lowest add-on costs. In conjunction with the WICP, an MBS technology is being developed which pursues the philosophy of high reciprocating cutting speeds (high stroke frequency combined with a long stroke) and increased cutting forces. This results in cutting rates which are higher by

Fig. 30. SILSO® wafers ready for shipment.

an order of magnitude than those obtained with 'conventional' MBS sawing (Werner 1982). The slicing of the preshaped columns of $10\times10\,cm^2$ cross-section is accomplished by this upgraded MBS version. In a pilot equipment with high reciprocating cutting speeds and increased feed forces a throughput of at least 6 wafers/minute can be projected for a three-shift fabrication from present experiences (Ermer et al. 1984, 1985). The square-shaped slices, as shown in fig. 30, are well qualified for various solar cell preparation techniques.

6. Material characterization

6.1. Crystallographic structure and defects

In multicrystalline silicon samples a preferred orientation of individual crystallites (texture) is observed. Selective etching as well as X-ray diffraction have been applied to evaluate the orientation of crystallites with re-

Table 1
Texture analysis of multigrain silicon with etching technique and X-ray diffractometry (results are normalized to ⟨110⟩ grain orientation) (courtesy of H. Seiter, Heliotronic GmbH).

Crystallographic orientation (*hkl*)	Etching technique		X-ray diffraction		
	Fraction of grains		$I_{theor.}$	$I_{theor.}/I_{meas.}$	
	Sample A1	Sample B1	Isotropic distribution (Si-powder)	Sample B1	Sample B2
110	1.0	1.0	1.0	1.0	1.0
100	0.16	0.13	0.47	~	0.09
111	0.16	0.06	0.15	0.21	1.87
210	0.83	0.26	–	–	–
211	1.08	0.94	0.32	1.79	2.81
221	0.16	0.20	–	–	–
310	nm	nm	0.46	–	0.91
311	1.49	0.47	0.39	0.05	5.53
321	0.08	–	–	–	–
331	–	0.06	0.20	–	–
511	0.42	0.13	–	–	–
531	–	–	0.16	–	6.25

Sample A1 cut parallel to growth direction. Samples B1/B2 cut perpendicular to growth direction; B1 has 600 crystallites/cm^2, rocking angle = ±5°; B2 has 150 crystallites/cm^2, rocking angle = ±12.5°; nm = not measured.

spect to the growth direction. In table 1 the results are shown for three specimens oriented parallel to the main growth direction (sample A1) and perpendicular to it (samples B1 and B2).

The etched multicrystalline silicon samples (20×20 mm^2) were compared with single-crystal standards for relative surface patterns. (Etching is done in 20% KOH, 1% 'Coplex' at 75–80°C for 10 min.) Results on the fraction of grains for each of the ten orientations selected are normalized to the grains with ⟨110⟩ orientation. Only when the fraction of grains is greater than 0.75 the significance of a particular value can be assured. X-ray diffraction pole figure analysis by the Schulz method (Cullity 1956) was also used to describe the crystallographic orientation. The results [again normalized to the (110) reflection] are contrasted to the theoretical intensities of an isotropic distribution of orientations verified using ≦ 10 μm silicon powder.

The results of both techniques are in agreement within their limit of precision (±10°). The results indicate that the growth direction for the long axis of the crystallites is generally parallel to ⟨311⟩ with significant components of ⟨211⟩ and ⟨110⟩. The preferred direction perpendicular to the growth direction is a mixture of ⟨211⟩ and ⟨110⟩. This is consistent with the ⟨311⟩ growth direction. Such results can be explained as follows: the selection of preferential and annihilation of non-preferential grains is governed by different crystallization rates in different crystallographic orientations. For the overgrowth of non-preferential grains the growth rate perpendicular to the fibre axis seems to be of higher importance than that parallel to it.

Cast silicon is characterized by low-energy coherent grain boundaries (fig. 31) and twin planes (figs. 32). The coherent grain boundaries appear as trajectories to the solid/liquid interface. Individual crystallites have different dislocation densities; the mean dislocation density is about 10^3–10^4/cm^2. In some cases dislocations accumulate in monocrystalline areas to form a sub-boundary (fig. 33; compare also fig. 47). A further extended

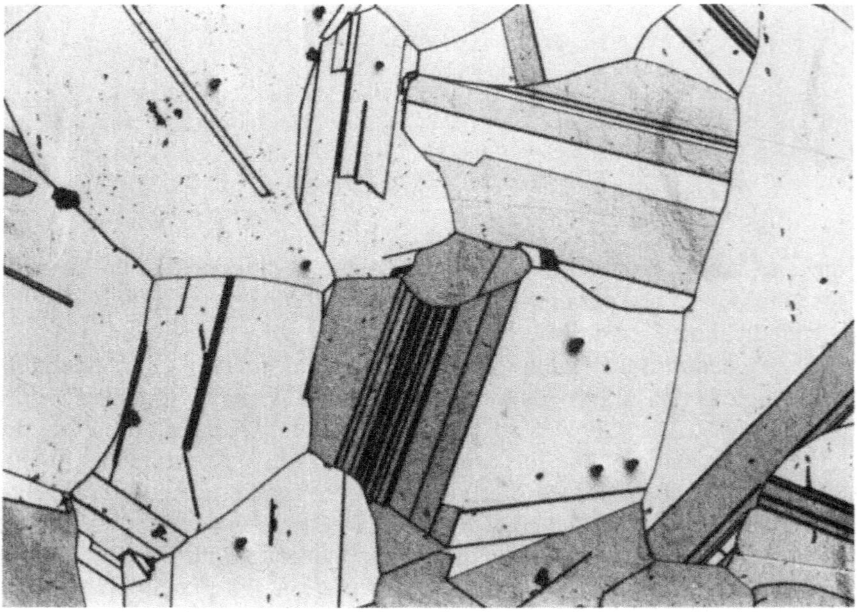

Fig. 31. Optical micrograph of a multigrain silicon wafer with coherent grain boundaries and twin boundaries.

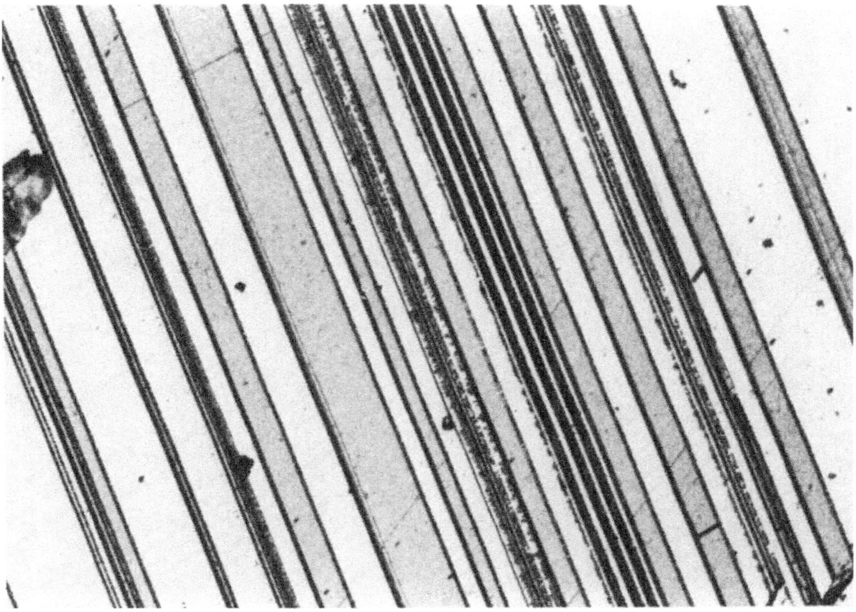

Fig. 32. Optical micrograph of an area with twin boundaries.

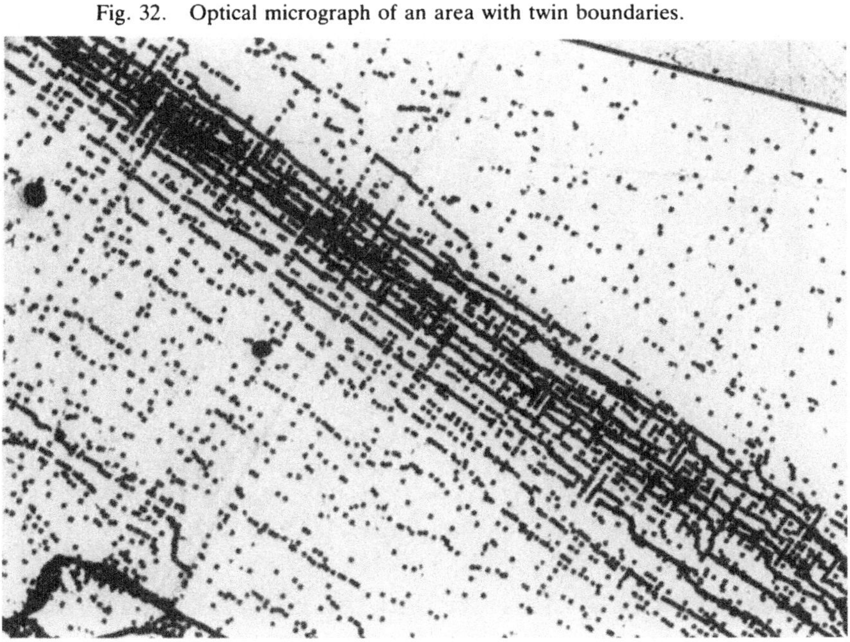

Fig. 33. Optical micrograph of an area with accumulated dislocations (sub-boundaries).

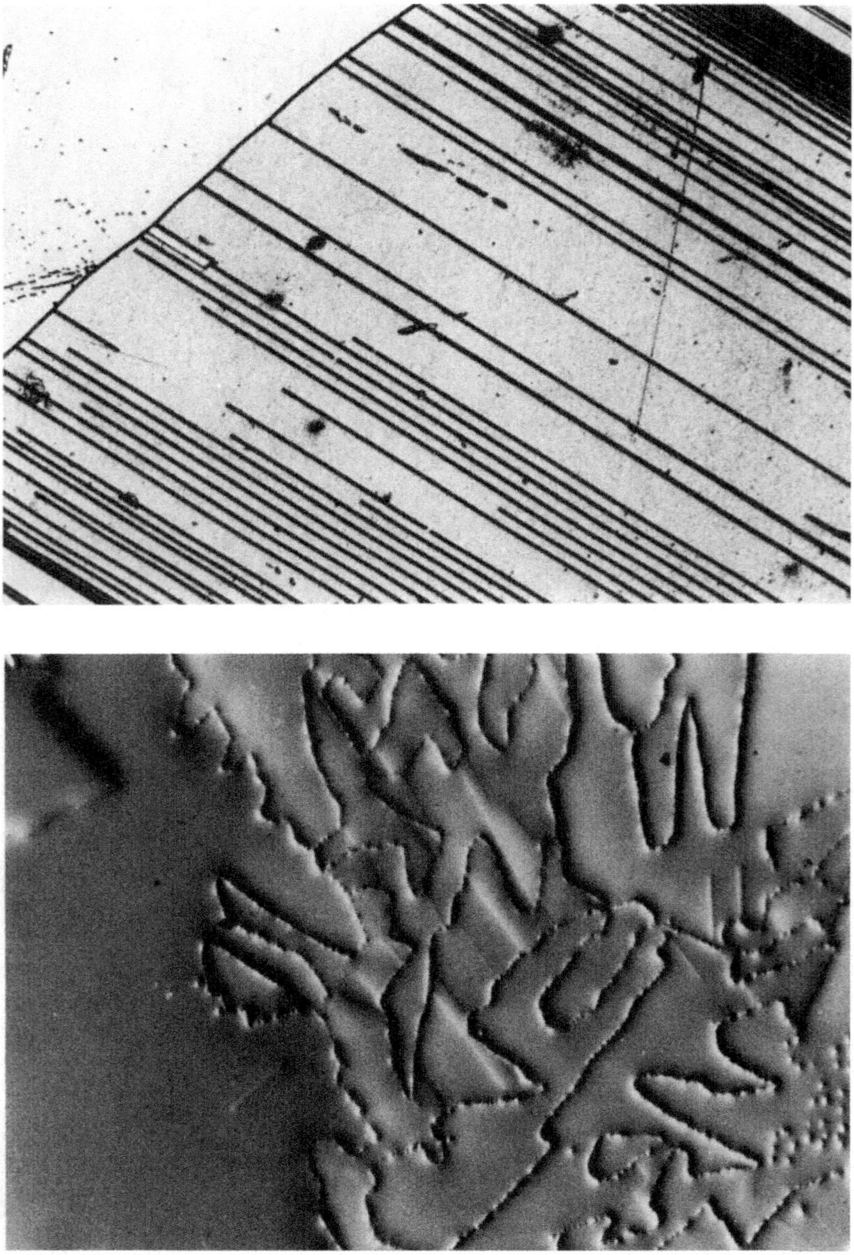

Fig. 35. Incoherent grain boundaries with SiC precipitates. Micrograph of an as-polished multigrain silicon wafer (ref. E1).

crystallographic defect – the stacking fault – mainly occurs through (partial) release of thermally induced stress (fig. 34).

High-energy incoherent grain boundaries appear in areas with carbon concentrations above the saturation limit of about $2\text{–}3 \times 10^{17}/\text{cm}^3$. They are decorated with SiC precipitates (fig. 35) and incorporate a lot of other impurities (compare fig. 42). As a consequence the conversion efficiency (CE) is strongly degraded in this region.

In contrast to coherent grain boundaries, twin boundaries and stacking faults which have a minor influence on CE intragrain defects considerably affect the photovoltaic properties of the material (see also Regnault and Yoo 1984, Leung and Iles 1984). With appropriate melting and crystallization parameters the concentration of intragrain defects like dislocations, precipitates (e.g. SiC or SiO_2 in the case of high carbon resp. oxygen contents) and complexes can be kept at a low level to guarantee CE's similar to those obtained with single crystalline silicon grown by the Czochralski method.

6.2. *Impurities*

As the picture of solidification can vary between extremes, segregation may occur likewise in widely variant patterns. In addition to the segregation coefficient mechanical, chemical, and thermal factors influence the degree and direction of solute migration via the additional characteristic data on diffusivity and saturation concentration. Two different kinds of segregation can be differentiated. Impurities with high diffusivity and favorable segregation coefficients preferredly accumulate as separate phases at the grain boundaries (Helmreich and Ast 1980, Pizzini et al. 1982, Kazmerski 1984, Saito et al. 1984). For typical (substitutional) doping elements directional solidification leads to macro segregation (Helmreich and Ast 1980). In fig. 36 critical impurity levels in a simulated meltstock are calculated from threshold values for terrestrial PV application (Dietl 1983). The assumption of macroscopic segregation characteristics similar to those observed in Czochralski technology has been verified.

A marked macroscopic segregation is found for the slow-diffusing elements like dopants (P, As, Sb, as well as B, fig. 37) and light elements (C, O) (figs. 37, 38). The individual concentration profiles are best fitted by the simple formula $C_S = k_{eff}C_0(1\text{-}g)k_0 - 1$ where g, the fraction of solidified melt, is taken as $g = 0.9$ for the calculation of the curves given in fig. 36. At low growth rates k_{eff} approaches k_0.

Neutron activation analysis (NAA) showed that there is no detectable macroscopic segregation for fast-diffusing elements (fig. 37). These elements rather show 'microscopic segregation' into grain boundaries and crystallographic defects. The different types of 'microsegregation' which can be expected in multicrystalline silicon are schematically shown in fig. 39. The evaluation of impurity distribution is favorably done by autoradiography because this evaluation technique is most sensitive to those elements which show a preferred 'microsegregation'.

Autoradiographic evaluation of a wafer prepared from a partially re-

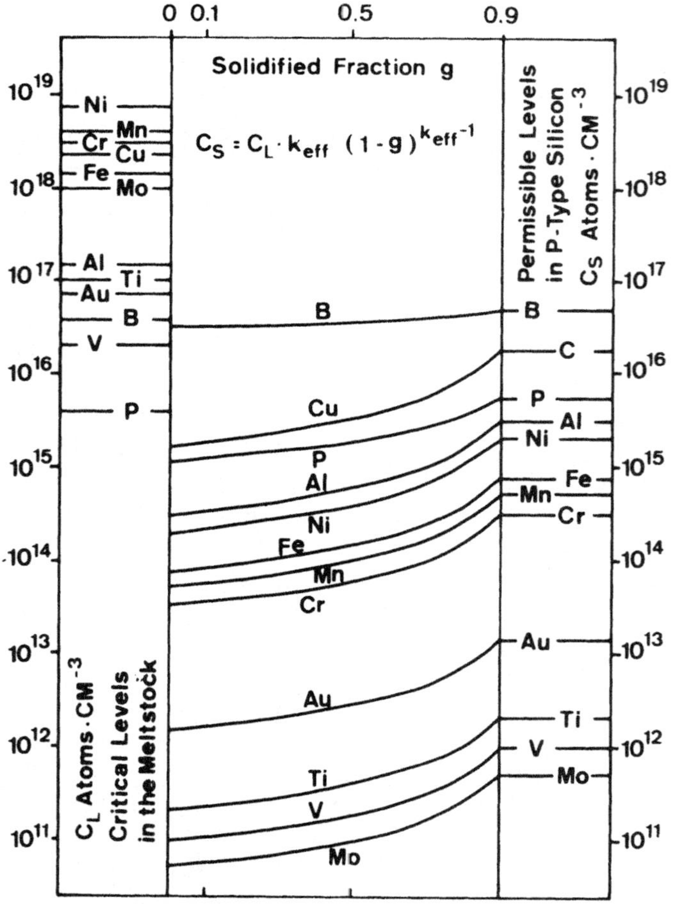

Fig. 36. Critical feed-stock impurity levels calculated from threshold values for solar cell degradation (Dietl 1983).

Fig. 37. Macroscopic segregation of impurities in cast multigrain silicon. Concentration of metallic impurities have been measured by neutron activation analysis (NAA); carbon/oxygen contents have been determined by IR absorption (Helmreich and Seiter 1979).

Element	Diffusion coefficient (cm^2/sec)	k_0 Segregation coefficient	Concentration (cm^{-3})			
			Detection limit	B*	M*	T*
As	2.5×10^{-13}	0.3	5×10^9	4.5×10^{11}	1.3×10^{12}	2.8×10^{12}
Au	6×10^{-6}		1×10^8	6.7×10^9	1.6×10^9	8.1×10^8
Co			2×10^{11}	1.6×10^{11}	4.8×10^{10}	6.3×10^{11}
Cr			6×10^{11}	2.4×10^{12}	1.1×10^{12}	1.1×10^{12}
Cu	7×10^{-6}		8×10^{10}	8.6×10^{13}	5.9×10^{13}	7.7×10^{13}
Fe	7×10^{-6}		5×10^{13}	1.7×10^{14}	$<2.2 \times 10^{13}$	$<3.5 \times 10^{13}$
Ni			2×10^{13}	2.2×10^{13}	4.5×10^{13}	1×10^{14}
Sb	3.6×10^{-13}	0.023	4×10^9	9.5×10^{10}	3.3×10^{11}	3×10^{12}
Zn	5×10^{-7}		5×10^{13}	4.6×10^{12}	9.5×10^{11}	4.5×10^{12}
C	2×10^{-11}	0.05				
O	1.6×10^{-10}	1.25				

* B, M, T: bottom, middle and top of ingot, respectively.

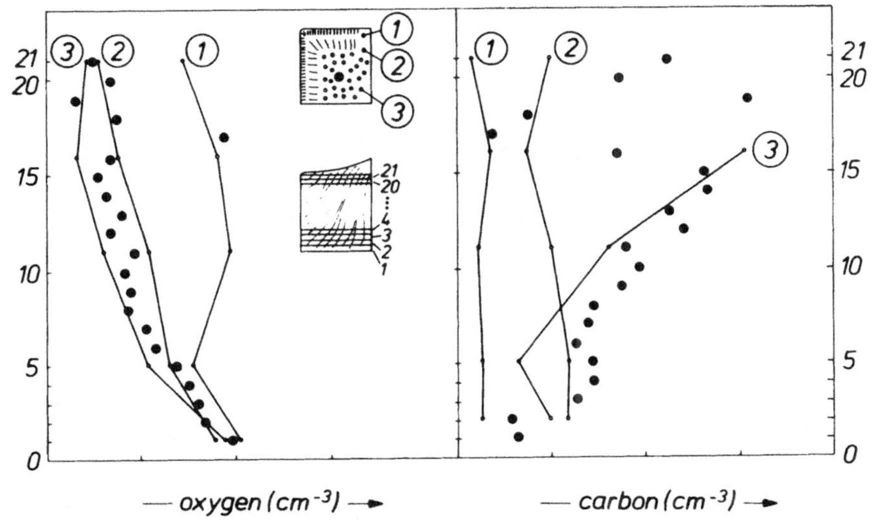

Fig. 38. Macrosegregation of carbon and oxygen in a cast silicon ingot. Inset shows locations of measuring points (Helmreich and Seiter 1979).

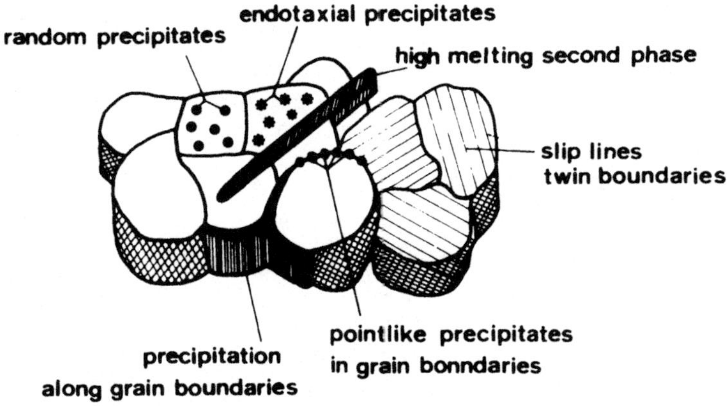

Fig. 39. Different types of microsegregation observed in multigrain silicon (Helmreich and Ast 1980).

fined metallurgical-grade silicon shows enrichment of metallic impurities (especially Fe, Co, and Ti) at the peripheral chill zone of the wafer and in areas characterized by the so-called 'grit'-structure (fig. 40). The enrichment factor is different for the various impurities and ranges from a few ten and hundred for doping elements (Sb, As) to a few thousand for heavy metals. This increased impurity level is paralleled with an increased level of SiC precipitates. A higher magnification of the so-called transcrystallization zone (fig. 41) exhibits a strong accumulation of such impurity elements in incoherent, high-energy grain boundaries. It demonstrates that grain boundaries of this kind act as efficient gettering centers thus reducing the impurity content within individual grains. Figure 41 also corroborates the inhomogeneous distribution of impurities along a given grain boundary. This has also been found by others (Aubert et al. 1980, Kazmerski and Russell 1982, Kazmerski 1984). Low-energy grain (twin) boundaries are usually clean and undecorated. A close-up view of a monocrystalline area shows that in the case of high impurity levels (especially carbon) this region is predestined for a high concentration of precipitates with accumulated impurities (fig. 42).

Due to the characteristics of the crystallization procedure in WICP a further specific pattern in defect and impurity distribution is found (Helmreich 1984). In a vertical cut of an ingot etched with Secco etch (fig. 43), striations are found which are parallel to the solid/liquid interfaces. The striations, which are schematically drawn in fig. 44, are not affected by grain boundaries.

Fig. 40. Autoradiography (a) of a multigrain silicon wafer (b) taken from an ingot prepared from partially refined metallurgical silicon (Helmreich and Ast 1980).

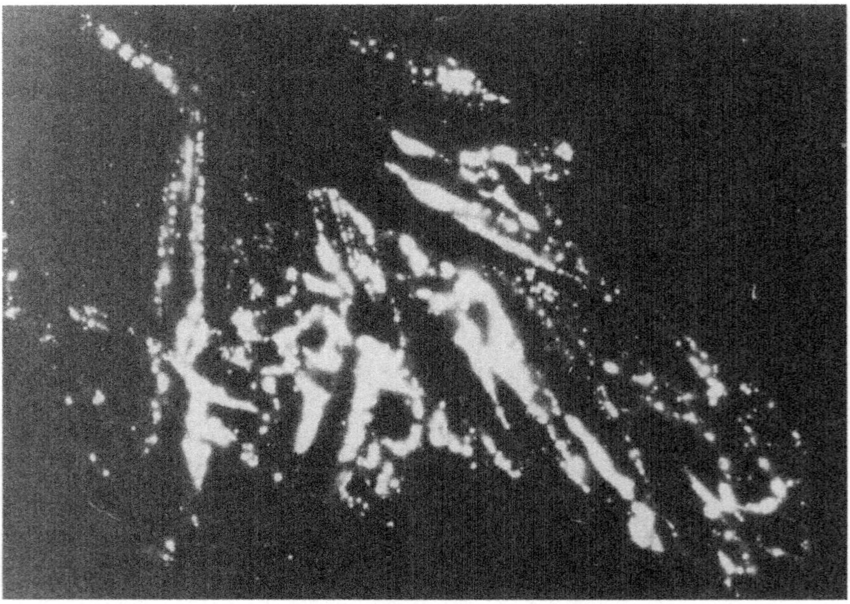

Fig. 41. Close-up view of the autoradiography given in fig. 40. Grain boundary segregation is found in the transcrystallization zone (Helmreich and Ast 1980).

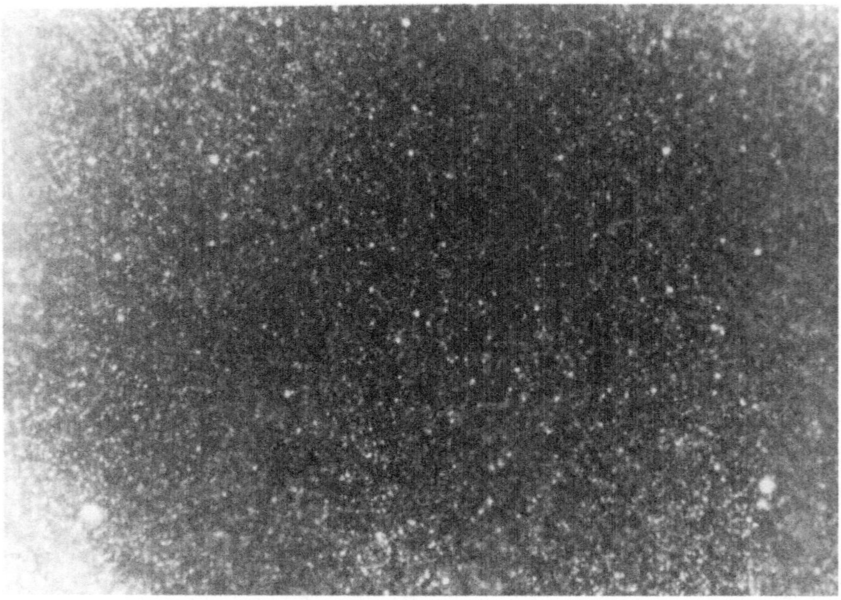

Fig. 42. Close-up view of the autoradiography given in fig. 40. Precipitates are found within an individual grain (Helmreich and Ast 1980).

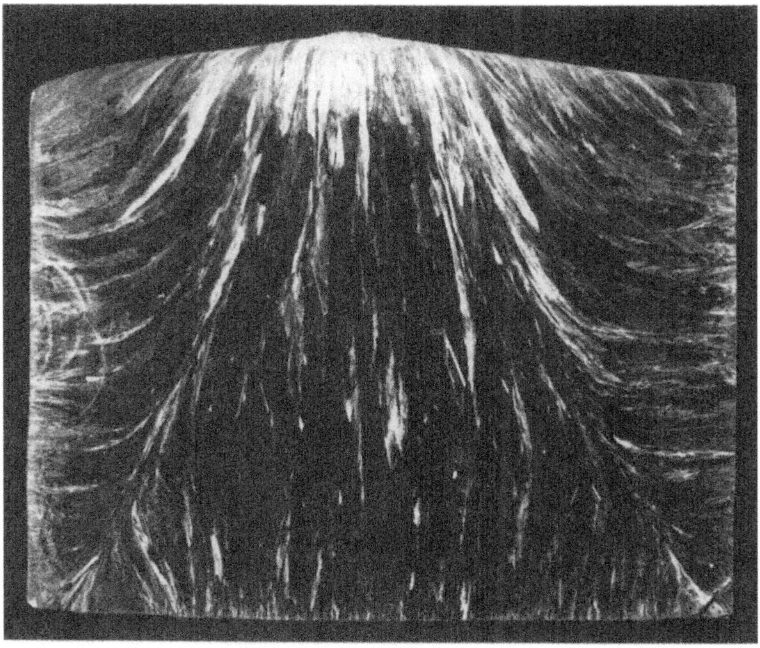

Fig. 43. Vertically cut wafer of an ingot intentionally prepared for investigation of striations (Helmreich 1984).

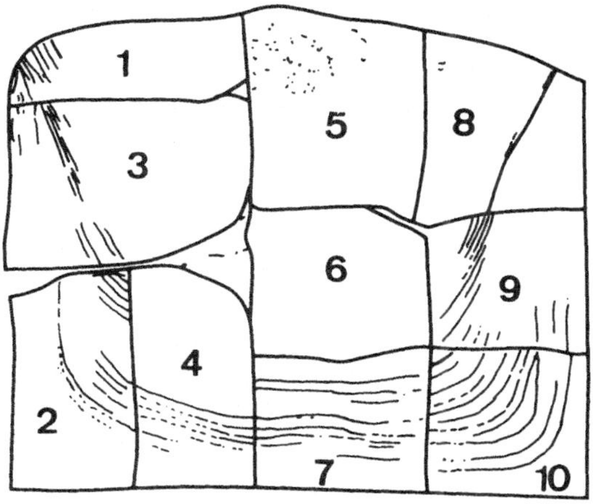

Fig. 44. Schematic graph of striations found in the wafer given in fig. 43 (Helmreich 1984).

The three different types of striations are presented in figs. 45–47:

(1) The appearance of steps on an etched surface of a vertical cut indicates that doping inhomogeneities are present in the material. Etching effects due to doping fluctuations are well known in silicon single crystals. They are caused by repeated partial remelting due to the divergences between the temperature field and the geometrical coordinate system of the rotating mold. The doping striations are confirmed by autoradiography. Only 23 days after neutron activation the short-living radio isotopes are predominant. The pattern of striations in fig. 45 is due to P and/or Sb.

(2) Striae with a mat appearance which are found on an etched shining surface consist of a striated accumulation of etch pits. These etch pits are caused by dislocations (fig. 46).

(3) Mat striae caused by point-like defects (fig. 47).

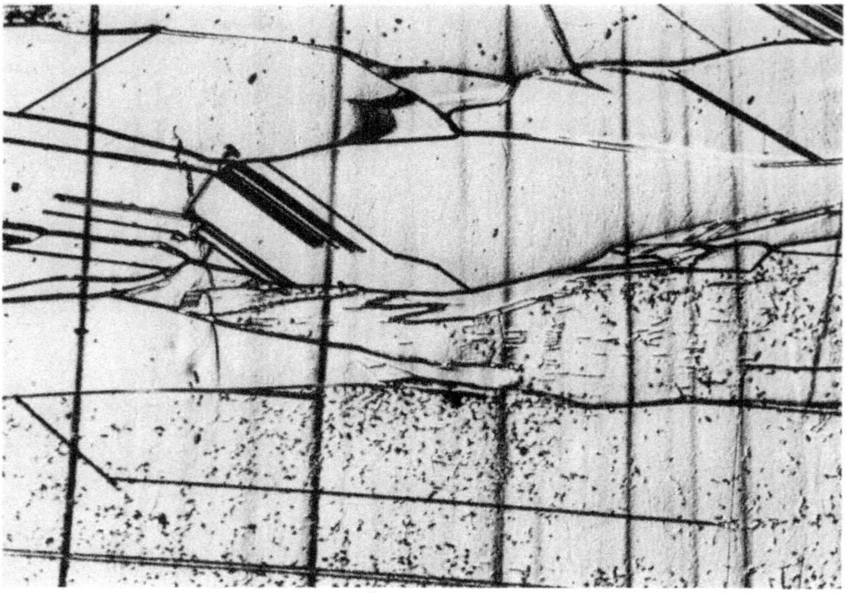

Fig. 45. Striations caused by differences in dopant concentration. Magnification 60× (Helmreich 1984).

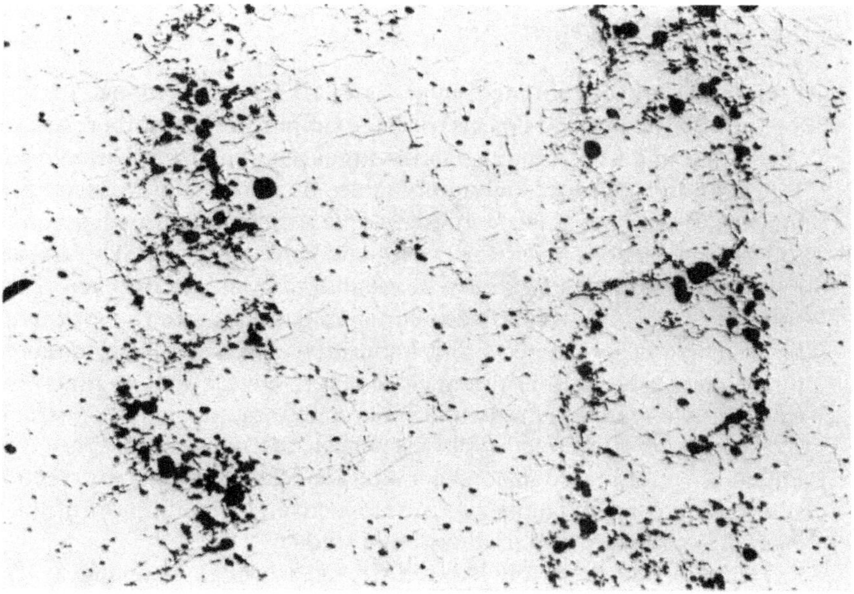

Fig. 46. Accumulation of dislocations in striated form. Magnification 60× (Helmreich 1984).

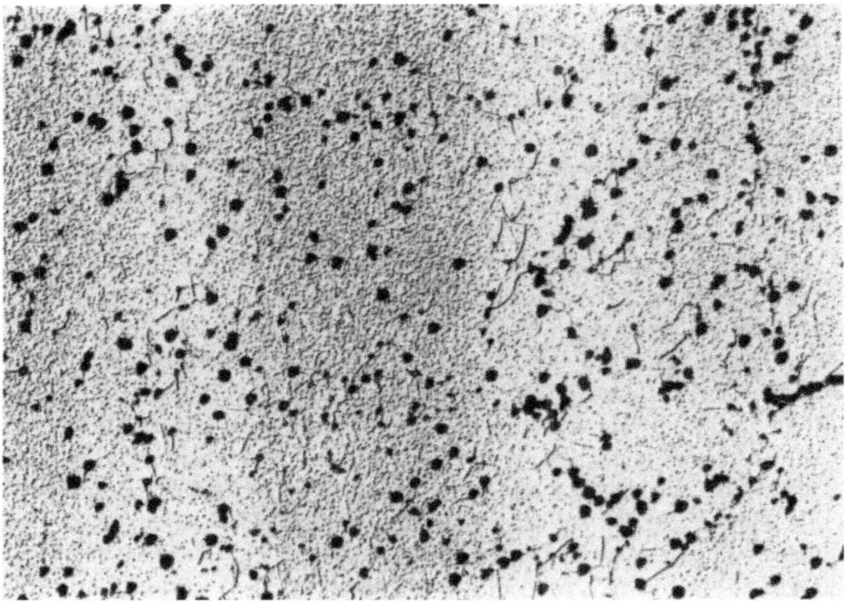

Fig. 47. Accumulation of point defects in striated form. Magnification 60× (Helmreich 1984).

6.3. *Temperature effects*

It has already been outlined, in sections 4.3 (crystallization), 6.1 (crystallographic structure and defects) and 6.2 (impurities), that there are special defect and impurity situations in the ingot resulting from variations in the WICP. In the course of high-temperature treatment during device production this situation cannot be preserved, it is changed through a number of solid-state reactions occurring in the solid medium silicon. Those reactions have rarely been investigated in detail and, thus, are far from any thorough comprehension. In order to develop better insight into the synergistic effects of the various defects and impurities, space-resolved measurement techniques, especially in the microscopic range, as well as time-resolved experiments are a very valuable tool. A few examples for illustration of those temperature effects will be given in the next paragraphs. It is not intended to fully account on temperature-related solid-state reactions in silicon. (In order to emphazise the respective effects, in most of the cases especially prepared material has been studied.)

The release of thermally induced stresses (through annealing at 1200°C / 20 h) is accompanied by the formation of slip lines which are decorated with newly formed dislocations. Slip lines and dislocations are mostly arranged along well-defined crystallographic orientations.

Annealing experiments at 1250°C / 1 h showed that the isotropic distribution of already present dislocations can be transformed into a net-work system. Such a picture is preferredly found in a region with higher impurity content (especially carbon and oxygen).

Due to their strainfield, SiC precipitates act as sinks for impurities and dislocations. Such aggregation phenomena are diffusion-controlled processes which can be influenced by variation of annealing programs. The accumulated impurities and dislocations in the surroundings of SiC precipitates are more likely to act as recombination centers than the SiC precipitates themselves (Helmreich et al. 1982). If the solubility limits of carbon (Yatsurugi et al. 1973) and oxygen (Kolbesen and Mühlbauer 1982) are exceeded, precipitation of both impurities occurs. Annealing of samples with an (intentionally) high carbon and oxygen content at 1250°C / 1 h results in the rearrangement and precipitation of both impurities. The IR spectra of an untreated and an annealed sample (fig. 48) demonstrate the appearance of SiC and SiO_x precipitates through the respective spectral bands. The increase of (interstitial) oxygen concentration through annealing shows that there is an obvious redistribution of formerly precipitated oxygen. The oxygen concentration found after annealing is identical with the solubility limit at the annealing temperature.

Normally, the precipitation of SiC needs temperatures near the melting point of silicon. The SiC-precipitation in C/O-rich samples at temperatures as low as 1250°C obviously takes advantage from the presence of oxygen, which seems to act as a catalyst. Significant precipitation of oxygen can

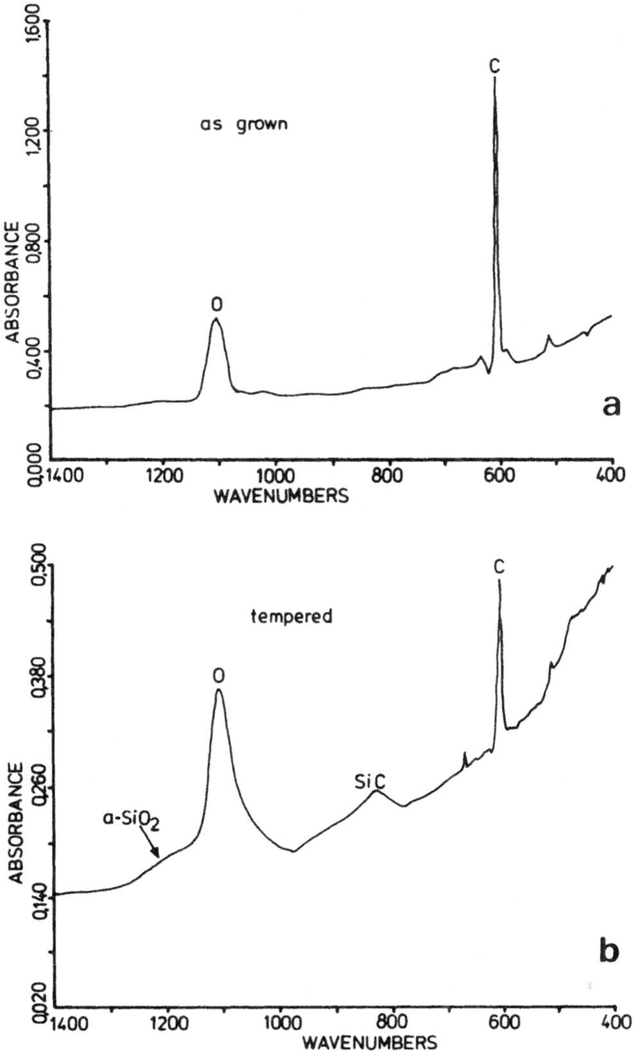

Fig. 48. IR spectrum of a wafer with high carbon and oxygen content: (a) sample in the as-grown state; (b) sample annealed at 1250°C/1 h (Helmreich et al. 1982).

already be expected at temperature conditions which are reached during the phosphorus diffusion (850°C / 1 h). Here again the SiO_x-precipitation is closely related with carbon and it seems to be confirmed that the coexistence of both carbon and oxygen is necessary for microdefect formation (Kishino et al. 1979, Matsushita 1982, Kolbesen and Mühlbauer 1982). In the immediate vicinity of SiC-precipitates the concentration of substitutional carbon may be reduced below a threshold level which inhibits SiO_x-precipitation (Reffle 1982).

The formation (at 1250°C / 1 h) of denuded zones around high-energy incoherent grain boundaries (fig. 49) is initiated through dissolution of SiO_x precipitates in the bulk and diffusion of oxygen to the grain boundaries (ref. E1, Redfield 1981, 1982). Low-energy (coherent) grain boundaries are much more less effective, simple twin boundaries were never found to develop denuded zones. Denuded zones show a quite peculiar photoelectric behaviour: the photocurrent in the core of a denuded zone exceeds that of the extended field of the grain boundary, sometimes the level is even higher than that found in defect-free grains. The more pronounced the denuded zone, the higher the intensity of the photocurrent

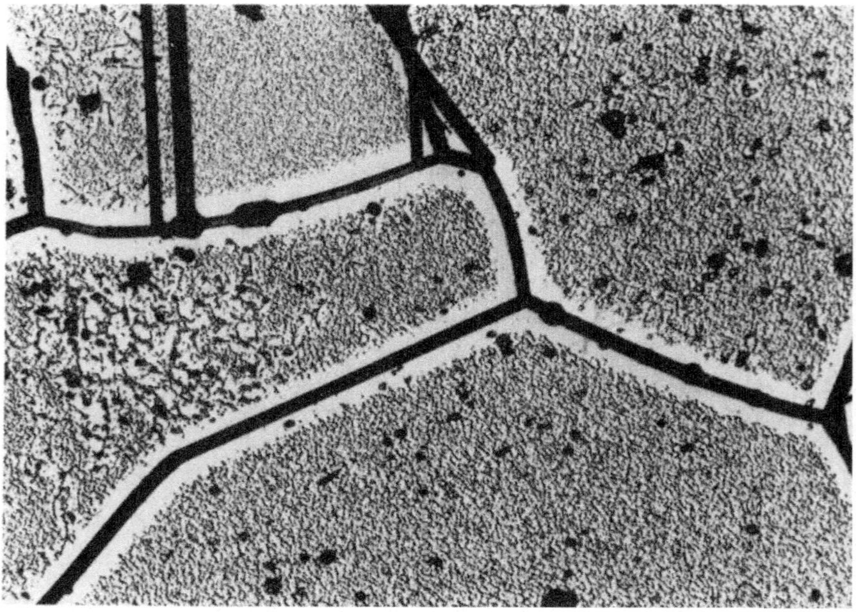

Fig. 49. Multigrain silicon wafer with high oxygen and carbon content annealed at 1250°C/ 1 h. Big dots: dislocations; small dots: precipitates (Helmreich et al. 1982).

in the core (see figs. 50–52). It seems likely that this photoelectric behaviour is correlated with the state of impurities (oxygen) in the grain boundary.

Although the list of temperature effects is very incomplete, a few conclusions can be drawn from such observations. Thermal treatment, as it happens already during crystallization as well as afterwards during solar cell preparation, transforms the defect and impurity situation into one with positive and/or negative aspects. In a high-purity material with good PV properties in the as-grown state low-temperature solar cell preparation techniques are recommended to preserve the positive characteristics of the material. In the case of a solar-grade starting material with low impurity content, the defect situation in the bulk can be used advantageously as an internal gettering system. Alternatively, it is possible to neutralize recombination centers through complexing, thus creating harmless aggregates. Each user has to deal with the pros and cons of this material, but the measures have to fit in the complete line. As has been stated recently (Sirtl 1984), the catch word 'solar grade' not only applies to the definition of the starting material for crystallization; it rather has to cover both material

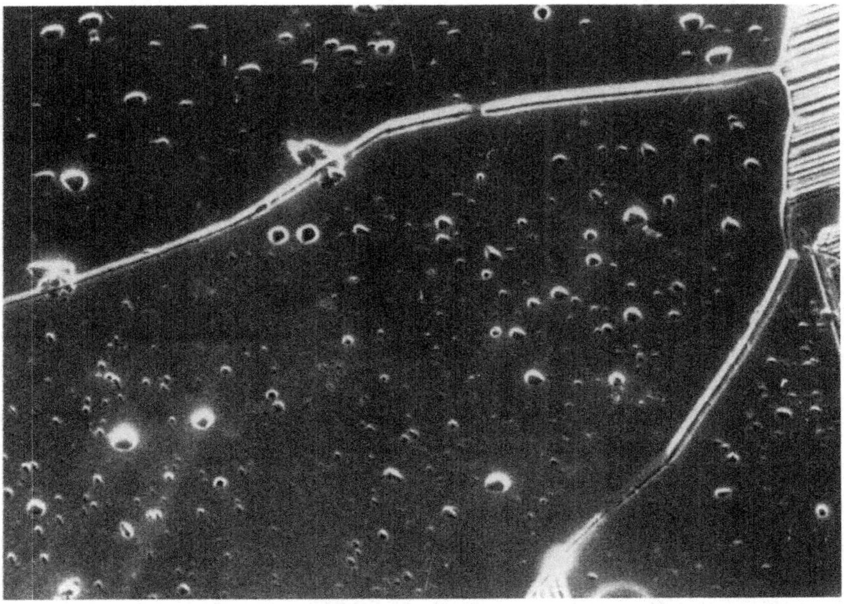

Fig. 50. REM picture of a multigrain silicon sample, heavily doped with oxygen and carbon and annealed at 1250°C/1 h. Magnification 66× (Helmreich et al. 1982).

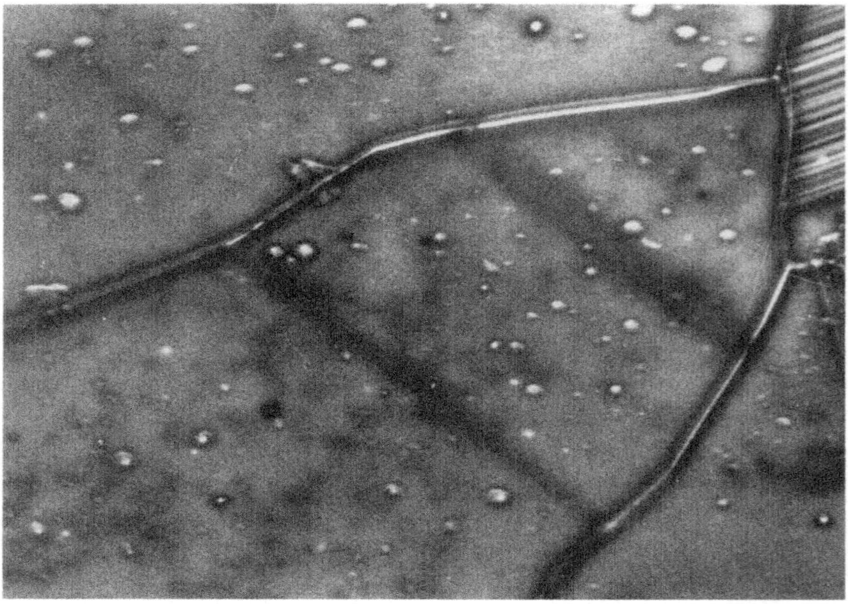

Fig. 51. EBIC picture of the sample given in fig. 50 (Helmreich et al. 1982).

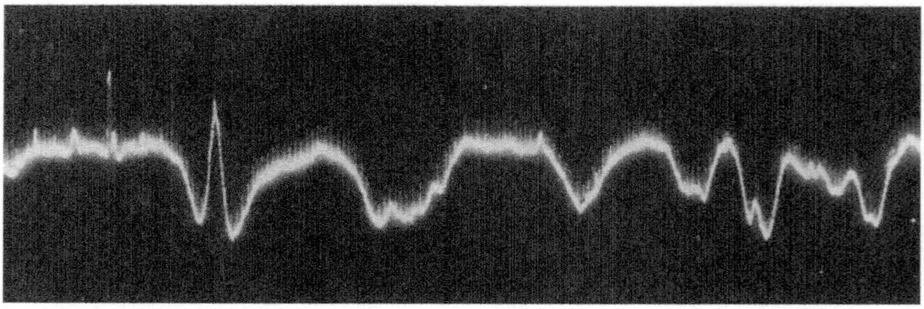

Fig. 52. EBIC line scan of the sample shown in fig. 50 (Helmreich et al. 1982).

characteristics and processing actions in the whole sequence from the basic material to the system.

6.4. *Photovoltaic applications*

During the last several years more than 10 million $10 \times 10\,cm^2$ SILSO® wafers have been processed successfully by a great number of appliers into solar cells with a variety of processing techniques and cell structures. Procedures and processing steps are described in detail in the pertinent literature [see key references (section 8) given at the end of this chapter]. Conventional solar cell preparation techniques which apply diffusion and evaporation techniques (Roy 1982) are equally qualified as unconventional ones relying mainly on screen-printing techniques (Rasch et al. 1982, Cheek et al. 1982, Kimura 1984). Ion implantation has been applied (Siffert 1982) with success as a preparation means as well as a technique based on flash heating (e.g., through high-power lasers; Fogarassy and Siffert 1983, Van der Leeden et al. 1983). MIS technology has proved very qualified in the case of p- and n-type basis material (Green and Godfrey 1977, Ghosh et al. 1980). So far the solar cells for commercial application have been prepared from wafers with semiconductor-grade silicon as the starting material. Within the coming years, solar-grade silicon will be introduced as starting material. This certainly requires the consideration of a new defect and impurity situation.

The prevailing experiences indicate that SILSO® is a material that can be handled without difficulties and that shows sufficient potential similar to what has been experienced with single crystalline silicon prepared by the Czochralski technique.

7. Summary

In this chapter a survey is given on the status of the ingot casting process developed and performed at Wacker. After a sketch of the historical development of this ingot technique, the fundamentals of this process are described in detail. The technological aspects of the individual processing steps are also discussed. As an efficient ingot technology has to be complemented by qualified cutting/wafering techniques one section (section 5) is dedicated to such techniques under development at Wacker.

Material characterization has to focus on those parameters which are important for terrestrial photovoltaic application.

Acknowledgement

I would like to express my gratitude to my co-workers and many colleagues for their engaged cooperation during the procedure of completing this article. Particular thanks go to C. Gessert and D. Miller for their distinguished R&D work in ingot casting, and to H. Seiter and P. Wagner for their excellent work in evaluation and characterization of the prepared materials. I am indebted Prof. E. Sirtl for his continuous invaluable advice in this R&D program and his comprehensive support in the preparation of this article. This study has been supported by the Bundesministerium für Forschung und Technologie of the Federal Republic of Germany under contract number 03E-4506-B.

8. Key references

The following key references are marked by symbols in order to facilitate further listing of the vast literature in the photovoltaic work:

A11. Proc. 11th IEEE Photovoltaic Specialists Conf., Scottsdale, AZ, 1975 (IEEE, New York).

A12. Proc. 12th IEEE Photovoltaic Specialists Conf., Baton Rouge, LA, 1976 (IEEE, New York).

A13. Proc. 13th IEEE Photovoltaic Specialists Conf., Washington, DC, 1978 (IEEE, New York).

A14. Proc. 14th IEEE Photovoltaic Specialists Conf., San Diego, CA, 1980 (IEEE, New York).

A15. Proc. 15th IEEE Photovoltaic Specialists Conf., Kissimee, FL, 1981 (IEEE, New York).

A16. Proc. 16th IEEE Photovoltaic Specialists Conf., San Diego, CA, 1982 (IEEE, New York).

A17. Proc. 17th IEEE Photovoltaic Specialists Conf., Kissimee, FL, 1984 (IEEE, New York).

B1. Proc. 1st EC Photovoltaic Solar Energy Conf., Luxembourg, 1977, (Reidel, Dordrecht).

B2. Proc. 2nd EC Photovoltaic Solar Energy Conf., Berlin, Fed. Rep. Germany, 1979, eds R. van Overstraeten and W. Palz (Reidel, Dordrecht).

B3. Proc. 3rd EC Photovoltaic Solar Energy Conf., Cannes, France, 1980, ed. W. Palz (Reidel, Dordrecht).

B4. Proc. 4th EC Photovoltaic Solar Energy Conf., Stresa, Italy, 1982, eds W.H. Bloss and G. Grassi (Reidel, Dordrecht).

B5. Proc. 5th EC Photovoltaic Solar Energy Conf., Athens, Greece, 1983, eds W. Palz and F. Fittipaldi (Reidel, Dordrecht).

B6. Proc. 6th EC Photovoltaic Solar Energy Conf., London, England, 1985, eds F.C. Treble and W. Palz (Reidel, Dordrecht).

C1. Proc. Symp. Electronic and Optical Properties of Polycrystalline or Impure Semiconductors and Novel Silicon Growth Methods, St. Louis, 1980, eds K.V. Ravi and B. O'Mara (Electrochemical Society, Pennington).

C2. Proc. Symp. Materials and New Processing Technologies for Photovoltaics, Hollywood, 1981, eds J.A. Amick, E. Sirtl, P. Ray-Choudhury, J.P. Dismukes (Electrochemical Society, Pennington).

C3. Proc. Symp. Materials and New Processing Technologies for Photovoltaics, Montreal, 1982, eds J.P. Dismukes, E. Sirtl, P. Ray-Choudhury and L.P. Hunt (Electrochemical Society, Pennington).

C4. Proc. Symp. Materials and New Processing Technologies for Photovoltaics, San Francisco, CA, 1983, eds J.A. Amick, V.K. Kapur and J. Dietl (Electrochemical Society, Pennington).

C5. Proc. Symp. Materials and New Processing Technologies for Photovoltaics, New Orleans, LA, 1984, eds V.K. Kapur, J.P. Dismukes and S. Pizzini (Electrochemical Society, Pennington).

D1. Technical Digest 1st International Photovoltaic Science and Engineering Conf., Kobe, Japan, 1984 (Nippon Press, Tokyo).

For comprehensive information on photovoltaic materials and preparation techniques for semiconductor silicon, see:

E1. Dietl, J., D. Helmreich and E. Sirtl, 1981, 'Solar' Silicon, in: Crystals: Growth, Properties and Applications, Vol. 5, ed. J. Grabmaier (Springer, Berlin).

E2. Gessert, C., and E. Sirtl, 1983, Elektronische Halbleitermaterialien, in: Chemische Technologie Band 3, Anorganische Technologie II, eds Winnacker, Küchler (Hanser, Münich).

E3. Zulehner, W., and D. Huber, 1982, Czochralski-grown Silicon, in: Crystals, Vol. 8, ed. J. Grabmaier (Springer, Berlin).

References

Anderson, J.R., 1980, Wire Saw for Low Damage, Low Kerf Loss Wafering, in: **A14**.

Aubert, J.J., J.J. Bacmann, A. Bourret, J. Daval, J. Desseaux and A. Rocher, 1980, Silicon Bicrystal: Growth and Charaterization, in: **B3**.

Authier, B.A., 1974, Neue Siliciumkristalle für Solarzellen, Final Report Research Program RV 11-V 70/74-TO-BB 74, BMFT/DFVLR.

Authier, B.A., 1975, German Patent (DOS) 25 08 803.

Authier, B.A., 1978, Polycrystalline Silicon with Columnar Structure, in: Advances in Solid State Physics, Vol. XVIII, ed. J. Treusch (Vieweg, Braunschweig).

Bradshaw, S.E., and G.K. Sargeant, 1955, Improvements in or Relating to the Processing of Silicon, Austral. Patent 200 405.

Cheek, G., R. Janssen, M. Leempoels, L. Frisson, R. Mertens and R. van Overstraeten, 1982, Polycrystalline Silicon Solar Cells Utilizing an Integral Screen Printing Technique, in: **B4**.

Ciszek, T.F., G.H. Schwuttke and K.H. Yang, 1979, J. Cryst. Growth **46**, 527.

Cullity, B.D., 1956, Elements of X-ray Diffraction (Addison–Wesley Reading, MA) p. 290.

Daud, T., J.K. Liu and G. Fiegl, 1981, Economics of Ingot Slicing with an Internal Diameter Saw for Low-Cost Solar Cells, in: **A15**.
Dietl, J., 1983, Advances in the Refining of Metallurgical-grade Silicon, in: **C4**.
Ermer, W., D. Helmreich, D. Regler and D. Seifert, 1984, Current Aspects of Silicon Cutting Techniques at Wacker, in: **D1**.
Ermer, W., D. Helmreich, A. Moritz and K. Mühlbauer, 1985, Ingot Cutting for Silicon Wafer Production, in: **B6**.
Fally, J., and C. Guenel, 1980, Study of the Elaboration of Semicrystalline Ingots, in: **B3**.
Fiegl, G.F., and A.C. Bonora, 1980, Low Cost Monocrystalline Silicon Sheet Fabrication for Solar Cells by Advanced Ingot Technology, in: **A14**.
Fischer, H., 1978, Solar Cells Based on Nonsingle Crystalline Silicon, in: Advances in Solid State Physics, Vol. XVIII, ed. J. Treusch (Vieweg, Braunschweig).
Fischer, H., and W. Pschunder, 1976, Low Cost Solar Cells Based on Large Area Unconventional Silicon, in: **A12**.
Fleming, J.R., S.C. Holden and R.G. Wolfson, 1979, Slicing of Silicon into Sheet Material, Final Report DOE/JPL contract no. 954 374-77/5.
Fogarassy, E., and P. Siffert, 1983, Laser Induced Diffusion for the Preparation of High Efficiency Silicon Solar Cells, in: **B5**.
Fujisaki, M., M. Geshi, T. Yamada, K. Tabata, N. Mizukami and T. Saito, 1984, Large Size Cast Silicon for Low Cost Solar Cells, in: **D1**.
Gessert, C., and D. Helmreich, 1985, Recent Advances in SILSO Casting, in: **B6**.
Ghosh, A.K., T. Feng and H.P. Maruska, 1980, SnO_2/Polycrystalline Silicon Solar Cells, in: **A14**.
Green, M.A., and R.B. Godfrey, 1977, The MIS and MISIM Solar Cell, in: **B1**.
Helmreich, D., 1980, Current Aspects in Silicon Casting, in: **C1**.
Helmreich, D., 1982a, Directional Solidification of "Solar" Silicon, in: **C3**.
Helmreich, D., 1982b, J. Phys. (France) **43**, Suppl. Nr. 10, C-289.
Helmreich, D., 1984, Kristallisationsverfahren für Solarsilicium, in: Statusbericht Photovoltaik (Deutsche Gesellschaft für Sonnenenergie e.V., München).
Helmreich, D., 1985, Experiences in Casting Silicon for Terrestrial PV-Application, in: **C5**.
Helmreich, D., and G. Ast, 1980, Microsegregation in Cast Multigrain Silicon, in: **C1**.
Helmreich, D., and H. Seiter, 1979, Polycrystalline Silicon and its Characterization, in: **B2**.
Helmreich, D., and F. Suk, 1982, Experiences with Open Hearth Casting of Multicrystalline Silicon, in: **C3**.
Helmreich, D., H. Seiter, D. Huber and R. Wahlich, 1982, Defect Interactions in "Solar" Silicon, in: **A16**.
Hwang, H.L., C.Y. Sun, C.H. Chu and L.R. Hwang, 1984, On the Growth of Polycrystalline Si Crystals by Gas Assisted Solidification, in: **D1**.
Kazmerski, L.L., 1984, Grain Boundaries in Polycrystalline Silicon Solar Cells: Impurities and Passivation, in: **B5**.
Kazmerski, L.L., and P.E. Russell, 1982, J. Phys. (France) **43**, Suppl. Nr. 10, C1-171.
Khattak, C.P., and F. Schmid, 1978, Directional Solidification of Crack-Free Silicon Ingots by Heat Exchanger Method, in: **A13**.
Kimura, K., 1984, Recent developments in Polycrystalline Silicon Solar Cells, in: **D1**.
Kishino, S., Y. Matsushita and M. Kanamori, 1979, Appl. Phys. Lett. **35**, 213.
Kolbesen, B.O., and A. Mühlbauer, 1982, Solid State Electron. **25**, 759.
Lauvray, H., A. Talpied and J.P. Basselere, 1980, New Wire Silicon Slicing Technology for Solar Cell, in: **B3**.

Leung, D., and P.A. Iles, 1984, Effect of Localized Polycrystalline Silicon Properties on Solar Cell Performance, in: **A17**.

Lindmayer, J., 1976, Semi-Crystalline Silicon Solar Cells, in: **A12**.

Matsushita, Y., 1982, J. Cryst. Growth **56**, 516.

Pizzini, S., L. Braicovich, L. Calliari, M. Gasparini, C.M. Mari, F. Redaelli and M. Sancrotti, 1982, Segregation of Impurities at Grain Boundaries and other Compositional Inhomogeneities in Chill-Cast Silicon Ingots, in: **B4**.

Pryor, R.A., 1980, Empirical Study of the Interaction of Silicon Substrate Thinness, Device Design and Solar Cell Processing, in: **A14**.

Pschunder, W., and H. Bebermeier, 1974, Experimentalstudie zur Verbesserung der Technologie und Elektrischen Lesitung von Silicium-Solarzellen, Final Report Research Program, RV 11-V 70/74 (59)-TO-BB 74, BMFT/DFVLR.

Rasch, K.-D., K. Roy, W. Schmidt and G. Wahl, 1982, Low Cost Processes for Cast Silicon Solar Cells, in: **B4**.

Redfield, D., 1981, Grain Boundaries in Silicon as Analogs of Surfaces, in: **A15**.

Redfield, D., 1982, Appl. Phys. Lett. **40**, 163.

Reffle, J., 1982, Private communication.

Regnault, W.F., and K.C. Yoo, 1984, Lifetime Enhancement in Semicrystalline Silicon by Structural Defect Density Reduction, in: **A17**.

Ross, J.R., 1980, Final Report DOE/JPL Contract no. 955 563.

Roy, K., 1982, Processing of Multicrystalline Silicon Into Solar Cells, in: **C3**.

Saito, T., M. Fujisaki, N. Mizukami, M. Geshi, T. Yamada and K. Tabata, 1984, Impurity Segregation to Grain Boundaries in Directional Solidification for Silicon Solar Cells, in: **D1**.

Schmid, F., C.P. Khattak, M.B. Smith and L.D. Lynch, 1982, Recent Development in Multi-Wire Fixed Abrasive Slicing Technique (FAST), in: **B4**.

Siffert, P., 1982, Possibilities of Ion Implantation in Silicon Solar Cell Manufacturing, in: **B4**.

Sirtl, E., 1984, The Status of (Crystalline) "Solar Grade" Silicon, in: **D1**.

Van der Leeden, G.A., R.T. Young, R.L. Sandstrom, R.F. Wood and R.D. Westbrook, 1983, Beam Processing for High-Efficiency Silicon Solar Cells, in: **B5**.

Werner, P.G., 1982, Advanced Slicing Techniques, in: **B4**.

Werner, P.G., E. Minke and W. Burkhardt, 1982, Abschlussbericht zu BMFT-Forschungsvorhaben, Bremer Institut für Angewandte Stahltechnik (BIAS), Bremen.

Wolf, M., 1975, J. Vac. Sci & Technol. **12**, 984.

Yatsurugi, Y., N. Akiyama, Y. Endo and T. Nozaki, 1973, J. Electrochem. Soc. **120**, 975.

Zauhar, H., B. Authier, R. Luptovits and L. Schmidhammer, 1979, Process and Apparatus for the Semicontinuous Production of Silicon Moldings, US-Patent 4 175 610.

CHAPTER 3

GROWTH OF SILICON INGOTS BY HEM FOR PHOTOVOLTAIC APPLICATIONS

C.P. KHATTAK and F. SCHMID

Crystal Systems, Inc.
27 Congress Street
Salem, MA 01970, USA

Silicon Processing for Photovoltaics II, edited by C.P. Khattak and K.V. Ravi

Contents

1. Introduction

Photovoltaic technology has been demonstrated for space applications. For terrestrial use, however, cost is the main obstacle to the commercialization of this alternate energy technology. Vigorous research and development programs in the USA, Europe and Japan have led to alternate approaches towards adaptation of photovoltaics for terrestrial use. Orders of magnitude cost reductions have been achieved in the price of photovoltaic modules; however, still further economies have to be achieved for widespread use of this alternate energy technology. One of the key areas of emphasis is still the cost of making silicon sheets as it constitutes about half the cost of the solar modules.

Present day modules use bulk silicon wafers for fabricating solar cells; these panels almost exclusively use wafers sliced from bulk single crystals grown by the Czochralski process from electronic-grade polysilicon produced by the Siemens process. This was the conventional approach used for space cells and economics have been achieved through development of existing technologies. In addition, a number of new technologies have been developed which, to date, have not seen widespread commercial application, but which offer potential for further cost reductions in large-scale manufacturing. One of these new approaches is the heat exchanger method (HEM) which will be discussed in detail; some of the other technologies are described elsewhere in this volume.

The HEM is a directional solidification technique which has been adapted for the growth of large, square cross-section silicon ingots from the melt. The process and its adaptation for silicon as well as development of alternate approaches in reducing total silicon sheet costs for terrestrial photovoltaic applications will be discussed.

2. Heat exchanger method (HEM) of ingot growth

The salient features of an HEM furnace for growth of material under reducing or neutral atmosphere are shown in fig. 1. The furnace consists of a graphite heat zone backed by layers of graphite insulation. This assem-

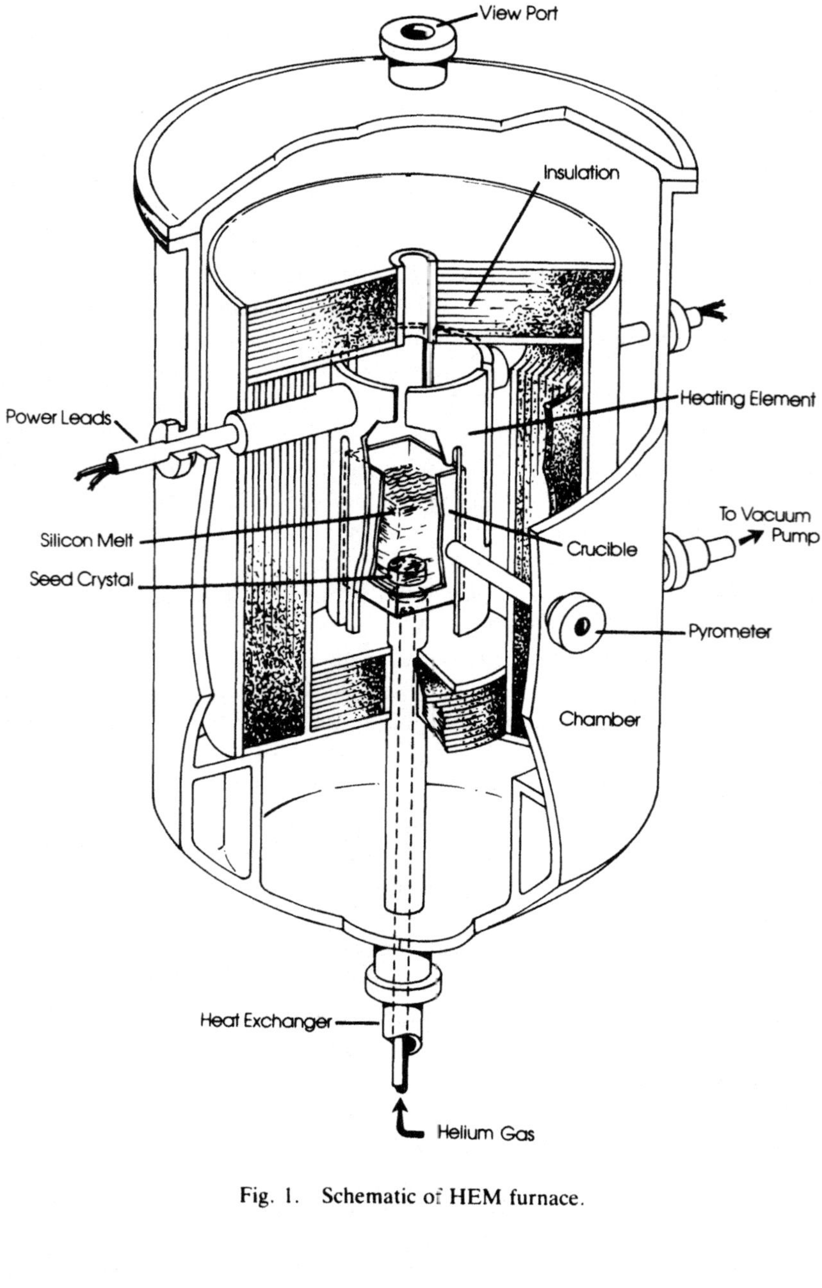

Fig. 1. Schematic of HEM furnace.

bly is placed in a vacuum-tight, water-cooled stainless steel chamber. Heat is supplied by a picket-fence type graphite heater resistively powered by an appropriate three-phase power supply. A high-temperature heat exchanger (Schmid and Viechnicki 1972) is inserted through the bottom of the chamber and heat zone. This heat exchanger is a closed end tube with an injection tube for flowing helium gas as coolant. There are no moving parts in the HEM furnace; this minimizes the seals required. Further, the solid–liquid interface is submerged below the melt hence only a small observation port is incorporated at the top of the furnace. Other ports in the furnace are for evacuation and for control and measurement pyrometers. These features allow a design of a well insulated heat zone. The control instrumentation is tied to a standard, dual-channel microprocessor which can be easily programmed for heat input as well as heat extraction.

The heat zone is designed such that with no coolant flow through the heat exchanger, there are no significant gradients built in the furnace. This is achieved with thermal symmetry, multi-layer insulation all around the heat zone and minimization of sight ports. Some natural temperature gradients are expected, for instance, at the edges of the heating element. The isotherms in an HEM furnace at various locations and under different conditions are shown qualitatively in fig. 2. Under these conditions the temperature along the crucible wall is nearly constant in the HEM furnace. This feature distinguishes HEM from gradient freeze techniques.

The heat exchanger method (HEM) (Schmid and Viechnicki 1970; Viechnicki et al. 1972, Schmid and Viechnicki 1973, Viechnicki and Schmid 1974, Khattak and Schmid 1978a,b, Schmid and Khattak 1981, 1983) has been developed to grow large high-quality crystals. A seed crystal is placed at the bottom of the crucible which is seated on a high-temperature heat exchanger. The silicon feedstock is then loaded into the crucible on top of the seed crystal. After evacuation, heat is supplied by the graphite heater and the charge is melted. The seed is prevented from melting by forcing minimal gaseous helium flow through the heat exchanger. After meltback of the seed, growth is progressed by increasing the flow of helium and thereby decreasing the heat exchanger temperature.

In essence, this method involves directional solidification from the melt where the temperature gradient in the solid is controlled by the heat exchanger, and the gradient in the liquid is controlled by the furnace temperature. After solidification is complete, the gas flow through the heat exchanger can be decreased to equilibrate the temperature throughout the crystal during the annealing and cooldown stage.

This technique is unique in that the liquid temperature gradient can be controlled independently of the solid gradient without moving the crucible, heat zone or ingot. The most significant feature is the submerged interface which is stabilized by the surrounding liquid. It is protected from hot spots, mechanical vibration and convention currents. Consequently, rotation of the crucible is not necessary to achieve thermal symmetry.

Growth with a submerged interface makes HEM ideally suited for low-

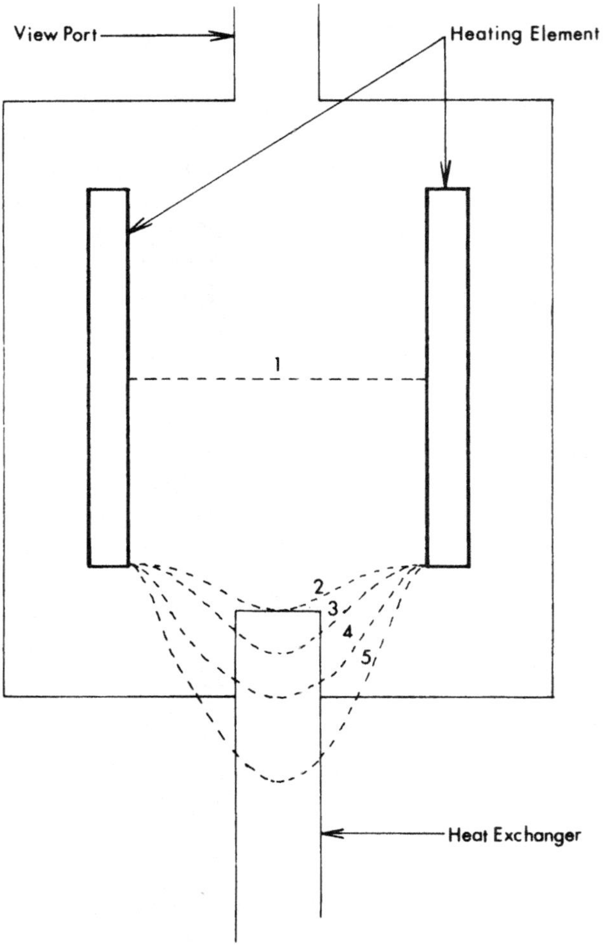

Fig. 2. Isotherms in an HEM furnace near the center of the heating element (1), and near the bottom of the furnace (2–5). Isotherms (2) through (5) show the increasing effect of the heat exchanger.

purity silicon where many of the second phase contaminants, such as carbides and oxides, tend to float on the surface of the melt, away from the growing interface. The melt acts as a buffer and protects the submerged solid–liquid interface during most of the growth cycle. Therefore, the temperature and concentration fluctuations at the interface are minimized in HEM because of the surrounding liquid. During growth the colder material is at the bottom and the hotter melt is on the top; this minimizes convection and therefore, growth occurs under stabilizing temperature gradients (Kim et al. 1972). The minimization of temperature and concentration fluctuations along with stabilizing temperature gradients minimize constitutional supercooling and promote uniform growth. This results in high crystal perfection and chemical homogeneity. This salient feature accounts for the unique capability of HEM to grow a nearly single-crystal ingot in one solidification, using commercially available metallurgical-grade silicon as meltstock (Schmid et al. 1980).

As ingot growth proceeds the size of the interface increases as shown in fig. 3 (Schmid 1975). Therefore, high growth rates are achieved with larger size ingots. As the distance of the interface from the heat exchanger increases linear movement of the interface is slowed; however, volume growth rates are still increasing because of the larger size of the interface. This feature is significant when low-purity meltstock is directionally solidified by HEM. As growth proceeds impurities are rejected to the liquid because of segregation effects; however, their effect is minimized because of the increasing size of the interface. As more and more impurities are piled up in the liquid, the slowing linear growth rate suppresses constitutional supercooling.

The shape of the solid–liquid interface is an important criterion in rejection of impurities during growth. Basically, there are three techniques in which the ingot is grown in a crucible and there is a somewhat submerged solid–liquid interface. These techniques are conventional casting, Bridgman or gradient freeze approach and the heat exchanger method (HEM). In conventional casting the first material to solidify is along the bottom and walls of the crucible. Preferential heat extraction through the bottom gives it a concave-shaped interface as shown in fig. 4a. The convective flow in the liquid is also shown. For Bridgman or gradient freeze approach, the heat extraction is also through the walls of the crucible and a similar pattern is achieved as shown in fig. 4b. In the HEM, ideally, to maintain single crystallinity throughout the entire ingot, growth has to take place off the solid–liquid interface. Heat flow has to be maintained such that all heat extraction occurs through the seed and no spurious

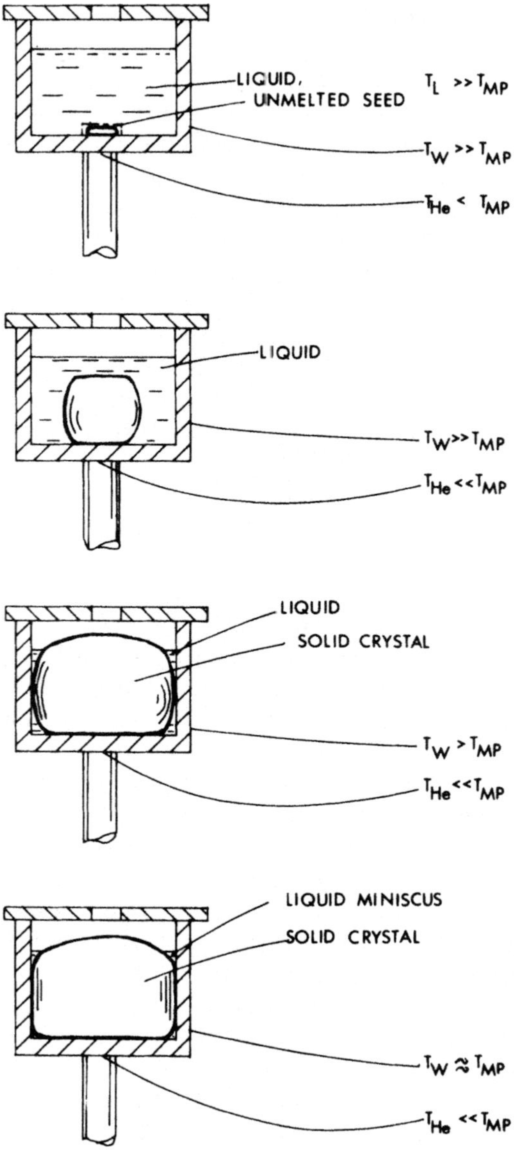

Fig. 3. Illustration of various stages of growth in HEM from 'seeding' (top) to fully solidified ingot (bottom).

nucleations occur off the crucible. A convex-shaped interface is, therefore, achieved in HEM [fig. 4c].

Silicon undergoes an expansion on solidification so that as growth proceeds the liquid level rises in the crucible. In directional solidification, therefore, the last material to solidify is at the highest point in the crucible. The shape of the top surface of the ingot will be determined by the shape of the interface. Figure 5 is a representation of the cross-section of ingots parallel to the growth direction. For conventional casting and Bridgman type growths with concave interface the central region is higher, as shown in fig. 5a, whereas for HEM growth with convex interface the edges are higher [fig. 5b]. Because of directional solidification, the impurities are rejected to the last material to freeze. For casting and Bridgman growth,

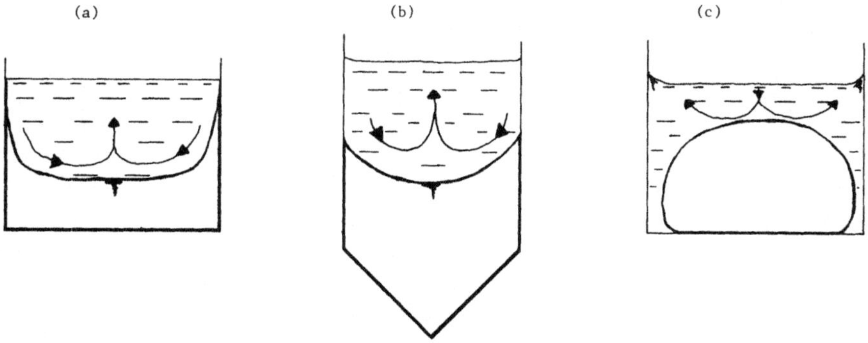

Fig. 4. Shape of the solid–liquid interface, convective flow in the liquid and rejection of impurities in: (a) conventional casting, (b) Bridgman or gradient freeze, and (c) heat exchanger method.

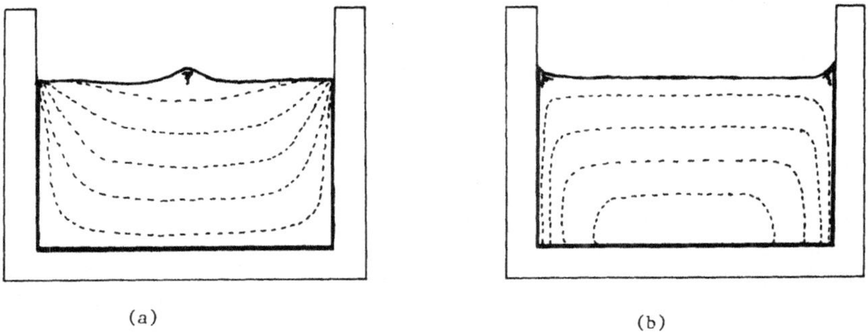

Fig. 5. Representation of shape of solid–liquid interface and final shape of the ingot for: (a) conventional casting or Bridgman, and (b) heat exchanger method.

this is in the top central region of the ingot, whereas for HEM it is near the top edges. In production, all crucible-grown ingots have to be trimmed to maintain dimensional accuracy. In the case of HEM ingots the trimmed material contains most of the impurities.

In HEM, the stability of the submerged solid–liquid interface is evident from the fact that, when particles were entrapped on the interface, growth progressed around the particle without breakdown in structure (Digges et al. 1976). The absence of high local gradients at the interface ensures the growth off the interface in preference to off the particle. This is contradistinct to the Czochralski process where such entrapment would cause spurious nucleation and, therefore, multi-crystalline growth (Voltmer and Padovani 1973).

The controllable heat exchanger of HEM allows precise control of the temperature and temperature gradients at the bottom of the crucible. This precise control of the interface also allows high growth rates under low temperature gradients. This reduces solidification stresses that cause defect formation. Further, in situ annealing of the ingot can be accomplished after growth is complete, since the boule does not move out of the heat zone during solidification. This is accomplished by reducing the temperature of the furnace just below the solidification temperature and then reducing the helium flow – the whole ingot can, therefore, be brought to high temperatures and then cooled uniformly at a controlled rate. This further reduces internal stresses, and a costly, separate annealing step is eliminated. This annealing and controlled cooldown prevents cracking due to thermal shock, thereby allowing large ingots to be produced.

The simple nature of HEM allows the use of the same furnace for a range of sizes and shapes of ingots. Silicon boules of 15 cm and 20 cm diameter as well as 10 cm and 20 cm square cross-section ranging from 3 kg to 20 kg were grown using the same HEM furnace with only minor modifications to the heat zone. Similarly 20 cm and 34 cm square cross-section and 40 cm diameter ingots ranging from 18 kg to 50 kg were solidified in the larger HEM furnace.

3. Growth of high-purity silicon ingots

HEM has been adapted for the growth of silicon ingots using high-purity silicon meltstock. Besides optimization of directional solidification parameters for silicon, some of the unique features necessary to maintain quality of the product were development of a high-purity silica crucible and techniques for processing under vacuum.

3.1. High-purity silica crucible

A significant development in the adaptation of HEM for silicon is the use of a high-purity silica crucible (Khattak and Schmid 1980). Silica is the container for silicon in the semiconductor industry because of its availability in high-purity form as well as its non-contaminant behavior in contact with silicon. The silica glass crucibles as they exist for the Czochralski process are not suitable for directional solidification of silicon because of the historical ingot cracking problem (Hino and Stauss 1952). Molten silicon reacts with silica to form SiO; on solidification a tenacious bond is formed (Khattak and Schmid 1978b) via SiO_x $(0 \leqslant x \leqslant 2)$. During cooldown, because of differential thermal expansion coefficients of silicon (Gibbons 1958, Maissel 1960) and silica (Powell et al. 1966), silicon is in tension and silica is in compression. Below 650°C, the ductile-brittle transition temperature for silicon (Pearson et al. 1957), the ingot is cracked to pieces. Above this temperature silicon flows plastically to strain relieve itself. Various types of crucibles were studied to prevent cracking of silicon ingots. Best results were achieved with a graded silica crucible (Khattak and Schmid 1978a, 1980). This is a high-purity, slip-cast silica crucible which is heat treated to develop a graded structure in density as well as phase composition. This crucible sinters to high density during the heat-up cycle and contains molten silicon; during cooldown stage the graded structure causes delamination of the crucible and thereby prevents cracking as well as stresses in the ingot. The non-contaminant, high-purity silica crucible is an important part of HEM for containment of molten silicon and retention of its high-purity state. In addition, as is discussed in section 4.3, this crucible aids in refinement when low-purity silicon meltstock is used. Such crucibles have been used to solidify 50 kg, 32 cm square cross-section silicon ingots (Schmid et al. 1983).

3.2. Vacuum processing

An important feature of HEM processing of silicon is operation under vacuum. After loading the furnace, the chamber is evacuated by a mechanical pump, to typically 0.1 Torr, and the meltdown, solidification and cooldown cycles are carried out under vacuum conditions. Molten silicon reacts with the silica crucible (Erasmus and Persson 1949) to yield a gaseous product, SiO

$$Si + SiO_2 \rightarrow 2SiO. \qquad (1)$$

This reaction is a strong function of pressure and temperature (Schmid et al. 1979). Under vacuum conditions the SiO is continuously removed, thereby driving the reaction to the right; this leads to erosion of the crucible. In the Czochralski process with convection in the crucible and continuous stirring of the melt reaction (1) is enhanced. Considerable grooving of the crucible is, therefore, observed in the area of contact with melt after a typical Czochralski crystal growth, even when the chamber pressure is maintained at about 20 Torr. In the HEM system under 0.1 Torr pressure even when molten silicon was contained in a silica crucible for 24 hours no appreciable grooving reaction was observed. This is attributed to stabilizing temperature gradients and absence of stirring. Under these conditions a boundary layer of SiO may be trapped between the molten silicon and silica crucible which impedes further erosion of the crucible.

An additional feature observed with vacuum operation was the contamination of silicon by silicon carbide. A detailed thermodynamic analysis of various reactions as a function of temperature and pressure led to the origin of SiC impurities (Schmid et al. 1979). Under vacuum the following reactions between the graphite retainer and silica crucible led to the formation of CO,

$$C + SiO_2 \rightarrow SiO + CO, \tag{2}$$

$$3C + SiO_2 \rightarrow SiC + 2CO. \tag{3}$$

The CO, so generated, was transported to the molten silicon to form SiC,

$$2Si + CO \rightarrow SiO + SiC. \tag{4}$$

Introduction of a barrier, such as a refractory metal sheet, between the crucible and retainer prevented the formation of CO and thereby reduced SiC contamination.

3.3. Single-crystal ingots

Initial emphasis in the adaptation of HEM for silicon was on growth of large ingots with a single-crystal structure. During early stages 15 cm diameter, 3 kg ingots were produced. Scale-up in size of ingots was continued to 10 cm, 15 cm and 20 cm square ingots in the same furnace. Figure 6 shows the structure of a slab sectioned parallel to the growth direction; a 20 cm square cross-section ingot with rounded corners is also shown in this figure. This was the largest size of the ingot which could be solidified in the experimental furnace. In the regions where single-crystallinity was

Fig. 6. A 20 cm square cross-section ingot solidified by HEM along with a typical cross-section showing crystal structure.

Fig. 7. A view of a 45 kg, 32 cm square cross-section ingot grown by HEM.

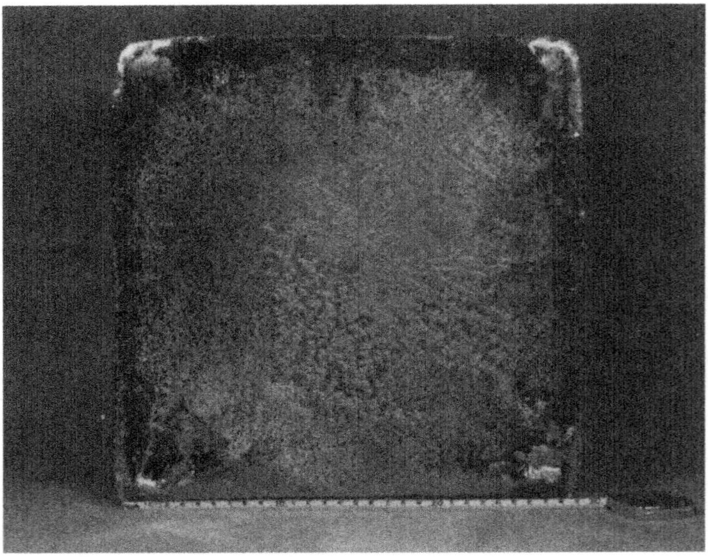

Fig. 8. A view of the top surface of the ingot shown in fig. 7.

not maintained very large-size columnar grains were formed.

After it was established that large silicon ingots can be grown by HEM and that the material showed a solar cell performance (Khattak and Schmid 1980) similar to Czochralski silicon, efforts were directed for further scale up of the size of the ingots. Emphasis was placed on producing 32 cm square cross-section ingots. With large cross-section the size of the interface is enlarged and, consequently, high growth rates are achieved. The 32 cm square cross-section is useful for sectioning nine 10 cm square bars or other convenient shapes to yield rectangular or square wafers in order to achieve high packing density of solar cells in photovoltaic modules.

Figure 7 shows a 45 kg, 32 cm square cross-section silicon ingot grown by HEM. A view of the top surface of such an ingot is shown in fig. 8. The corners of the ingot are rather sharp and the sides are flat; this is important to achieve high yields of rectangular/square wafers.

During early stages of 32 cm square ingots, single crystallinity was achieved in the central region with large, columnar grain structure on the sides. Efforts to prevent breakdown in single crystallinity showed improvement, as shown in fig. 9. In the latter case spurious nucleation of grains was limited to the periphery of the melted-back seed with the crucible. As

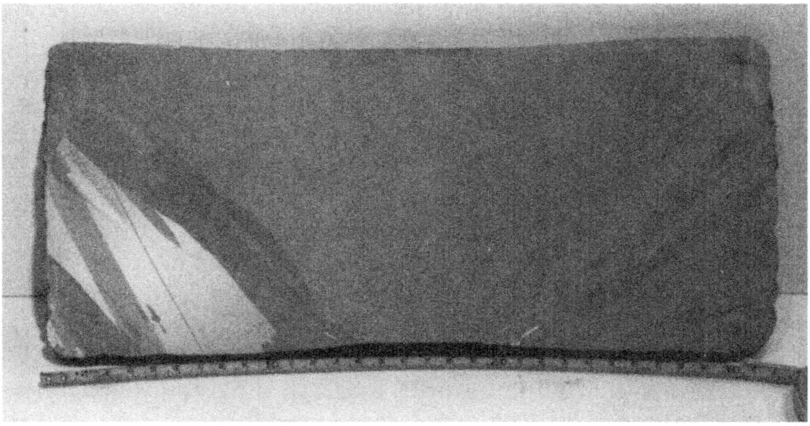

Fig. 9. Typical cross-section of a 32 cm square ingot showing a nearly single-crystal central region and very large grains near the edges.

growth progressed, these grains expanded to a few cm size. Two of the initial 35 kg, 32 cm square ingots grown under non-optimized conditions were characterized in detail for solar cell applications (Dumas et al. 1981). It was found that the useable material averaged over the entire ingot showed an 85% solar cell performance as compared to the control cells. In comparison with other sheet growth technologies the HEM material has shown the best performance (Yoo and Liu 1980, Leung et al. 1984); the average performance of HEM test cells was 15% (AM1) efficiency with the best cell at 15.5%.

3.4. Multicrystalline ingots

During investigation of solar cell performance of HEM silicon it was demonstrated that there was no significant variation between the single-crystal and multicrystalline material (Dumas et al. 1981). For large-scale production operation it may be desirable to maintain uniform quality for various processing steps. For example, it is desirable to use a non-preferential etch for multicrystalline wafers (Roy and Pschunder 1980). In view of such practical parameters, emphasis was placed on solidifying ingots with controlled, large-size columnar grains. The adoption of this practice resulted in elimination of the 'seeding' step and thereby incorporated more automation in the processing cycle. As a result faster meltdown and higher growth rates were achieved with less labor supervision. Silicon ingots of 50 kg,

32 cm square cross-section were directionally solidified; The resultant material was more uniform in quality (Schmid et al. 1983).

Another important feature in multicrystalline growth is the shape of the solid–liquid interface. This shape controls the grain size and orientation of the grain boundaries. In conventional casting and Bridgman growth, a concave-shaped interface is formed (figs. 4a,b) which results in very small grains being formed near the top central region towards the end of growth cycle (Roy et al. 1980). In these situations growth occurs from the bottom of the crucible as well as the sides. The control of the interface is minimal and it results in impinging interfaces (Helmreich 1980). These regions yield poor performance in solar cells (Legros et al. 1980, Schmidt et al. 1982). In HEM growth a convex interface is formed (fig. 4c) which favors enlargement of grain size as growth progresses. Further, the heat extraction is only through the heat exchanger; hence, there is no spurious nucleation along the walls of the crucible. The high-purity silica crucible in HEM growth does not contaminate the silicon melt, hence, it is not necessary to form a chilled silicon layer on the surface of the crucible to prevent reactions with the crucible. The combination of these features allow formation of nearly vertically oriented, columnar, grain boundaries with no impinging interfaces.

The controllable heat exchanger of HEM allows precise control of the temperature and temperature gradients at the bottom of the crucible. This feature allows flexibility of nucleation with respect to the size and type of grains as they relate to the purity of the meltstock. Large grains can, therefore, be nucleated which expand in size as growth proceeds. Heat extraction from the melt can be optimized such that the orientation of the grain boundaries is nearly vertical. Figure 10 shows the structure of a 32 cm square cross-section ingot where large grains of uniform size and vertical grain boundaries were produced using multicrystalline growth procedures. A section of such an ingot perpendicular to the growth direction shows that a few-centimeter-size grains are formed (fig. 11). The uniformity of material is also demonstrated by the narrow range of resistivity values for different areas in the 32 cm square cross-section, 40 kg ingot, as shown in fig. 10.

The potential of HEM silicon using high-purity meltstock has been demonstrated for photovoltaic applications. The material has been evaluated by a number of solar cell manufacturers as well as research laboratories (Cheek et al. 1982, Schmidt et al. 1982). Based upon these results, HEM silicon has been suggested as a strong candidate for near-term, widespread use of low-cost photovoltaics as an alternate energy technology (Barnett

	1	2	3	4	5	6	7	8	9	10
A	0.82	0.92	1.09	1.16	1.15	1.20	1.10	1.15	0.92	0.92
B	1.06	1.18	1.31	1.43	1.51	1.32	1.34	1.35	1.29	1.04
C	1.36	1.44	1.54	1.53	1.61	1.55	1.56	1.55	1.44	1.15
D	1.50	1.60	1.61	1.66	1.78	1.76	1.75	1.64	1.58	1.47
E	1.82	1.76	1.85	1.71	1.81	1.71	1.88	1.71	1.75	1.72
F	1.87	1.84	1.97	1.90	1.86	1.79	1.81	1.87	1.91	1.94

Fig. 10. Vertical cross-section of a multi-crystalline HEM ingot, showing uniform grain size and vertical orientation of the grain boundaries. Resistivity mapping of the surface shows a very narrow range of resistivity values (Ω cm, p-type).

1982). Figure 12 shows an HEM silicon ingot sectioned into nine 10 cm × 10 cm bars, ready for wafering and solar cell fabrication.

4. Growth of lower-purity silicon ingots

In order to make photovoltaics a reality for terrestrial applications it is necessary to reduce the cost of the silicon meltstock used for ingot growth. Present-day electronic-grade silicon is too expensive, and commercially available metallurgical-grade silicon contains a lot of impurities. A number of intermediate grades are under development, and some of these have been evaluated to study the potential of using HEM as well as understanding the purification of impurities.

Fig. 11. Cross-section of a 32 cm square ingot perpendicular to growth direction, showing very large grain size and uniformity.

4.1. *Solar-grade silicon*

There are only two types of commercially available silicon meltstocks. The first is the electronic-grade (EG) silicon which, except for carbon and oxygen, contains impurities in the ppb levels. Most of this EG silicon is produced by the Siemens process, involving reduction pyrolysis of trichlorosilane. The second grade of silicon available is metallurgical-grade (MG). In comparison to EG silicon, MG is a very dirty form of silicon. It is produced by the carbothermic reduction of quartz in a submerged electrode arc furnace. The purity of MG silicon is 98–99%. While EG silicon is acceptable for photovoltaic applications, it is expensive with today's commercially practiced technologies. On the other hand, MG silicon contains a lot of impurities which need to be refined before it is useful for solar cells.

In order to bridge the gap between the two extremes of silicon meltstocks, EG and MG, the US Department of Energy coined a new

Fig. 12. A 32 cm square ingot sectioned into nine 10 cm × 10 cm bars ready for slicing into wafers.

term: 'solar-grade' (SG) silicon. At the onset, SG silicon was expected to be low-cost and of a purity which was inferior to EG; however, the purity was high enough to produce high-efficiency solar cells. A number of approaches were pursued (Lutwack 1980, Dietl et al. 1981, Pizzini 1982) to develop SG silicon. Meanwhile, research was pursued towards defining of SG silicon by studying the effect of various impurities on the solar cell performance (Davis et al. 1980, Rohatgi et al. 1981). A data base was developed on impurity effects, thermal stability of impurities (Hanes et al. 1981) and tolerance level in single versus multicrystalline wafers (Rohatgi et al. 1982). It was soon recognized that the tolerance of impurities in meltstock varied with different crystallization techniques.

Two main courses were pursued in the development of SG silicon; i.e., chemical route and metallurgical route. While significant progress was made in the chemical approach, the results have been demonstrated on portions of the process in the laboratory stage. Therefore, rather limited quantities have been available for testing. Large pilot plants need to be built to demonstrate the complete sequence of this approach.

The metallurgical approach was initially adopted at Dow Corning, and labelled Direct Arc Reactor (DAR) silicon (Hunt et al. 1977, 1978, Dosaj and Hunt 1981). Their process involved use of high-purity raw materials in an electric arc furnace followed by directional solidification to produce SG silicon. This SG silicon was intended to be used with further crystallization technologies to produce silicon sheet for solar cells. Other metallurgical approaches pursued at Heliotronic (Sirtl 1980, Dietl et al. 1981, 1982, Dietl 1983) were a combination of hydrometallurgical and pyrometallurgical operations as well as aluminothermic reduction technique. A recent review (Amick et al. 1983) by Exxon/Elchem details their study using the DAR approach. This group has also evaluated the use of purified rice hulls (Amick 1982) as the silica source. Other attempts at upgrading MG silicon have been by dissolving it followed by precipitation from solution (Kotval and Strock 1978, Dawless 1981a,b).

While all these processes have been pursued to various stages, none of them has reached commercialization.

4.2. Commercially available MG silicon

Metallurgical-grade (MG) silicon is produced worldwide with a capacity of approximately 600 000 tons per year. Silicon is tapped from a submerged electrode arc furnace after carbonaceous reduction of quartz. The molten metal is poured into a ladle where gas blowing is carried out followed by pouring into shallow cast iron molds for rapid solidification. The technology for MG silicon production is driven more by experience than by science. The purity practices are typical of the metallurgical practices which is very much unlike semiconductor operations.

Commercially available MG silicon is generally of 97–99% purity. In most plants each 'heat' is analyzed for Fe, Al, Ti and Ca and the output is graded, based on these results. The major impurities are, typically, Al and Fe concentrations of a few tenths of one percent, and other metallic impurities such as Mn, Ti and V at concentrations of several hundred parts per million.

MG silicon, as commercially produced today, is a rather inhomogenous material with considerable scatter in impurities content from supplier to supplier as well as from batch to batch for the same producer. This is not surprising because the quality of the MG silicon is dependent on the purity and mix of the raw materials, operating and tapping parameters of the arc furnace, and handling and solidification procedures used in the MG silicon plant.

Samples of commercially available MG silicon were obtained from 11 suppliers, worldwide. In some cases a number of batches were obtained;

in all, 23 batches of MG silicon were received. There samples were analyzed for impurities using spark source emission spectroscopic techniques. Because of the inhomogenous nature of MG silicon as well as multi-impurity problems, there was considerable scatter in the data. Typical impurities for MG silicon from six different sources are listed in table 1; in one case three different batches were analyzed.

4.3. *Processing of MG silicon by HEM*

Directional solidification experiments of commercially available MG silicon has been carried out by HEM (Schmid et al. 1980, Khattak et al. 1981,

Table 1
Spark source emission spectrographic analysis of commercially available MG silicon from six different sources. Carbon analyses are by fusion technique. All values are in ppmw.

Element	KBI[a]	UC[b]	HMC[c]	Brazil[d]	S.Africa[e]	India[f]		
						#1	#2	#3
B	38	<5	<10	<5	<10	<10	<5	<10
Mn	4	55	85	20	170	190	15	5
Mg	6	12	<1	17	<1	<1	2	5
Fe	2000	2800	1150	560	1500	1450	430	480
Al	1300	1600	1000	540	1300	360	54	2100
V	170	320	270	<1	<3	<3	<1	<2
Ti	310	360	170	62	190	98	45	43
Cu	64	58	42	28	10	14	15	235
Ni	104	98	170	<1	38	<1	1	<1
Ca	23	32	100	86	14	350	19	62
Cr	8	160	<1	2	11	<1	<1	<1
Ba	<1	1	7	2	<1	36	1	6
P	<50	<50	<50	<50	<50	<50	<50	<50
Mo	14	4	17	<5	<10	<10	<5	<4
C	130	200	270	30	90	40		

[a] Kawecki Berylco Industries, P.O. 1462, Reading, PA 19603, USA.
[b] Union Carbide Corp., P.O. 72, Marietta, OH 45720, USA.
[c] Hanna Mining Co., 100 Erieview Plaza, Cleveland, OH 44114, USA.
[d] Ligas de Aluminio S.A., Liasa, Sao Saulo, Escritoriohua, Sud Menucci 234, Brazil.
[e] Silicon Smelters, Ltd., P.O. 657, Pietersburg 0700, South Africa.
[f] Indian Metals & FerroAlloys, Ltd., Bomikhal, P.O. Rasulgarh, Bhubaneswar-751010, Orissa, India.

1982, Khattak and Schmid 1983). In these experiments the as-received material, without any cleaning, was used as meltstock and ingots of 15 cm square cross-section were solidified. In all cases a single-crystal structure was achieved in the central area with breakdown in crystallinity near the edges of the ingot. The extent of the single crystal in the boule varied with the source of the meltstock. In some cases nearly the entire ingot was single crystal (fig. 13).

The unique result of achieving single-crystal structure with one directional solidification of MG silicon is attributed to the following processing features of HEM:

(i) The high-vapor pressure impurities are volatilized under vacuum operation.

(ii) The lighter impurities float to the surface of the melt away from the interface.

(iii) The absence of stirring and motion combined with stabilizing temperature gradients minimizes sweeping of impurities to the interface once they are segregated.

(iv) The submerged solid–liquid interface is stable such that even if a

Fig. 13. Single-crystal structure achieved in one single directional solidification of MG silicon by HEM.

SiC particle is entrapped at the interface, the silicon grows around the particle instead of causing breakdown of structure.

(v) The convex solid–liquid interface rejects impurities to the outside of the ingot.

(vi) As impurities are segregated to the melt, the size of the interface increases thereby minimizing their effect.

(vii) As growth proceeds and the melt is enriched with impurities the distance of the interface from the heat exchanger increases; the lateral movement of the interface is slowed so that purification is still achieved with segregation.

(viii) The high-purity silica crucible does not contaminate the molten silicon.

Of all the various sources specified in table 1, MG silicon from India, South Africa and Brazil was found to be of superior quality (Khattak and Schmid 1983). The best solar cell performance obtained was 7.2% for diffused cells (Schmid et al. 1980) and 12% for epitaxial cells (Khattak et al. 1982).

DAR silicon was also directionally solidified by HEM (Khattak et al. 1980). Considerable purification was achieved, and 12.3% efficiency diffused solar cells were fabricated on twice directionally solidified HEM material.

A common problem identified with arc-furnace-derived silicon was the high carbon content of the meltstock. In the submerged arc furnace, silicon is produced by the carbothermic reduction of quartz through an intermediate step – the formation of silicon carbide (Dosaj et al. 1978). Incomplete reactions could result in carbon and silicon carbide in the silicon product. Commercially available MG silicon, therefore, contains carbon levels above those of the solid-solubility limit. During heating of meltstock in the HEM furnace prior to directional solidification silicon carbide particles are formed. These particles, depending on their morphology, are present at various positions in the melt. The larger particles sink to the bottom of the crucible because they have a higher density than molten silicon, and the smaller particles tend to float on the surface of the melt because of surface tension; some of the intermediate sizes are suspended in the melt. Figure 14 shows a heavy concentration of large silicon carbide particles on the surface of the melted-back seed and at the bottom section of the boule. The stability of the submerged interface allows growth to proceed around these particles (Digges et al. 1976). The convex shape of the HEM interface is useful in pushing the suspended particles ahead of the interface till such time that they agglomerate and form larger particles.

Such particles are then entrapped in the solid and usually form twins in the crystal as evidenced in figs. 15a,b. The last material to solidify is along the wall of the crucible, and near the top corners of a square ingot; consequently, all the impurities and particles are dumped near the skin of the ingot. It was observed that most of the impurities were segregated in a small volume at the top of the four corners of the square ingot; the particles were rejected in heavy concentrations on the sides and top of the ingot.

The silicon carbide particles are probably complexes of silicon, carbon, oxygen and other impurities. The strain field around these particles getters other impurities during directional solidification. A significantly higher concentration of impurities has been observed in the vicinity of the particles. When wafers containing such particles are processed into solar cells, severe shunting has been observed (Khattak et al. 1981). This behavior is associated with impurities surrounding the particles, because substrates fabricated from electronic-grade silicon and containing SiC precipitates are processed into solar cells without shunting problems.

Therefore, the carbon level in the meltstock has to be correlated with other impurities in determining the tolerance levels for directional solidification. For elements that have low segregation coefficients considerably high concentrations in the meltstock can be refined provided the carbon

Fig. 14. Structure of an ingot after HEM solidification of MG silicon, showing heavy concentration of SiC particles near the bottom and on the melted-back seed surface.

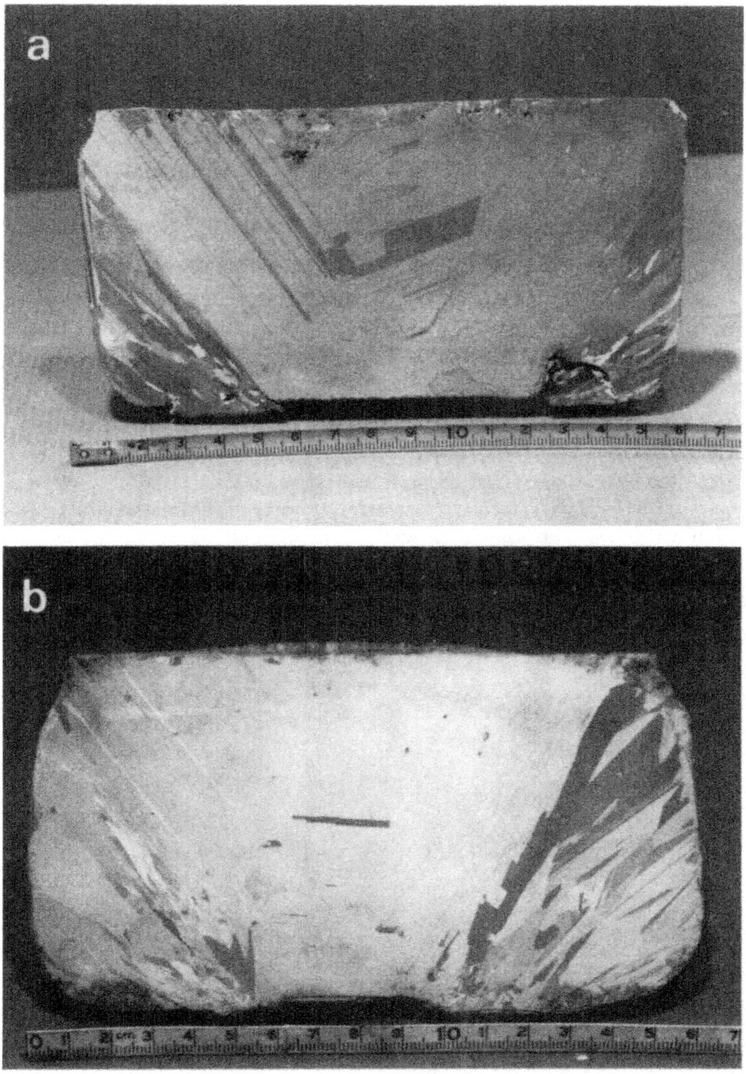

Fig. 15. Silicon carbide particles trapped in the bulk of the ingot (a) as well as acting as a source for formation of twins (b).

level is low, preferably lower than the solid-solubility limit. Under these conditions SiC particles would not be formed, and the gettering of impurities in the vicinity of the particles is prevented. If the carbon level in the meltstock is high, as is the case with the arc-furnace-derived silicon, then the tolerance level of metallic impurities is severely limited unless the particles are considerably reduced by rejection, segregation or reaction.

Directional solidification of arc-furnace-derived silicon has been processed by HEM (Schmid et al. 1980, Khattak et al. 1980, 1981, 1982). Even with commercially available MG silicon of 98–99% purity, the metallic impurities in the bulk of the material were significantly reduced by directional solidification effects. Purifications by in situ processing, i.e. gas blowing, slagging, moist argon blowing, were therefore evaluated in the

Table 2
Spark source mass spectrographic analysis of in situ refined HEM-solidified ingots using same batch of MG silicon as meltstock. Sample A corresponds to material near the middle of the ingot and sample B is towards the end of solidification. All values are in ppmw.

Element	Meltstock	In situ purification									
		DS [1]		Argon blowing		Slagging		DDS [2]		H_2O + Argon + Slag [3]	
		$\varrho = 0.18\ \Omega$ cm		$\varrho = 0.40\ \Omega$ cm		$\varrho = 0.15\ \Omega$ cm		$\varrho = 0.07\ \Omega$ cm		$\varrho = 0.07\ \Omega$ cm	
		A	B	A	B	A	B	A	B	A	B
B	5	10	10	6	15	5	20	5	50	0.8	2
Mn	50	1	50	<1	<1	<1	<100	<5	500	0.02	0.04
Mg				1	20					0.2	0.5
Fe	M	200	M	15	70	50	M	50	M	1	4
Al	M	200	200	5	145	20	M	20	500	1	1
V	2	<1	50			<1	100	<1	1000	0.02	<0.02
Ti	100	2	M	2	2	<1	M	<1	M	<0.3	<0.3
Cu	20	20	50	2	8	2	200	20	M	1	1
Ni	20	20	50			2	200	10	M	<2	<2
Ca	M	5	10	7	40	<1	2	2	10	0.3	3
Cr	2	<1	50			<1	100	<1	1000	0.04	0.04
Ba	20	<2	10	<1	5	<2	<2	<2	10	0.5	0.5
P	20	5	10	5	20	2	5	2	2	1	3

[1] Directional solidification, resulting in a bulk resistivity ϱ as given.
[2] Double directional solidification, resulting in a ϱ as given.
[3] Moist argon blowing slagging, resulting in a ϱ as given.
[4] M denotes major.

HEM for further refinements (Khattak and Schmid 1983). The effect of two solidifications was also studied. Table 2 shows the level of impurities after these processing steps. Simultaneous incorporation of the in situ processing techniques yielded material with each impurity reduced to below 1 ppm level.

The in situ processing was followed by directional solidification; hence, all the additional steps show an improvement in purification. The level of impurities was reduced significantly by directional solidification and these levels were reduced further on double directional solidification. A purification procedure employed was addition of high-purity silica powder to the charge in the crucible. Silica reacts with the impurities in MG silicon such as Ca, Ti, B, etc., to form a slag and thereby purify the melt. Even in the absence of silica addition to the charge, the high-purity silica crucible aids in refining of impurities.

The argon blowing through the melt increases the reaction rates and exposes more melt to the vacuum in the furnace, thereby aiding evaporation of impurities. Addition of moisture to the argon helps oxidation of impurities and, therefore, facilitates their incorporation in the slag in preference to molten silicon. Even impurities which have low segregation coefficient, such as B, P, Al, are reduced with the in situ processing by HEM. The addition of slag, and moist argon blowing through the melt did not affect the structure of the ingot. Figure 16 shows the crystallinity achieved

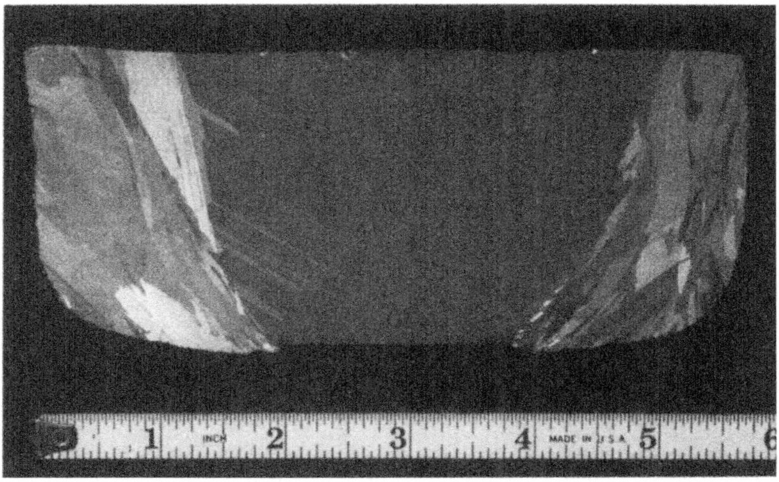

Fig. 16. Structure of a 15 cm square ingot after in situ purification of MG silicon and directional solidification by HEM.

in a slab sectioned parallel to the growth direction of a 15 cm square cross-section silicon ingot solidified with simultaneous use of all these purifications.

Bulk resistivity measurements were carried out on slabs sectioned parallel to the growth direction of HEM solidified ingots. In table 2 the values for the central portion of these slabs are shown. The data show that except for argon blowing the resistivity of ingots decreased with in situ purifications. An interesting feature observed with moist argon blowing was the change in resistivity with position in the ingot. The initial material to solidify was 0.24 Ω cm (p-type) resistivity. This value increased till near the edge of the ingot where low values (0.1–0.2 Ω cm) n-type resistivity were observed. The moist argon blowing through the melt, therefore, affects the 3+ and 5+ type impurities. During directional solidification the melt is enriched in P because it has a lower segregation coefficient than B. Continuous rejection of P could change the resistivity of the last material to solidify to n-type; however, this will depend on the amount of the 3+ and 5+ impurities as well as their ratio. Such behavior has not been observed for all types of MG silicon.

5. Summary

HEM has been developed into a very effective technique for processing silicon for photovoltaic applications. Silicon ingots of 50 kg with 32 cm square cross-section have been produced using high-purity silicon meltstock. The development of a high-purity graded silica crucible allows solidification of crackfree ingots and prevents contamination of silicon. Very large, columnar grains of uniform size and distribution have been produced because of the flexibility of the heat exchanger. The growth of multicrystalline ingots is considerably faster than single-crystal material. HEM silicon, therefore, combines the high-purity, large size, square shape, multicrystalline structure for use in high-efficiency, close-packed modules at low cost.

The HEM furnace has also been demonstrated as a holding furnace for in situ refinements of low-purity meltstocks prior to directional solidification. Nearly single-crystal structure has been produced with one single directional solidification of commercially available metallurgical grade (MG) silicon. The incorporation of in situ refinements reduced the level of each impurity to about 1 ppm level in the HEM solidified ingots. The high level of carbon in MG silicon as it exists today, limits its use for photovoltaic silicon. However, HEM is very effective in reducing metallic impurities of low-purity silicon meltstocks.

References

Amick, J.A., 1982, J. Electrochem. Soc. **129**, 864.

Amick, J.A., J.P. Dismukes, R.W. Francis, L.P. Hunt, P.S. Ravishankar, M. Schneider, K. Matthei, R. Sylvain, K. Larsen and A. Schei, 1983, in: Proc. Symp. Materials and New Processing Technologies for Photovoltaics, San Francisco, CA, 1983 (Electrochemical Society, Pennington, NJ) p. 67.

Barnett, A.M., 1982, in: Proc. XVI IEEE Photovoltaic Specialists Conf., San Diego, CA, 1982 (IEEE, New York) p. 1165.

Cheek, G., R. Janssens, M. Leempoels, L. Frisson, R. Mertens and R. Van Overstraeten, 1982, in: Proc. IV EC Photovoltaic Solar Energy Conf., Stresa, Italy, 1982 (Reidel, Dordrecht) p. 926.

Davis Jr, J.R., A. Rohatgi, R.H. Hopkins, P.D. Blais, P. Rai-Choudhury, J.R. McCormick and H.C. Mollenkopf, 1980, IEEE Trans. Electron Devices **ED-27**, 677.

Dawless, R.K., 1981a, US Patent No. 4246249.

Dawless, R.K., 1981b, US Patent No. 4256717.

Dietl, J., 1983, in: Proc. Symp. Materials and New Processing Technologies for Photovoltaics, San Francisco, CA, 1983, eds J.A. Amick, V.K. Kapur and J. Dietl (Electrochemical Society, Pennington, NJ) p. 52.

Dietl, J., D. Helmreich and E. Sirtl, 1982, Proc. IV EC Photovoltaic Solar Eniergy Conf., Stresa, Italy, 1982 (Reidel, Dordrecht) p. 941.

Digges Jr, T.G., M.H. Leipold, K.M. Koliwad, G. Turner and G.D. Cumming, 1976, in: Proc. XII IEEE Photovoltaic Spec. Conf., Baton Rouge, LA, 1976 (IEEE, New York) p. 120.

Dosaj, V.D., and L.P. Hunt, 1976, US Patent No. 4247528.

Dosaj, V.D., L.P. Hunt and A. Schei, 1978, J. Met. **30**, 8.

Dumas, K.A., C.P. Khattak and F. Schmid, 1981, in: Proc. XV IEEE Photovoltaic Spec. Conf., Orlando, FL, 1981 (IEEE, New York) p. 954.

Erasmus, H.D., and J.A. Persson, 1949, Trans. Electrochem. Soc. **95**, 316.

Gibbons, D.F., 1958, Phys. Rev. **112**, 136.

Hanes, M.H., R.H. Hopkins, A. Rohatgi, P. Rai-Choudhury, R.B. Campbell and H.C. Mollenkopf, 1981, in: Proc. XV IEEE Photovoltaic Spec. Conf., Kissimee, FL, 1981 (IEEE, New York) p. 530.

Helmreich, D., 1980, in: Proc. Symp. Electronic and Optical Properties of Polycrystalline or Impure Semiconductors and Novel Silicon Growth Methods, St. Louis, MO, 1980 (Electrochemical Society, Pennington, NJ) p. 184.

Hino, J., and H.E. Stauss, 1952, J. Met. **4**, 656.

Hunt, L.P., and V.D. Dosaj, 1979, in: Proc. II EC Photovoltaic Solar Energy Conf., Berlin, W. Germany, 1979 (Reidel, Dordrecht) p. 98.

Hunt, L.P., V.D. Dosaj, J.R. McCormick and L.D. Crossman, 1977, in: Proc. XII IEEE Photovoltaic Spec. Conf., Baton Rouge, LA, 1977 (IEEE, New York) p. 125.

Hunt, L.P., V.D. Dosaj, J.R. McCormick and A.W. Rauchholz, 1978, in: Proc. XIII IEEE Photovoltaic Spec. Conf., Washington, DC, 1978 (IEEE, New York) p. 333.

Khattak, C.P., and F. Schmid, 1978a, in: Proc. XIII IEEE Photovoltaic Spec. Conf., Washinghton, DC, 1978 (IEEE, New York) p. 137.

Khattak, C.P., and F. Schmid, 1978b, Am. Ceram. Soc. Bull. **57**, 609.

Khattak, C.P., and F. Schmid, 1980, in: Proc. XIV IEEE Photovoltaic Spec. Conf., San Diego, CA, 1980 (IEEE, New York) p. 484.

Khattak, C.P., and F. Schmid, 1983, in: Proc. Symp. Materials and New Processing Technologies for Photovoltaics, San Francisco, CA, 1983, eds J.A. Amick, V.K. Kapur and J. Dietl (Electrochemical Society, Pennington, NJ) p. 478.

Khattak, C.P., F. Schmid and L.P. Hunt, 1980, in: Proc. Symp. Electronic and Optical Properties of Polycrystalline or Impure Semiconductors and Novel Silicon Growth Methods, St. Louis, MO, 1980 (Electrochemical Society, Pennington, NJ) p. 223.

Khattak, C.P., M. Basaran, F. Schmid, R.V. D'Aiello, P. Robinson and A. Firester, 1981, in: Proc. XV Photovoltaic Spec. Conf., Kissimee, FL, 1981 (IEEE, New York) p. 1432.

Khattak, C.P., F. Schmid, P.H. Robinson and R.V. D'Aiello, 1982, in: Proc. XVI IEEE Photovoltaic Spec. Conf., San Diego, CA, 1982 (IEEE, New York) p. 128.

Kim, K.M., A.F. Witt and H.C. Gatos, 1972, J. Electrochem. Soc. **119**, 9.

Kotval, P.S., and H.B. Strock, 1978, US Patent No. 4 124 410.

Legros, B., G.R. David, M.C. Boissy, J. Lebailly, T.P. Ngo and E. Fabre, 1980, in: Proc. III EC Photovoltaic Solar Energy Conf., Cannes, France, 1980 (Reidel, Dordrecht) p. 593.

Leung, D., P.A. Iles, S. Hyland and A. Kachare, 1984, in: Proc. XVII IEEE Photovoltaic Spec. Conf., Kissimee, FL, 1984 (IEEE, New York) p. 264.

Lutwack, R., 1980, in: Proc. III EC Photovoltaic Solar Energy Conf., Cannes, France, 1980 (Reidel, Dordrecht) p. 220.

Maissel, L., 1960, J. Appl. Phys. **31**, 211 (L).

Pearson, G.L., W.T. Reed Jr and W.L. Feldman, 1957, Acta Metall. **5**, 1981.

Pizzini, S., 1982, Sol. Energy Mater. **6**, 253.

Powell, R.W., C.Y. Ho and P.E. Liley, 1966, Thermal Conductivity of Selected Materials (National Bureau of Standards, Washington, DC) **NBS-8**, Category 5, 99.

Rohatgi, A., R.H. Hopkins and J.R. Davis, 1981, IEEE Trans. Electron Devices **ED-28**(1), 103.

Rohatgi, A., J.R. Davis, R.H. Hopkins and P. Rai-Choudhury, 1982, in: Proc. XVI IEEE Photovoltaic Spec. Conf., San Diego, CA, 1982 (IEEE, New York) p. 411.

Roy, K., and W. Pschunder, 1980, in: Proc. II EC Photovoltaic Solar Energy Conf., Cannes, France, 1980 (Reidel, Dordrecht) p. 263.

Roy, K., K.-D. Rasch and H. Fischer, 1980, in: Proc. XIV IEEE Photovoltaic Spec. Conf., San Diego, CA, 1980 (IEEE, New York) p. 897.

Schmid, F., 1975, US Patent No. 3 898 051 (Aug. 5, 1975).

Schmid, F., and C.P. Khattak, 1981, Opt. Spectra **15** (5), 65.

Schmid, F., and C.P. Khattak, 1983, Laser Focus **19** (9), 147.

Schmid, F., and D. Viechnicki, 1970, J. Am. Ceram. Soc. **53**, 528.

Schmid, F., and D. Viechnicki, 1972, US Patent No. 3 653 432 (April 4, 1972).

Schmid, F., and D. Viechnicki, 1973, Solid State Technol. **16**, 45.

Schmid, F., C.P. Khattak, T.G. Digges Jr and L. Kaufman, 1979, J. Electrochem. Soc. **126**, 935.

Schmid, F., M. Basaran and C.P. Khattak, 1980, in: Proc. III EC Photovoltaic Solar Energy Conf., Cannes, France, 1980 (Reidel, Dordrecht).

Schmid, F., C.P. Khattak and P.A. Isles, 1983, in: Proc. V EC Photovoltaic Solar Energy Conf., Athens, Greece, 1983 (Reidel, Dordrecht) p. 1019.

Schmidt, W., K.-D. Rasch and K. Roy, 1982, in: Proc. XVI IEEE Photovoltaic Spec. Conf., San Diego, CA, 1982 (IEEE, New York) p. 537.

Sirtl, E., 1980, in: Proc. III EC Photovoltaic Solar Energy Conf., Cannes, France, 1980 (Reidel, Dordrecht) p. 236.

Viechnicki, D., and F. Schmid, 1974, J. Cryst. Growth **26**, 162.
Viechnicki, D., F. Schmid and J.W. McCauley, 1972, J. Appl. Phys. **43**, 4508.
Voltmer, F.W., and F.A. Padovani, 1973, in: Semiconductor Silicon – 1973, eds H.R. Huff and R.R. Burgess (Electrochemical Society, Pennington, NJ) p. 75.
Yoo, H.I., and J.K. Liu, 1980, in: Proc. III EC Photovoltaic Solar Energy Conf., Cannes, France, 1980 (Reidel, Dordrecht) p. 548.

CHAPTER 4

PROGRESS IN DEVELOPMENT OF EFG PROCESS CONTROL IN SILICON RIBBON PRODUCTION FOR PHOTOVOLTAIC APPLICATIONS

J.P. KALEJS

Mobil Solar Energy Corporation
16 Hickory Drive
Waltham, Massachusetts 02254, U.S.A.

Silicon Processing for Photovoltaics II, edited by C.P. Khattak and K.V. Ravi

Contents

1. Introduction

The objectives for silicon sheet growth for terrestrial photovoltaic applications are achievement of throughput rates considerably in excess of those available in conventional silicon melt growth systems in combination with a sheet, e.g., ribbon geometry that obviates the need for wafering. In order to attain these goals, the Edge-defined Film-fed Growth (EFG) technique embodies a number of unique aspects not found in the practice of more conventional melt growth, such as float-zone or Czochralski silicon crystal production. Development of EFG systems for both ribbon and closed-shape sheet geometries is being actively pursued. This chapter examines the relationships between EFG process parameters and sheet properties which arise because of special constraints placed on operating conditions by the need to fulfill the above requirements. These have stimulated development of models to study means to control growth phenomena and material characteristics not previously examined in detail for silicon crystal growth from the melt.

Fundamental aspects of meniscus-controlled melt growth, particularly ribbon for photovoltaics, have been thoroughly treated in a number of papers (Swartz et al. 1975, Surek et al. 1977, Ravi 1977, Tatarchenko and Brener 1980). Comprehensive reviews of shaped crystal growth (Rudolph 1982) and of the EFG technique as applied to silicon sheet growth for photovoltaics (Wald 1981) are also available. Only recently, models for the EFG process have been developed to enable extensions of the qualitative concepts to be examined within the framework of real-system behavior. This has been made possible by application of finite-element techniques to analyze combined heat and mass transfer (Kalejs et al. 1983a), coupled solutions for heat transfer and capillarity (Ettouney et al. 1983), and plastic deformation effects on sheet stress distributions (Lambropoulos et al. 1983). Modeling real-system geometries allows investigation of relationships between process variables and system response, as manifested in material properties, at quantitative levels. Furthermore these models make it possible to study phenomena associated with impurity transport, which has led to progress in understanding the, generally, complex relationships among operating parameters, material properties and solar-cell performance that arise from specific impurity contamination, particularly from carbon and oxygen.

The scope and requirements of models for EFG process behavior to be discussed become more evident from comparison of several key areas of EFG and Czochralski silicon growth, as is illustrated in table 1. A cross-sectional schematic of the interface region for sheet growth is shown in fig. 1. Several of the differences listed in table 1 are a direct consequence of the use of the EFG die, made of graphite, to shape the crystal into its sheet geometry. The die contributes carbon to the melt and confines segregated impurities to the meniscus volume at the die top; it also isolates the interface region from the bulk (crucible) melt. This provides an effective segregation coefficient of approximately unity in practice for all impurities except carbon and oxygen. The latter two may further be transported out of or into the meniscus volume by melt–die and ambient–meniscus melt reactions.

The achievement of high pull rates in the vertical sheet growth mode is aided by the large surface-to-volume ratio of the sheet geometry and by maintaining interface temperature gradients much greater than those typical of Czochralski systems (table 1). The influence of these factors on stress and on impurity-related defect formation mechanisms is examined separately here. Stress manifestations to be discussed are creep phenomena at high temperatures, residual stress that may promote fracture, and buckling. The high cooling rate impacts additionally on mi-

Table 1
Comparison of growth variable ranges for CZ silicon and EFG silicon (sheets).

Variable	CZ Silicon	EFG Silicon
Metallic impurities	$\leq 10^{14}$ at/cm^3	$\sim 10^{16}$ at/cm^3
k_{eff}	≤ 0.01	~ 1
Substitutional carbon C_s	$\leq 1 \times 10^{17}$ at/cm^3	$\sim 1 \times 10^{18}$ at/cm^3
Interstitial oxygen O_i	$\sim 1 \times 10^{18}$ at/cm^3	$\leq 1 \times 10^{17}$ at/cm^3
O_i/C_s ratio	10	0.1
Growth speed V_s	0.1 cm/min	2 cm/min
Interface temperature gradient G	200°C/cm	1000°C/cm
Interface cooling rate	20°C/min	2000°C/min

crodefect formation processes that are associated with impurities and point defects. These processes have been extensively studied in Czochralski silicon growth, particularly those involving participation of oxygen and carbon. A new situation is encountered in EFG materials. The oxygen/carbon ratio can be decreased from that in the Czochralski system by a factor of one hundred or more by use of graphite crucibles to contain the bulk melt. Sources of these impurities associated with ambient gases reactive with the

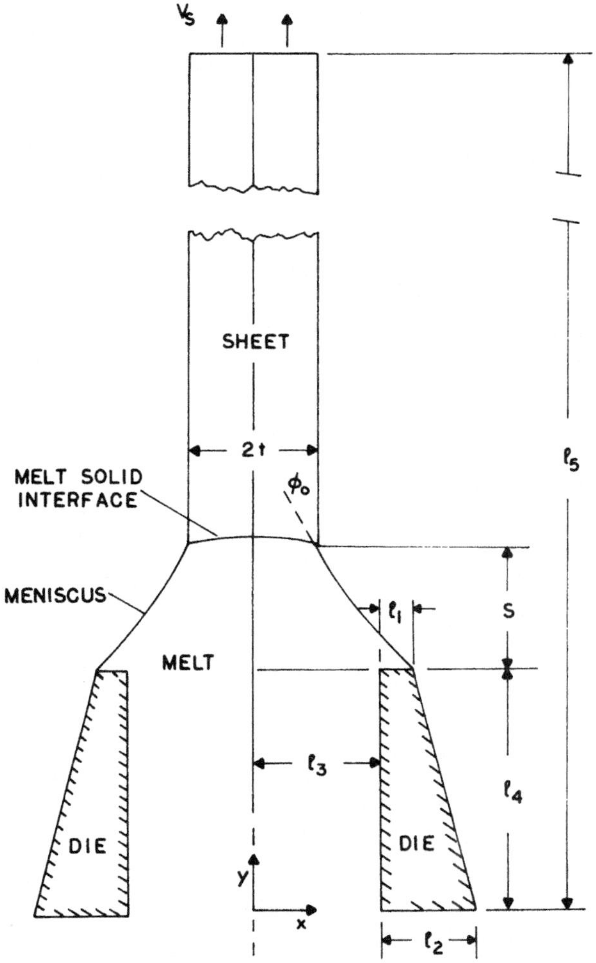

Fig. 1. A cross-sectional schematic of an EFG system for ribbon growth. Die and calculation domain dimensions are given in table A1.

silicon melt then must be additionally considered for EFG sheets. Ambient influence is known to produce inhomogeneities around the periphery of the Czochralski boule. However, they can become a dominant influence on sheet properties, which need to be controlled over sheet thicknesses of the order of the inhomogeneity depth in the surface region of the Czochralski boule.

In the following sections of this chapter the effects of operational constraints on EFG process control are examined as they influence material properties in three general areas: sheet dimensional uniformity and interface shape, stress, and impurity effects that impact on solar cell performance. Section 2 discusses heat and mass transfer models. These models interrelate process variables such as temperature field parameters and die geometry to sheet thickness, interface location and shape. Representations of interface melt flow fields are included to permit calculation of impurity segregation. Mechanisms responsible for generating thermoelastic stress are discussed in section 3. The effects of creep on stress arising during high-speed sheet growth are studied with a finite-element scheme for calculating sheet stress distributions with plastic deformation. In section 4, special relationships between growth conditions that impact on impurity incorporation and on material behavior during cell processing are examined. These point to the need to assume a dominant role for carbon in processes that govern recombination response in EFG solar cells.

2. Transport phenomena analysis

The location of the EFG die in close proximity to the growth interface complicates the analysis of heat and mass transfer phenomena for the EFG system. The transport equations need to be solved for three distinct regions: for the die, for the capillary and meniscus melts, and for the sheet. These solutions must be consistent with those for the Young–Laplace equation, which relates the die geometry to meniscus height and sheet thickness.

Initial heat transfer calculations for the EFG system (Chalmers et al. 1975, Ciszek 1976, Swartz et al. 1975) used one-dimensional approximations. This is generally satisfactory for representing heat transfer in the sheet, but it does not properly account for meniscus and die influence. Two-dimensional models, both in planes through the sheet thickness and across its width, are adequate for study of most phenomena because of the extreme aspect ratio (width $\gg$ thickness) of the sheet geometry. These have been used to model convection influence on impurity redistribution in the sheet width dimension (Kalejs 1978) and the effects of die asymmetry on interface shape and impurity segregation through the sheet thickness (Kalejs et al. 1983a). None of these early models included capillarity effects, and thus were limited to studying transport phenomena at a qualitative level.

Only recently have numerical schemes been developed for solving coupled systems of heat transfer and capillary equations (Ettouney 1983, Ettouney et al. 1983). These allow all parameters in meniscus-controlled growth to be included in relating process variables to operating conditions and interface shape. This has provided the first opportunity to study at a quantitative level the influence of external temperature fields and process variables, such as growth speed and ambient temperature, on sheet thickness, on its temperature distribution and on the melt/solid interface shape.

Heat and mass transfer modeling with capillarity included are discussed in section 2.1 below. Predictions for process variable relationships and for operating curves for EFG silicon sheet growth are examined in section 2.2. Application of the modeling to study real-system behavior is the subject of section 2.3. Interface shape and impurity segregation effects are considered in sections 2.4 and 2.5, respectively.

2.1. Model description

The calculation domain with the three regions of different phases is shown in fig. 1. The discussion of the two-dimensional transport and capillary equations, and boundary conditions employed can be found in the paper by Ettouney et al. (1983). Relevant material constants, modeling variables and system geometrical considerations are given in the Appendix. The heat and mass transport phenomena and associated interface shape behavior in the cross-sectional plane through the sheet thickness shown in fig. 1 have the most important bearing on material properties and on process variables related to sheet thickness control. In the sheet width dimension, the only significant effects arise from convective redistribution of impurities associated with die capillary geometry, rather than with heat transfer effects on interface shape. These redistribution studies are discussed separately in section 2.5. A model to describe heat transfer and capillarity effects at the ribbon edge is not yet available. These considerations do not enter process control description in closed shape sheet growth by EFG, where edges are eliminated (Taylor et al. 1981). The thickness-plane modeling now is all that is required to obtain a complete description of this process.

Process description is simplified here for purposes of facilitating discussion to consideration of four control variables initially. There are two temperatures, T_∞ (ambient) and T_0 (die base), the capillary rise height, h_{eff}, and the sheet pull speed V_s. All other parameters are held constant at values given for the baseline case described in table A3 unless changes are explicitly mentioned. The baseline temperature configuration itself (see table A3) does not represent practical growth configurations for which environment temperature distributions are generally nonuniform and gradients are reduced on average in order to cope with thermoelastic stresses. Temperature profiles encountered in practical EFG ribbon systems are discussed in section 2.3 and in section 3 in connection with modeling of stresses in sheet growth. The dependent variables in this model are the sheet thickness $2t$, the meniscus height s, and the temperature fields in the die, melt and sheet, which determine the interface configuration.

The relationship between the crystal thickness $2t$ and meniscus height s is determined by capillary statics and by the wetting angle ϕ_0 at the sheet–

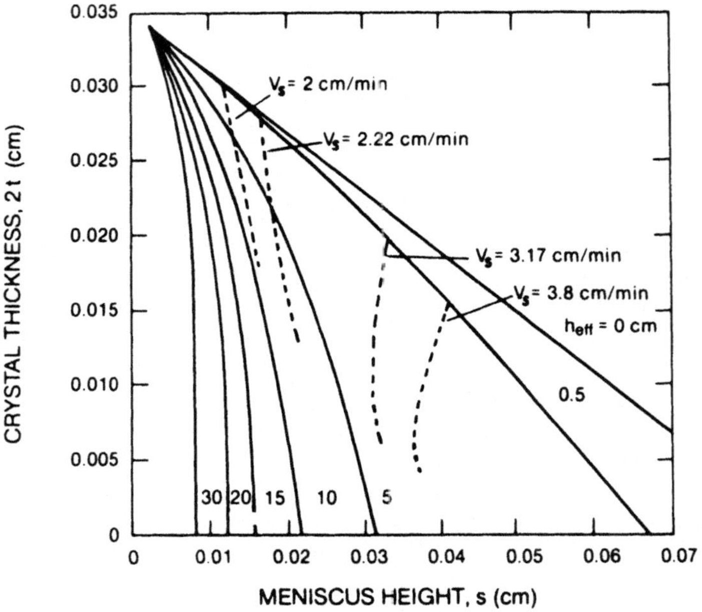

Fig. 2. Relationship between crystal thickness $2t$ and meniscus height s for solutions of the Young–Laplace equation (solid lines) compared to operating paths at fixed growth speed (dashed lines) predicted from thermal-capillary solutions, as in fig. 5.

melt interface. This is illustrated by the solid curves in fig. 2 for $\phi_0 = 11°$, the condition relevant to silicon (Surek and Chalmers 1975). However, the particular values of $2t$ and s attained during growth depend on heat transfer established through the choice of the environmental temperature field, represented here by T_∞ and T_0. Allowed operating paths calculated for the present model at fixed growth rates are superimposed on the capillary solutions as dashed lines in fig. 2.

An additional constraint on operating conditions arises from the choice of die tip geometry because the shape of the meniscus is influenced by the wetting conditions at the line of contact between the melt and die. The EFG die is made of graphite for silicon sheet growth, for which the equilibrium contact angle of melt is 30°. The apparent contact angles θ_c made by the meniscus with the horizontal plane at the edge of the die are shown in fig. 3 for the die of fig. 1, which has a die top angle of 60° with the horizontal, and for the meniscus shapes corresponding to those of fig. 2. Dashed lines in fig. 3 represent the Gibbs' criteria for the pinning of a wetting line along a sharp geometrical edge of a solid surface. Thus, the

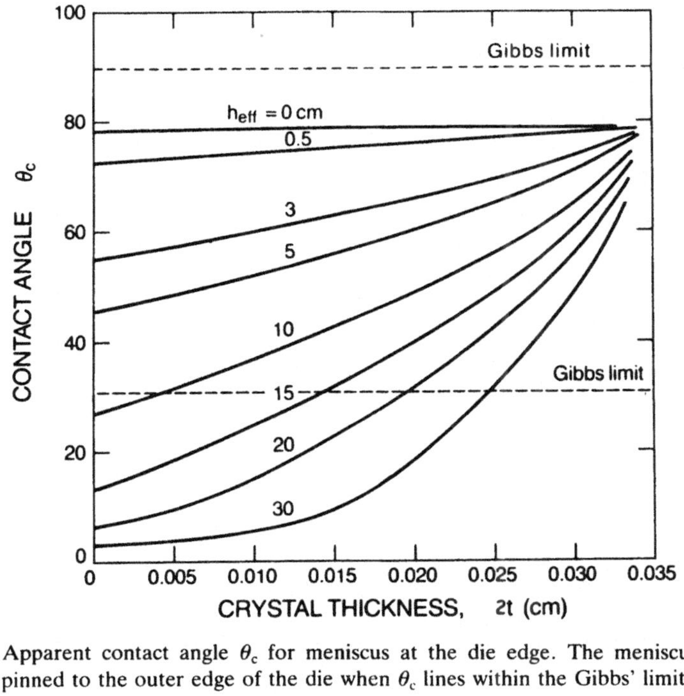

Fig. 3. Apparent contact angle θ_c for meniscus at the die edge. The meniscus remains pinned to the outer edge of the die when θ_c lines within the Gibbs' limits.

meniscus either advances over the die edge and down the slanted face if θ_c exceeds values represented by the upper dashed line, or retracts from the die edge and moves inward if θ_c is reduced below 30° (lower dashed line). Growth is interrupted in either case. Solutions of the heat and mass transfer models need to be examined to delineate those cases that cannot be sustained in practice because of die top contact angle incompatibilities.

2.2. *Operating variable relationships*

Typical operating region diagrams for sheet thickness response generated from solutions of the coupled heat transfer and capillary equations are given in figs. 4 and 5. The temperature T_0 parameterizing the t–V_s curves in fig. 4 is associated with a control thermocouple temperature used to monitor growth conditions in practice and to set the interface location. For the given T_∞, a range of T_0 values can be deduced that prevent the sheet from freezing at the die top [e.g., as happens for curves (a) and (b)] during

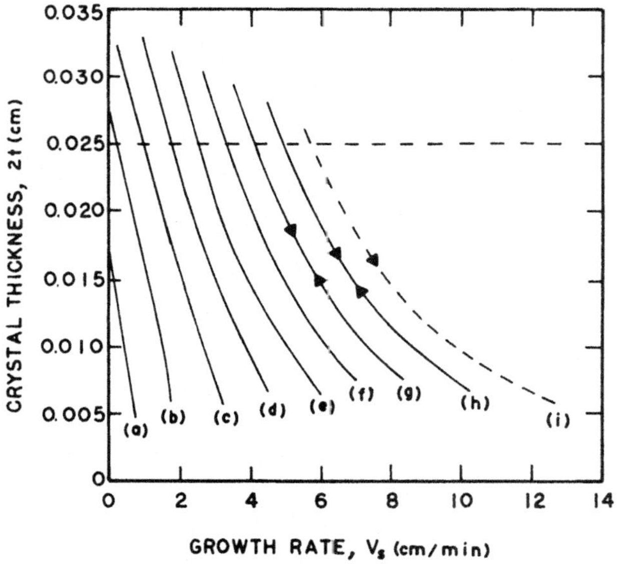

Fig. 4. Operating diagrams of crystal thickness as a function of growth speed for nine die base temperatures T_0 and $h_{eff} = 0.5$ cm. Dashed curves indicate combinations of T_0 and V_s for which the melt is thermally supercooled. The arrows give the direction of changing V_s for decreasing the interfacial temperature gradient in the melt. T_0 varies from: (a) 1793 K down to (i) 1713 K, with 10 K decrements.

seeding ($V_s = 0$). Thickness dependence on T_0 was shown previously to dominate the EFG system response (Ettouney et al. 1983). Measurements of the variation of t with T_0 have been made and give an approximately linear relationship with a slope of 9×10^{-4} cm/K. This is in the range of values that can be deduced from the operating curves for $T_0 = 1743$ K and 1753 K in fig. 4. Additional comparisons of model and experimental results are discussed below when more realistic environmental temperature profiles are treated.

The operating diagram of fig. 5 relates sheet thickness to the capillary parameter, h_{eff}, or static head, with V_s as the variable. h_{eff} is roughly the separation distance of the bulk (crucible) melt level and the die top (interface). Calculations indicate that increasing h_{eff} at constant growth speed brings the system operating point eventually to a critical value where very rapid decreases in t with small changes in h_{eff} occur. At each such 'limit' point, $\partial t / \partial h_{eff}$ tends toward infinity. The limit point occurs at progressively lower values of h_{eff} as V_s increases. The operating limit for $V_s = 2$ cm/min is shown in fig. 5, and the unstable domain is indicated by a dashed line. In fig. 2, the V_s = const. operating paths of fig. 5 are superimposed on the t–s curves obtained solely from solutions of the Young–Laplace equation.

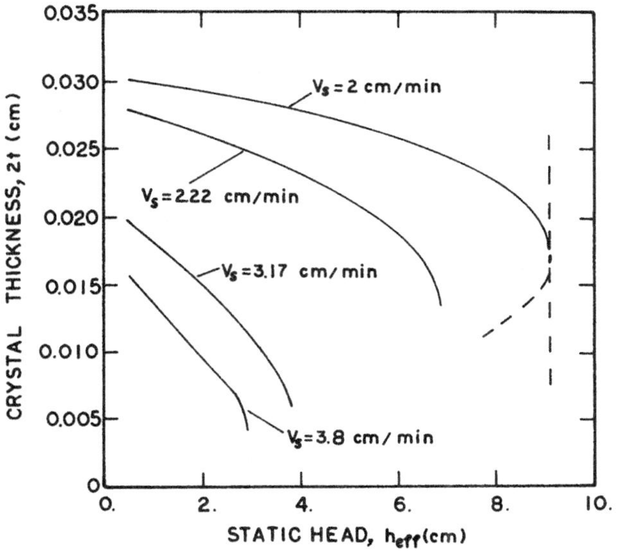

Fig. 5. Crystal thickness as a function of static head at several growth speeds, with $T_0 = 1753$ K. The vertical dashed line demarcates the limiting value of 9.2 cm for the static head at $V_s = 2$ cm/min.

Closer examination of heat transfer conditions associated with particular limiting regions of figs. 4 and 5 provides insight into the physical conditions associated with growth failure modes. Restrictions on the steady-state growth regime at the boundaries of the operating diagram represented by the curves in fig. 4 are illustrated with the help of calculated axial temperature profiles in the die top and capillary melts, and in the sheet, such as given in fig. 6. At constant $T_0 = 1763\,K$, for example, a growth rate increase leads to a progressively higher meniscus and to a thinner sheet, while the axial temperature gradient increases (fig. 6). The limit of zero thickness is thus approached without melt supercooling [e.g., as in curves (c) and (d) in fig. 4]. However, supercooling at higher growth rates will occur as the speed is raised at a constant sheet thickness. Such an operating path is indicated by the dashed horizontal line in fig. 4. The melt-temperature gradient decreases until thermal supercooling occurs on (i), at which point process control is lost. The curve (i) in fig. 4 represents an operating path at constant T_0 with supercooled melt conditions. Arrows on each of the adjacent curves indicate the direction of varying growth rate for decreasing the axial melt gradient. The calculations additionally show that

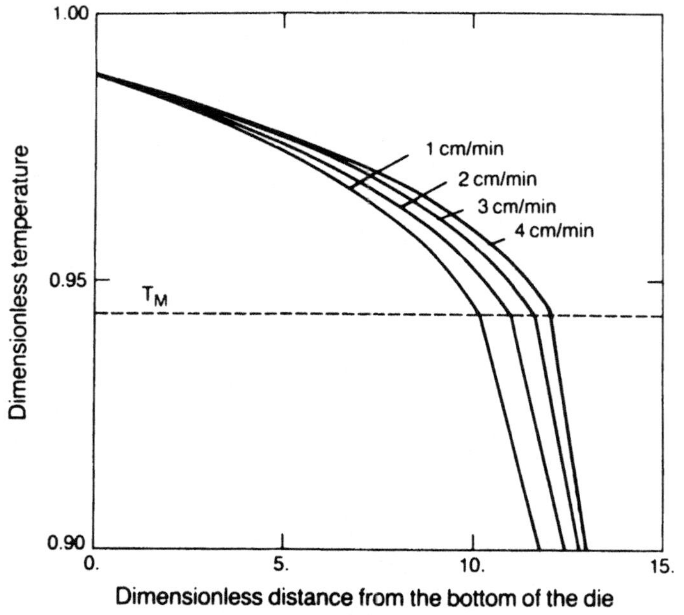

Fig. 6. Axial temperature profiles at four growth speeds for melt and sheet with $T_0 = 1763\,K$ and $h_{eff} = 0.5\,cm$. The axial coordinate is scaled with $2\ell_3 = 0.025\,cm$.

sheet freezing at the die top [curves (a) and (b) in fig. 4] occurs here, not because of supercooling but with positive melt temperature gradients.

The sensitivity of crystal thickness to h_{eff}, as represented by the slopes of the curves in fig. 5, is governed by the dominant heat transfer mechanism in the meniscus, which in turn depends on the sheet thickness. At $V_s = 2$ cm/min and $h_{eff} = 0.5$ cm conduction through the melt is greater than the radiative heat transfer to the environment from the meniscus surface by a factor of 20; at $V_s = 3.8$ cm/min this ratio is only about 4. Increasing h_{eff} has little effect on the meniscus height until the limit point for this speed is approached, where a rapid decrease of t with s occurs (see also fig. 2). This produces eventual growth failure because conduction can no longer carry away the latent heat generated at the interface. The limiting value of h_{eff} decreases as the growth speed increases because a greater latent heat contribution to the total heat flux thins the sheet and drives up the meniscus height more rapidly than at lower speeds, thereby severely constricting further conductive heat loss through the sheet.

The effects of die geometry on operating curves in the t–V_s diagram are illustrated in fig. 7. The die capillary opening, $2\ell_3$, is the parameter here.

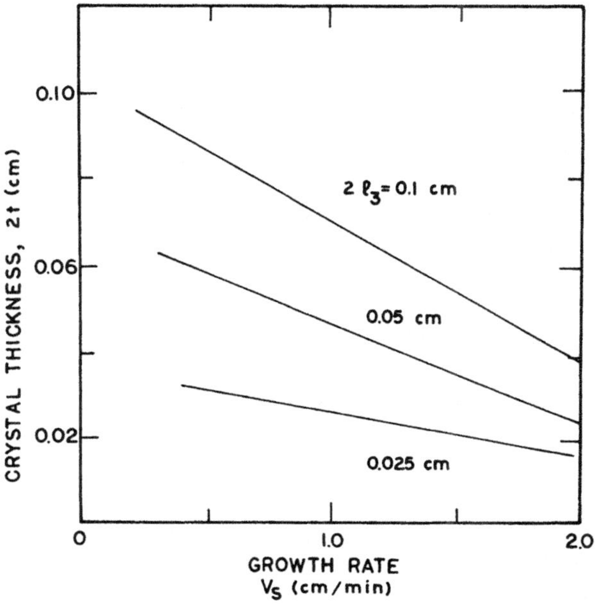

Fig. 7. Operating curves for crystals grown at three different die capillary thicknesses for $T_0 = 1773$ K and $h_{eff} = 0.5$ cm.

The operating curve slopes, such as represented by $\partial t/\partial V_s|_\ell$ in fig. 7 and $\partial t/\partial h_{\mathrm{eff}}|_{V_s}$ in fig. 5, are important in choosing operating regimes that optimize dimensional control. For example, operation with a wider die capillary leads to an increased sensitivity of t to V_s according to the results of fig. 7.

The treatment of coupled heat transfer and capillary effects with the help of numerical models described here is the first step toward developing understanding of stability of shaped crystal growth beyond the qualitative level of the early analytical studies of Surek et al. (1980), Sachs (1980), Sachs and Surek 1980, and Tatarchenko and Brener (1980). In each of these cases, simplified heat transfer models were used in order to examine specific aspects of stability. They are moreover limited in their predictive capabilities to a phenomenological description and cannot account for the full complexity of thermal and capillary parameter interactions. This complexity is illustrated by application of the present analysis to comparison of silicon and sapphire sheet growth in the work of Kalejs et al. (1983b). Full stability analysis of shaped crystal growth will require a combination of finite-element methods and computer aided techniques for nonlinear analysis, as described by Ungar and Brown (1983).

2.3. *Model application*

In order to obtain predictions at a quantitative level, additional considerations are required in applying the model presented in the preceding section to the analysis of experimental EFG systems. In general, the complexity of the system has required approximation of real-system conditions, and this has led to several approaches in dealing with heat transfer particularly, which are discussed below.

2.3.1. *Temperature fields*

Schemes have been developed to model heat transport in a high-speed EFG cartridge system for 10 cm wide ribbon (Kalejs and Bell 1984), which has been studied for its potential to produce low-cost sheet (Kalejs 1982). This system incorporates a water-cooled element, or 'cold shoe', about 0.1 cm above the growth interface in order to enhance the growth speed capacity, followed by an active afterheater to provide annealing so as to reduce sheet stresses to acceptable levels. Two aspects of modeling this system are examined here:

(1) heat transfer calculations of the sheet temperature profile, i.e., those above the growth interface; and

(2) application of the coupled capillary and heat transfer model examined in section 2.2 with a nonuniform environment temperature field.

The 10 cm cartridge system environment temperature profile along the growth axis and associated calculated ribbon profile are shown in fig. 8. A schematic of the ribbon growth enclosure is shown in the inset. T_A and T_C represent measured afterheater and cold shoe surface temperature distributions; T_B is representative of radiation from below the plane of the growth interface, and is treated as a design variable for purposes of sheet profile calculation. Interpolation is used to set enclosure temperatures in regions between T_A and T_C, and T_C and T_B, which represent cartridge openings, e.g., the interface view slot. The ribbon profile is calculated on the basis of radiative heat exchange between the plane ribbon and its enclosure walls (in three dimension).

A sensitivity analysis has included examination of the impact of variations in the controlled temperatures T_A and T_C, and the unknown T_B, and of some material constants on the first and second derivatives of the temperature profile at the growth interface. These characterize, respectively, the speed capability and the thermo-elastic stress for a sheet growth system.

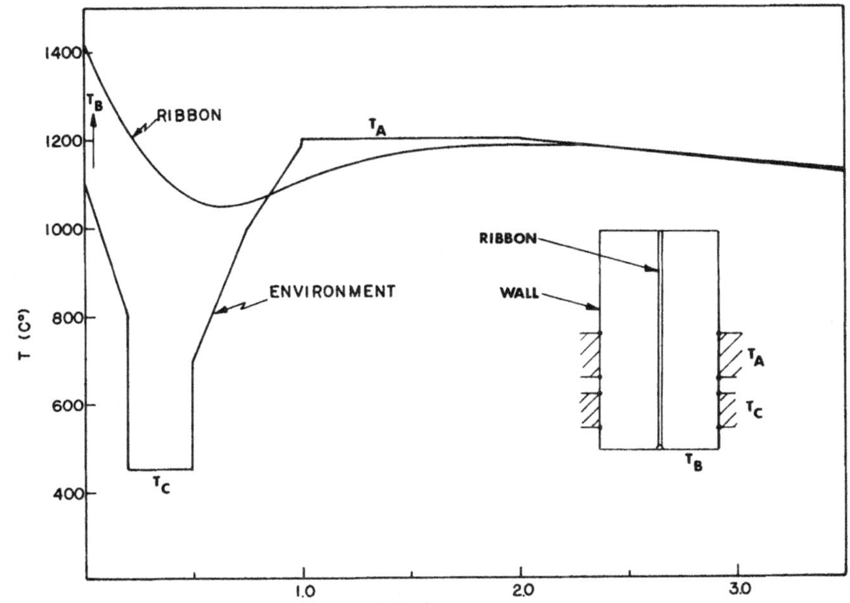

Fig. 8. Temperature distribution of the environment and in a 300 μm thick sheet for a 10 cm cartridge system. The base plate temperature is 1500°C.

The results of the study are presented in fig. 9 and tables 2 and 3. All calculations are for a 300 μm thick sheet. Variations in enclosure wall temperature distributions with the given model parameters do not have as large an impact on either the first or second temperature derivatives near the interface as does the change in absorption coefficient in the solid. The latter is related to uncertainties arising from literature values (Bell 1982). If the solid is transparent in the infrared, heat transfer by long-wavelength

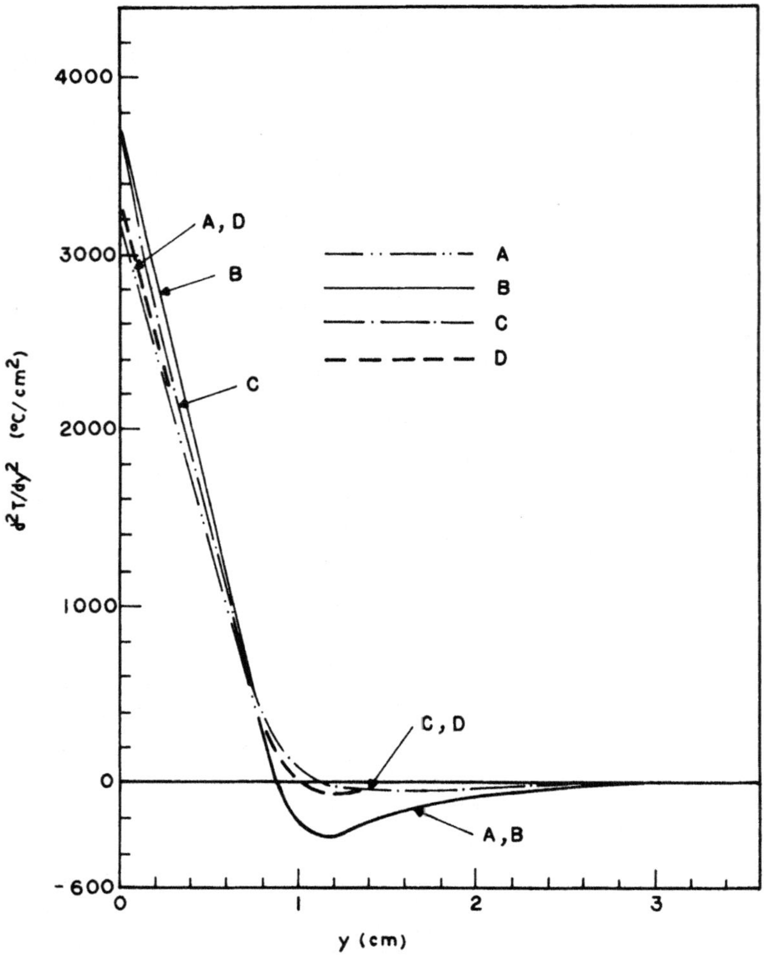

Fig. 9. Variation of second derivative of sheet vertical temperature profile with distance from interface for the cases reported in table 2.

radiation from the interface can augment axial heat transfer. However, the free-carrier absorption in silicon becomes very high near the melting point and makes the solid opaque, while it remains semi-transparent at lower temperatures. The lower absorption coefficient decreases the spectral emissivity, thus decreasing the heat flux from the face of the ribbon. The overall effect in silicon is to decrease the thermal gradient at the interface and significantly influence the second-order temperature derivative also, as is illustrated in table 3.

The above model treats only the region above the growth interface, and

Table 2
Thermal conditions used in modeling. Ribbon thickness is 300 μm, growth speed $V_s = 0$. All temperatures in °C.

	Curve			
	A	B	C	D
T_B, Base plate temperature	1500	1200	1200	1500
T_C, Cold shoe temperature	450	450	400	400
T_A, Afterheater temperature	1200	1200	960	960
$\left.\frac{dT}{dy}\right\|_{y=0}$, Vertical gradient	1310	1370	1490	1427

Table 3
Effect of optical absorption beyond the energy gap on first and second derivatives of the temperature. In the expression for α, T is in K, λ in μm and $E_g = 1.194–2.79 \times 10^{-4}$T. Radiation from the interface is neglected. Environmental temperature is that of case C in fig. 9 and table 2.

Case	α (cm^{-1}) for $E_\lambda < E_g$	$\frac{dT}{dy}$ (°C/cm)	$\frac{d^2T}{dy^2}$ (°C/cm^2)
Black body	∞	1610	4290
Sze (1969)	$1.36 \times 10^{-2} \lambda^2 T^2 \exp(-5803\, E_g/T)$	1490	3745
Ukhanoff (1962)	$2.71 \times 10^{-6} \lambda^2 T^3 \exp(-5803\, E_g/T)$	1365	3475
No free carriers	0	730	1460

thus cannot be used to predict operating variable relationships, such as is possible with the analysis of section 2.2. The latter has been extended (Kalejs 1983b) to include variable environment temperatures representative of the EFG system of fig. 8 and a temperature dependent thermal conductivity $k_s(T) = k_M T_M / T$. All other parameters are kept constant at the values given in table A3. Only two variables remain unspecified in this model: the die base temperature T_0, and the environment axial temperature distribution below the die top represented by $T_\infty = T_\infty(y)$, with $T_\infty^0 = T_\infty(0)$. In fitting of the model, a value is assigned to T_∞^0 and then the distribution between T_∞^0 and the known cold shoe temperature in fig. 8 is set by linear interpolation.

A comparison of model predictions and experimental t–V_s data for a 10 cm wide ribbon grown in the system of fig. 8 are shown in fig. 10. Also indicated are system responses to variations in T_0 and T_∞^0. This sensitivity

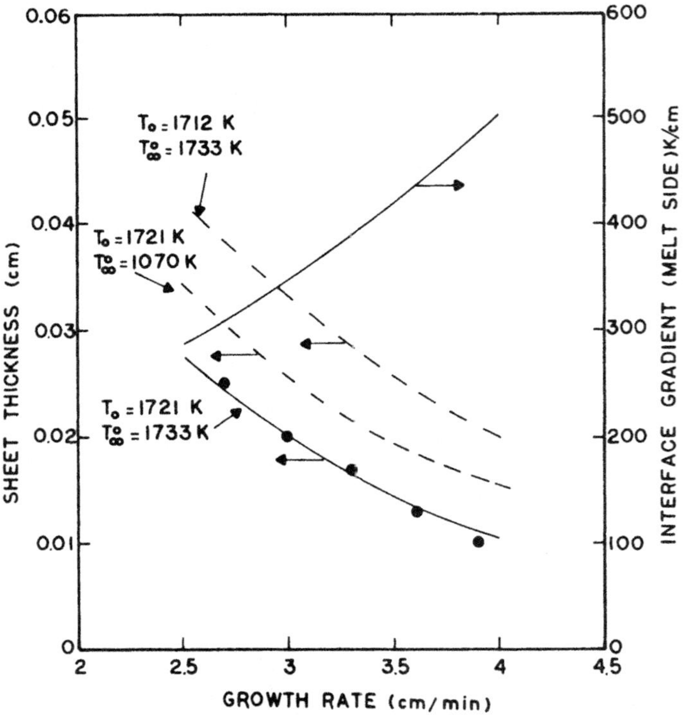

Fig. 10. Comparison of model predictions (–) and experimental data (•) for 10 cm cartridge system. Dashed curves illustrate sensitivity of model predictions to T_∞ and T_0. Solid gradients are for lower curve fit to data.

analysis shows that these operating curves are relatively insensitive to the choice made for T_∞^0 in comparison to T_0, so that T_0 becomes the dominant fitting variable. The fit of the model represented by T_0 in fig. 10 now is dependent primarily on the details of the heat transfer model in the solid. All the considerations for the sheet heat transfer used in the calculations described in the first part of this subsection can in principle be included at this point to more precisely fix T_0 and T_∞^0 within limits allowed by the data for the thermophysical constants of silicon. Additional extensions of this model to calculations of solute segregation and interface configurations are described in sections 2.3.2 and 2.4.2, respectively.

2.3.2. *Model extensions*

Two extensions of the finite-element model of section 2.2 illustrate behavior of parameters in experimental systems that need to be accounted for in some circumstances. Ettouney (1983) has examined the influence of ambient heat transfer at the meniscus and sheet surface using a much simplified EFG system representation to identify qualitative trends influencing the t–V_s relationship and interface shape. Ambient gas convection must be included in situations where helium gas is used to enhance speed capacity (Kalejs et al. 1980b). It is not found to be of importance when argon is utilized as the purge gas under normal operating conditions.

A second extension of the analysis by Ettouney and Brown (1984a) has examined the limitations inherent in the melt convection approximation, which includes only unidirectional (axial) convection. Solutions for the present model have been compared to those obtained from the full two-dimensional Navier–Stokes equations. It is found that the convection field approximation does not introduce significant error in predicted t–V_s relationships and interface configurations for silicon melt up to speeds of 10 cm/min. However, the full Navier–Stokes solutions are required to predict physically meaningful impurity segregation behavior because of the dominance of melt convection over interface curvature in influencing interface impurity redistribution. This relationship is examined in a later section.

2.4. *Interface configurations*

A primary aim of the transport process modeling described here is to understand factors that impact on the interface configuration, which is the single most important growth variable influencing material properties, both with respect to impurity segregation and structure. For the EFG system, the interface configuration is not only dictated by heat transfer considerations, as in conventional melt growth, but is also dependent on the capillary die parameters. Initial models for EFG did not include boundary conditions representative of real systems below the growth interface (Chalmers et al. 1975, Ettouney and Brown, 1982). Interfaces concave toward the melt were obtained for a wide range of growth variables. In later studies (Kalejs et al. 1983a, Kalejs 1983a), full two-dimensional treatments of heat transfer and melt convection were incorporated to examine the influence of both symmetric and asymmetric (displaced) die geometries through the sheet thickness on interface shape and on impurity segregation. For symmetric growth, the qualitative relationships between interface shape and process variables found for the earlier models were not changed with the added consideration of heat transfer in the die top capillary and meniscus melts. However, significant changes in impurity segregation are predicted for asymmetric die configurations. These are attributed to convective transport in the meniscus which dominates normal segregation arising from interface curvature. More recently, the model of Ettouney et al. (1983) presented in section 2.2 has, for the first time, examined relationships between process variables and the interface configuration obtained by including both capillary and heat transfer effects. The predictions of the models of Kalejs et al. (1983a) and Ettouney et al. (1983) on interface configurations and related experimental data are the topics of sections 2.4.1 and 2.4.2.

2.4.1. EFG die influence

The model of Kalejs et al. (1983a) represents the meniscus shown in fig. 1 by straight line boundaries extending from the die top to the liquid–solid interface. The melt–solid interface location and shape are not specified a priori but are derived from solutions of the heat and momentum transport equations for a fixed die geometry and sheet thickness. Although this approach does not lead to self-consistent solutions as with the model of section 2.2, it represents the only attempt to analyze asymmetric growth to date.

The interface configuration is parameterized by meniscus height, and by curvature and tilt variables. Curvature is expressed as a percentage of sheet half-thickness t using the interface function f referenced to the die tip as illustrated in fig. 14, and by defining a radius of curvature $R = t^2/2\Delta f$; Δf is the difference in f between the sheet cross-section centerline and its surface. A tilt angle θ is defined as the angle made with respect to the horizontal by a line drawn through the points of the interface intersecting the sheet surface.

Prescribed temperature boundary conditions were used on all the domain boundaries of fig. 1 to parameterize the temperature field solutions. The primary variables investigated were die tip temperatures T_1 and T_2, a die base melt inlet temperature T_i, the growth speed and sheet thickness. Die asymmetry was imposed by displacing one die top flat with respect to the other by a constant amount of 0.005 cm, viz., as shown in fig. 14. Prescribed temperatures at the sheet surface were taken to represent measured ribbon temperature profiles typical of the system of fig. 8 with (the 'high-speed' case) and without (the 'low-speed' case) the cold shoes. The solid interface temperature gradients in the former are of the order of 1500–2000°C/cm in the latter they are generally below 1000°C/cm.

The main results of the calculations are illustrated in figs. 11–14. All graphs are for the low-speed case. Interface curvature dependence on growth speed and sheet thickness for a symmetric die is shown in figs. 11 and 12. An independent parameter in this model is the die melt inlet temperature T_i. Curvatures typically remain of the order of 5% (fig. 11) over a wide range of values of the temperature and growth variables, and

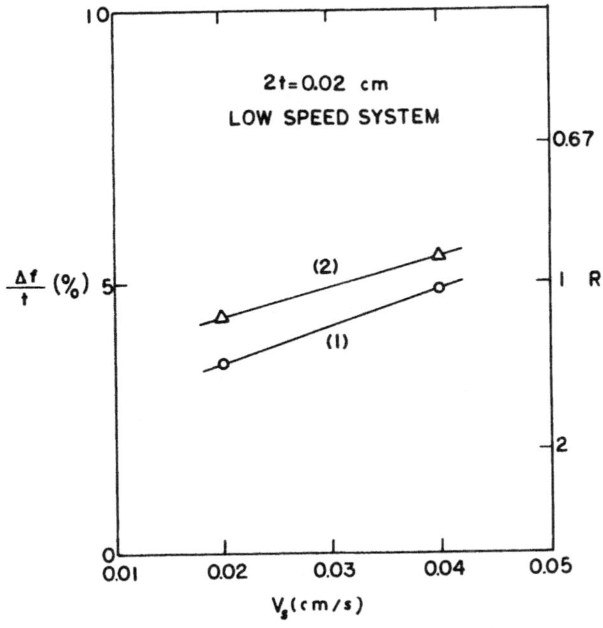

Fig. 11. Interface curvature dependence on growth speed at two melt inlet temperatures: (1) $T_i = T_M + 50°C$, and (2) $T_i = T_M + 150°C$.

are not strongly dependent on the growth speed or on sheet thickness. Thus, interface segregation for an impurity with $k_0 = 0.07$ and $D = 5 \times 10^{-5}\,\mathrm{cm^2/s}$, representative of carbon, as shown in figs. 12 and 13, also is relatively insensitive to operating parameters. The symmetric and asymmetric die configurations are compared in fig. 14 with T_1 and T_2 as variables. This illustrates the central finding of this analysis: the demonstration of the dominant impact of the die tip temperature on the interface configuration. Interface curvature itself is not significantly changed over the range of tilt angles examined. From a comparison of the results of an

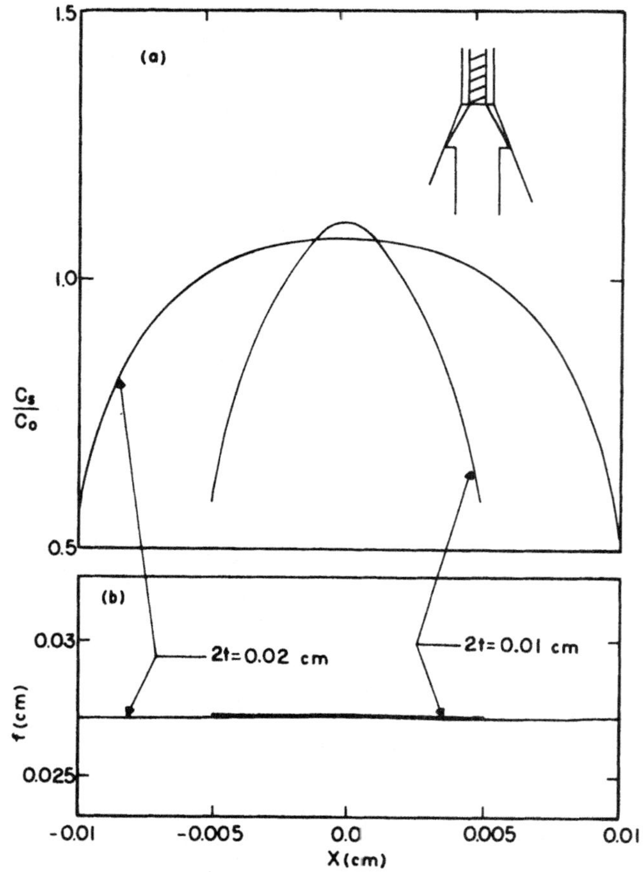

Fig. 12. Sheet interface impurity distributions (a) and interface configurations (b) at two thicknesses with $V_s = 0.04\,\mathrm{cm/s}$ in a low-speed symmetric system with $T_i = T_M + 150°\mathrm{C}$.

earlier study (Ettouney and Brown 1982) and those presented here, it is concluded that the introduction of the die and two-dimensional convection has little impact on the heat transfer that sets the interface curvature. However, the die asymmetry does lead to important changes in segregated impurity distributions because of convective transport variations, as is discussed in section 2.5.

While this analysis establishes relationships between prescribed system parameters and the interface configuration, its main limitations arise in that capillarity variables are not accounted for, and the meniscus shape is only approximately correct. Thus it cannot be used to relate interface configurations to operating diagrams such as given in figs. 4, 5 and 7. This can be done, however, with the model of section 2.2, and these results are described in the next section.

2.4.2. Process parameter impact

Interface shapes predicted by the model of section 2.2 for representative sets of operating variables considered there are shown in figs. 15 and 16.

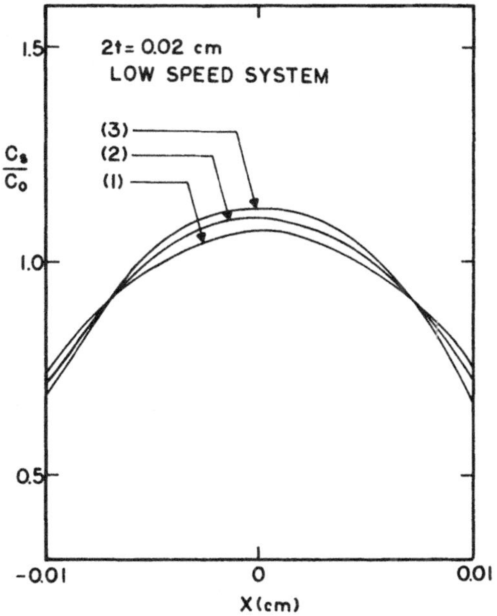

Fig. 13. Sheet interface impurity distributions for symmetric die at three speeds: (1) $V_s = 0.02$ cm/s (2) $V_s = 0.03$ cm/s and (3) $V_s = 0.04$ cm/s.

Only variables deviating from the baseline case given in table A3 are explicitly noted. Figure 15 illustrates interface shape dependence on growth rate, while the variation with die base temperature T_0 is shown in fig. 16. $h_{\rm eff} = 0.5$ cm for both cases, and the x-coordinate is scaled with $2\ell_3 = 0.025$ cm. The location of each interface at the sheet centerline has been translated to zero, and the height of the meniscus can be calculated

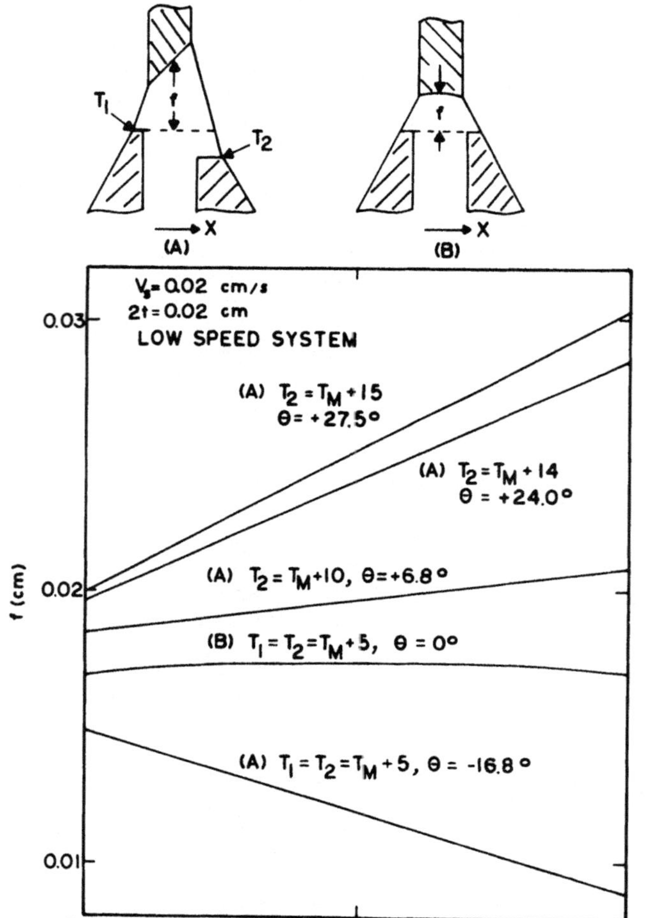

Fig. 14. Comparison of interface configurations for asymmetric die (A) and variable die tip temperature T_2 with symmetric die (B) in low-speed system.

by reading the sheet thickness from these figures and then using fig. 2 with $h_{eff} = 0.5$ cm to obtain s.

Interfaces convex toward the melt, arising when the sheet thickness decreases with increasing T_0, are obtained for the first time. These are not predicted from earlier models of Chalmers et al. (1975), Ettouney and Brown (1982), or Kalejs et al. (1983a), which do not properly account for heat transfer effects in the meniscus region. The importance of these effects is further illustrated in fig. 17, where interface shape is given at several values of the contact angle ϕ_0 of melt to the sheet. ϕ_0 is 11° for steady-state growth of silicon sheet. However, fluctuations about this steady-state value, such as during transients, could therefore lead to significant interface shape perturbations. Detailed heat transfer considerations and discussion of isotherm configurations are given by Ettouney et al. (1983) for these cases.

It is possible to account for all heat transfer effects on interface shape arising from process and material parameters with this model. Its limitation in the use of an approximation for convection was noted in section 2.3.2.

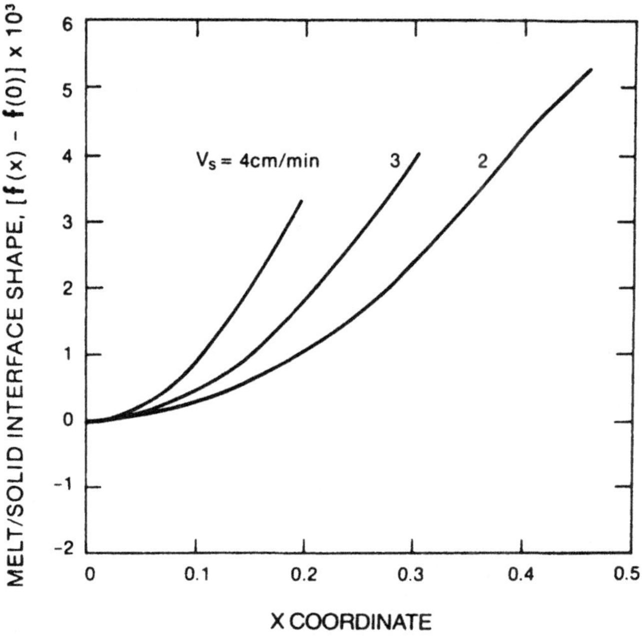

Fig. 15. Variations of melt–solid interface shape with growth speed at $T_0 = 1753$ K and $h_{eff} = 0.5$ cm.

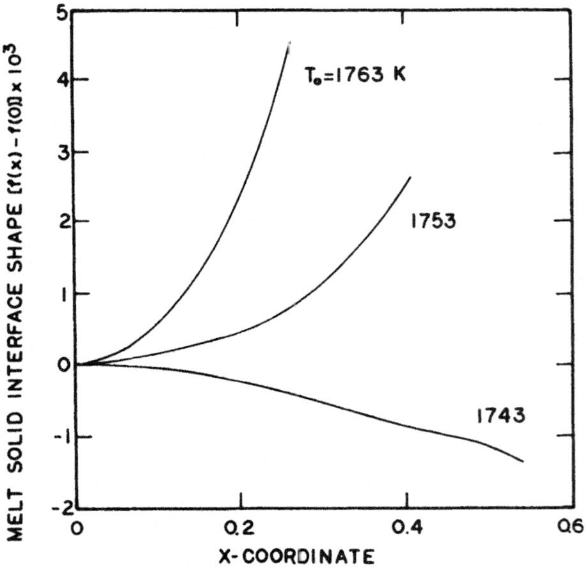

Fig. 16. Variations of melt–solid interface shape with die base temperature at $V_s = 3$ cm/min and $h_{eff} = 0.5$ cm.

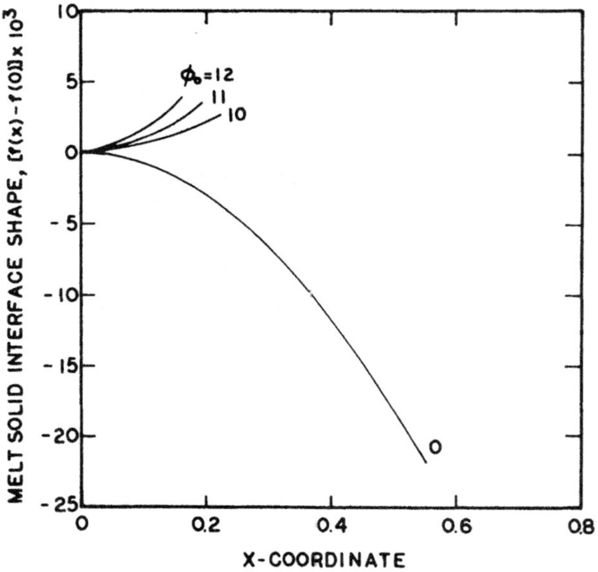

Fig. 17. Variations of melt–interface shape with melt–solid contact angle for $V_s = 4$ cm/min, $T_0 = 1763$ K and $h_{eff} = 0.5$ cm.

The work underway (Kalejs 1983b) to integrate full two-dimensional convection along with capillarity and heat transfer in this model will allow a self-consistent treatment of interface shape and impurity segregation to be achieved for sheet growth for the first time.

2.5. *Impurity transport*

Impurity segregation phenomena in EFG involve consideration of transport mechanisms in a different context from that usually treated in conventional melt growth. This is primarily due to the presence of the EFG die, which isolates the interface from the bulk melt. Consequently, new relations arise between convective and diffusive transport which can be parameterized by die geometry variables. An important consequence of the die influence is that interface curvature no longer dominates segregated impurity redistribution, but is usually secondary to convective transport in the meniscus, hence segregation depends on the meniscus configuration.

The study of impurity redistribution in EFG is facilitated by the extreme aspect ratio of the sheet geometry in the same manner as is the heat transfer, and hence impurity transport processes in the width dimension can be treated independently of those through the thickness to reduce the calculations to two-dimensional form. The extreme meniscus aspect ratio further makes it necessary to consider transport processes for oxygen and carbon associated with reactive ambient gases in contact with the meniscus surface, which are shown to impact on material properties. The special models developed for studying these transport processes and their experimental manifestations are discussed below.

2.5.1. *Die capillary convection*

The EFG die contains the capillary path by which silicon is transported between the crucible (bulk) melt and the interface (fig. 18). Solidification flow in the capillary is termed die capillary convection here because it is closely related to die geometry. The small meniscus volume effectively suppresses thermal convection; although the possibility exists for surface-tension driven flow to be vigorous, no evidence for it is observed (Kalejs 1978, Ettouney and Brown 1984b).

The wide, very thin aspect ($w \gg t$) of the sheet geometry justifies application of two-dimensional momentum and diffusion equations in a plane containing the growth axis and the sheet width to study impurity transport due to die capillary convection independently of transport in the thickness dimension. A schematic of a typical EFG die, and representative capillary cross sections in the sheet width plane are shown in fig. 18. Solutions of the Navier–Stokes equations have been obtained numerically for such capillary channels by Kalejs (1978) and used to calculate impurity redistribution ahead of the growth interface using the diffusion equation in the width plane. These results are parameterized by a die geometry variable w_d/w_c, the growth speed, and the melt diffusion and segregation coefficients of the impurity species.

Nonuniformity in the die top melt velocity field produces significant redistribution of segregated impurities with respect to the sheet width dimension. In steady-state growth at a speed V_s, an impurity boundary layer of thickness of the order D/V_s forms ahead of the growth interface. The interface melt concentration $C_I = C_0 k_0$, and thus depends on k_0 of the impurity species. Segregated melt impurity levels are uniform across the width of the sheet only for a flow field that is one-dimensional; i.e., directed normal to the interface. However, for the EFG die geometries where the melt flow channel connecting the bulk melt and the interface varies in cross sections (e.g., as in fig. 18) lateral melt flow parallel to the interface is necessary to supply melt to regions of the sheet not directly

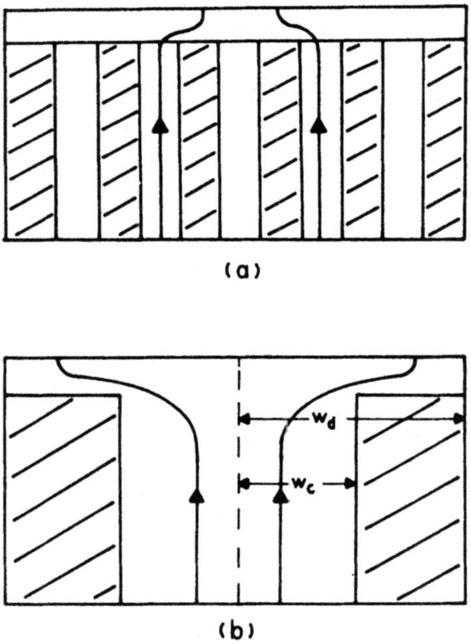

Fig. 18. Schematics of ribbon die cross sections showing several capillary arrangements: (a) multicapillary die, (b) central channel die.

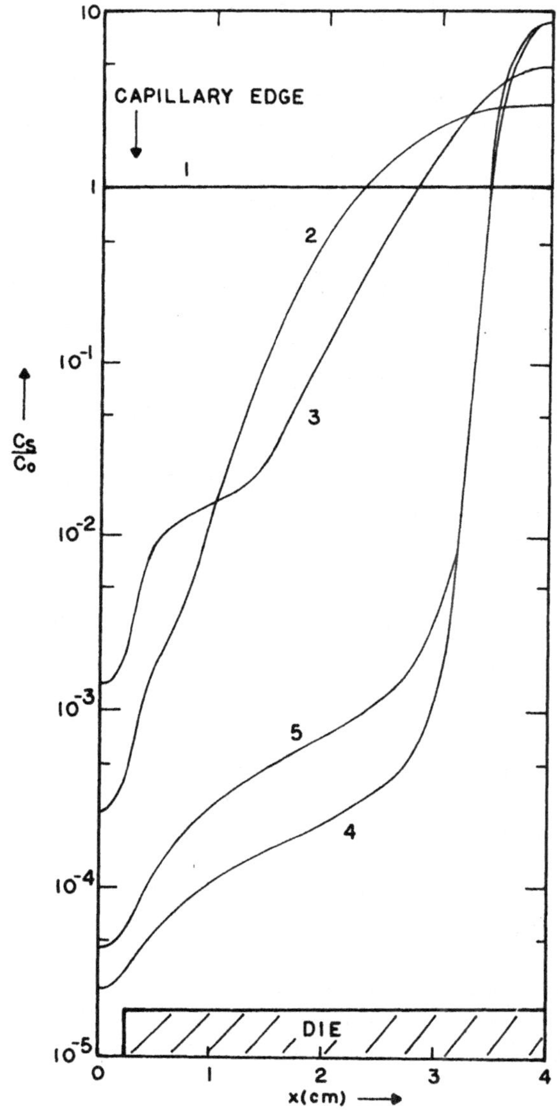

Fig. 19. Interface concentration ratio in the solid, C_s/C_0, along the sheet width for $w_d/w_c = 20$ and $k_0 = 10^{-5}$. Graph parameters: (1) $D = 10^{-5} cm^2/s$, $V_s = 0.042$, 0.125 cm/s (2) $D = 10^{-4} cm^2/s$, $V_s = 0.042$ cm/s (3) $D = 10^{-4} cm^2/s$, $V_s = 0.125$ cm/s (4) $D = 10^{-3} cm^2/s$, $V_s = 0.042$ cm/s (5) $D = 10^{-3} cm^2/s$, $V_s = 0.125$ cm/s.

above the capillary exit. Although a momentum boundary layer is also formed, overlap of interface impurity and momentum boundary layers allows the former to respond to nonuniformity in the lateral flow speed, and thus to produce lateral impurity redistribution.

The dependence of the impurity distribution across the sheet width on representative silicon impurity parameters and growth speed is illustrated in fig. 19 for $w_d/w_c = 20$ and the die design as in fig. 18b. An impurity with k_0 near unity, such as boron with $k_0 = 0.8$, has only minor redistribution. Order of magnitude variations arise for lower k_0, and these are also sensitively dependent on the magnitude of D. A minimum in the calculated C_s/C_0 occurs above the capillary exit for the geometry modeled in fig. 19. In general, as w_d/w_c decreases this minimum shifts also and remains above the capillary edge. The maximum enhancement in impurity concentration occurs in the region of the sheet farthest removed from the capillary exit.

Redistribution of aluminum in silicon ribbon EFG has been demonstrated (Kalejs et al. 1980a). Calculated aluminum impurity profiles are compared to measured ribbon resistivity profiles in fig. 20 for the given die geometry for a range of melt diffusion coefficients. Agreement is within anticipated experimental uncertainties, for D between $3\text{–}10 \times 10^{-4} \text{cm}^2/\text{s}$, which is in the range of the highest value of $5 \times 10^{-4} \text{cm}^2/\text{s}$ reported previously.

2.5.2. Asymmetric (displaced) die

Impurity redistribution has been studied in the sheet thickness dimension by assuming that transport phenomena at a given width location act essentially independently of die capillary convection across the sheet width. The validity of this approach at a qualitative level is supported by comparison of the predictions of the model of section 2.4.1 and experimental data obtained for ribbon grown with asymmetric, or displaced dies of the configuration shown in fig. 14 (Kalejs 1983a). Aluminum and undoped melts were used in this study.

Application of the model of section 2.4.1 to carbon and aluminum predicts the relationships between impurity distributions and interface configurations shown in fig. 21 in the case of displaced die growth. Resistivity profiles through the thickness for ribbon grown under conditions closely approximating those modeled are given in fig. 22. An interface tilt angle of the order of −17°, as for fig. 21a, is necessary to produce the average 20% aluminum surface concentration difference found experimentally. Although it has been possible to obtain agreement of model results and observed aluminum redistribution patterns using this approach, a compari-

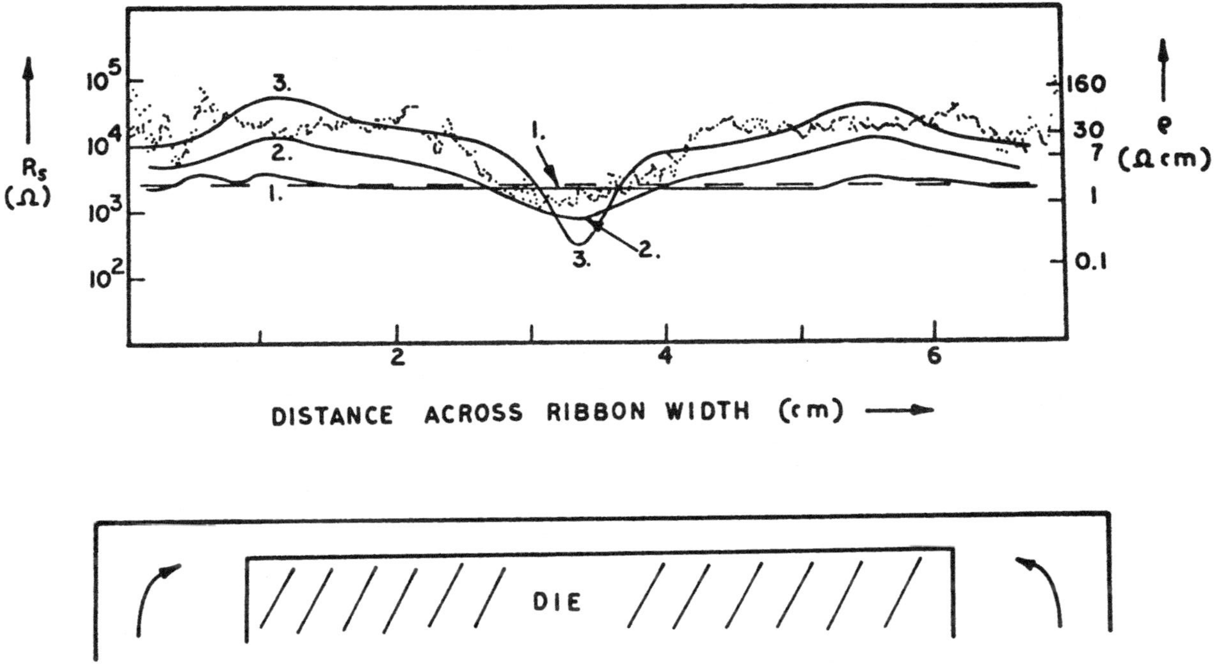

Fig. 20. Comparison of experimental data (dotted line) and calculated contribution of aluminum to resistivity of a doped ribbon for several values of the diffusion coefficient: (1) $0.24–0.37 \times 10^{-4} cm^2/s$, (2) $3 \times 10^{-4} cm^2/s$, and (3) $10 \times 10^{-4} cm^2/s$. Dashed line represents bulk melt resistivity level.

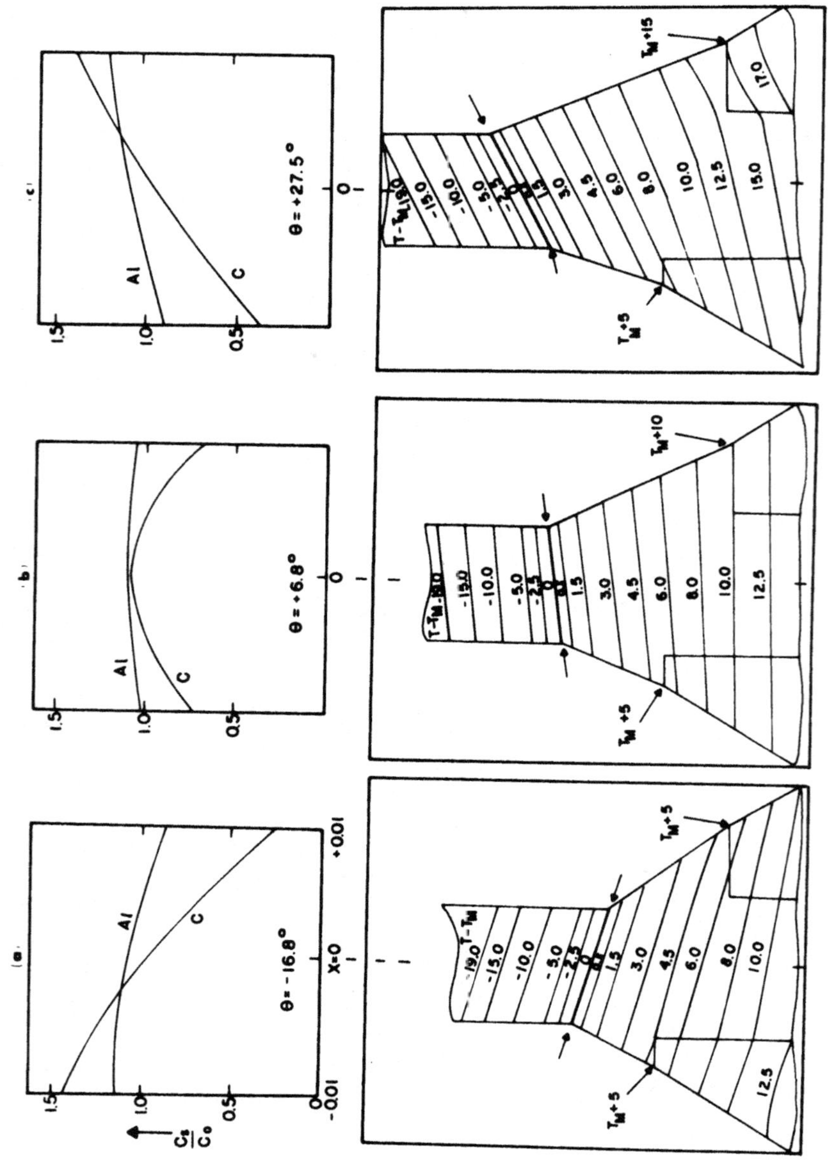
(a)
(b)
(c)
θ = -16.8°
θ = +6.8°
θ = +27.5°
Al
C
Cs/Co
X=0
-0.01
+0.01
T-TM
TM+5
TM+10
TM+15

son at a quantitative level is not meaningful because of the approximations used to treat capillarity. The most significant limitation arises because this formalism is constrained to describe only symmetric sheet configurations, i.e., ones for which the solid and the capillary centerlines correspond. This will only occur for certain values of capillary parameters (e.g., h_{eff}), and deviations from this symmetry may significantly contribute to producing segregated impurity asymmetries of the type observed in many situations. To model these effects, the analysis of section 2.2 including the full two-dimensional Navier–Stokes flow field needs to be used.

Ribbon surface asymmetries in silicon carbide density and in subsurface structure also occur with displaced die growth. The predicted carbon redistribution shown in fig. 21 can in principle account for the former. However, other factors must be considered in carbon transport in EFG (see sections 2.5.3 and 4.1). In addition, meniscus height differences result with this die asymmetry, and these are known to influence sheet SiC particle densities. The development of subsurface structure is more clearly related to asymmetry in interface shape and associated impurity profiles in the results shown in fig. 23. Experiments suggest that the frequency of incidence of the subsurface boundaries in evidence here rises with increasing ribbon thickness, and they are more often observed in ribbon grown with cold shoes (i.e., the more extreme temperature profile modeled in section 2.3.1) than without them (Rao et al. 1980). The effects of die asymmetry on location of the subsurface structure are in evidence in fig. 23 in the preferred appearance of these deviations nearer the ribbon face growing from above the non-displaced die side. It is not possible to distinguish between interface shape and associated impurity segregation phenomena in the melt and solid-state processes as causes for the subsurface structure on the basis of the experimental data. Carbon enhancement has been suggested as one factor contributing to its formation by promoting polygonization during post-growth cooling of the sheet at segregated carbon inhomogeneities within the cross section. However, resistivity discontinuities are observed to correspond to locations of the subsurface structure in aluminum doped ribbon (Kalejs 1983a). These most probably originate from redistribution at the growth interface and suggest competing proces-

---

Fig. 21. Calculated aluminum (Al) and carbon (C) segregation profiles through the sheet cross section for three interface configurations. The results are for a low-speed system with a die displacement of 0.0125 cm, a sheet thickness of 0.020 cm, and a growth speed of 0.02 cm/s.

ses may arise from grown-in dislocations that are focused toward regions of high local interface curvature.

Changes in impurity segregation that increase the incidence of subsurface structure are of particular concern when cold shoes are introduced to enhance the speed capability. The predictions of the model for segregation dependence on growth speed for an impurity typical of carbon are given in fig. 24 for several asymmetric growth situations. The practical aim of using the asymmetric die in this situation is to attempt to displace subsurface structure away from the sheet midplane and perhaps prevent its formation altogether. Large differences in the radial segregation are not predicted between growth at 0.02 cm/s (1.2 cm/min) with the low-speed sys-

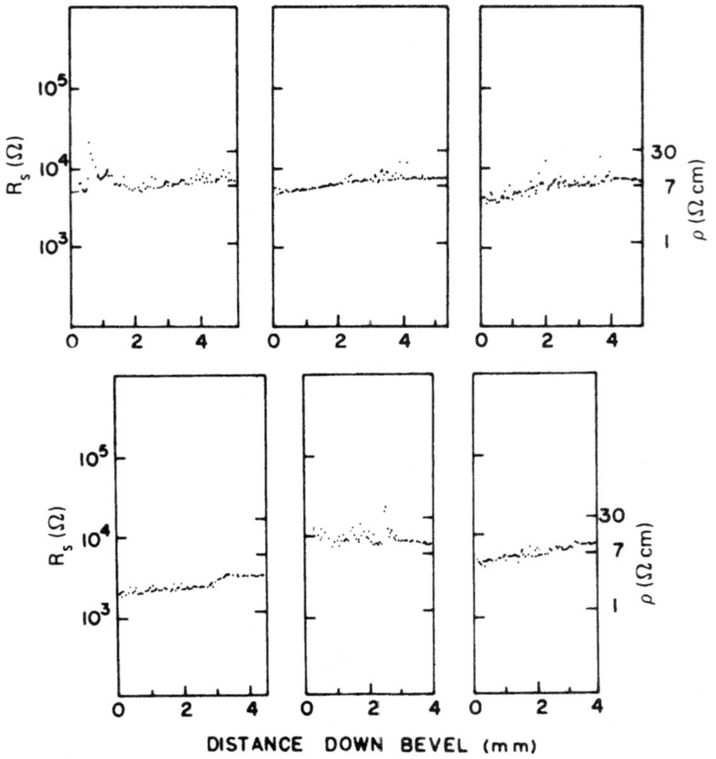

Fig. 22. Spreading resistance (resistivity) profiles for an aluminum doped ribbon grown in a low-speed system. The displaced die side is on the right. Ribbon thicknesses corresponding to the traces shown from left to right are: 0.38, 0.28, 0.26, 0.24, 0.22 and 0.20 mm, respectively.

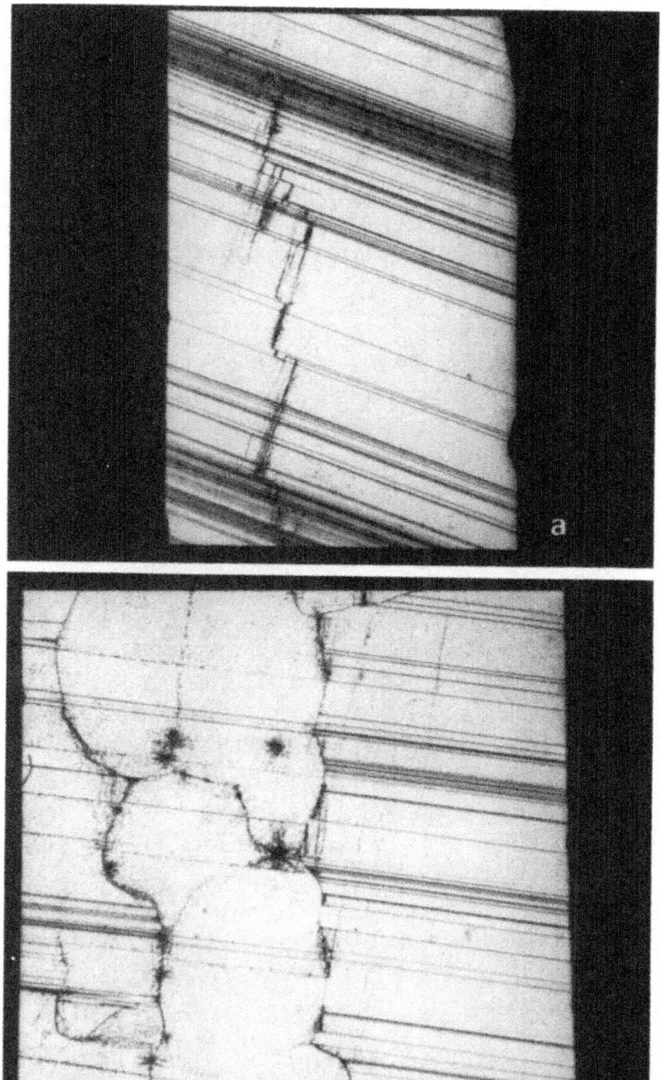

Fig. 23. Subsurface structure at two thicknesses for ribbon grown at a constant speed of 0.06 cm/s in a high-speed system: (a) $2t = 0.026$ cm, and (b) $2t = 0.040$ cm. Magnification is 185 ×. The growth direction is normal to the plane of the photographs and the displaced die side is on the right.

tem cooling profile and at 0.06 cm/s (3.6 cm/min) with the high-speed cooling profile. Interface curvatures for the two cases also do not differ appreciably. Detailed features of melt convection patterns rather than interface curvature impact most significantly on impurity segregation.

These impurity transport models for silicon ribbon EFG make it possible to identify dominant processes affecting interface impurity redistribution through the sheet thickness. In particular, the calculations that lead to the distributions of figs. 12, 13 and 21 show that melt convection dominates impurity redistribution over normal interface curvature-related effects

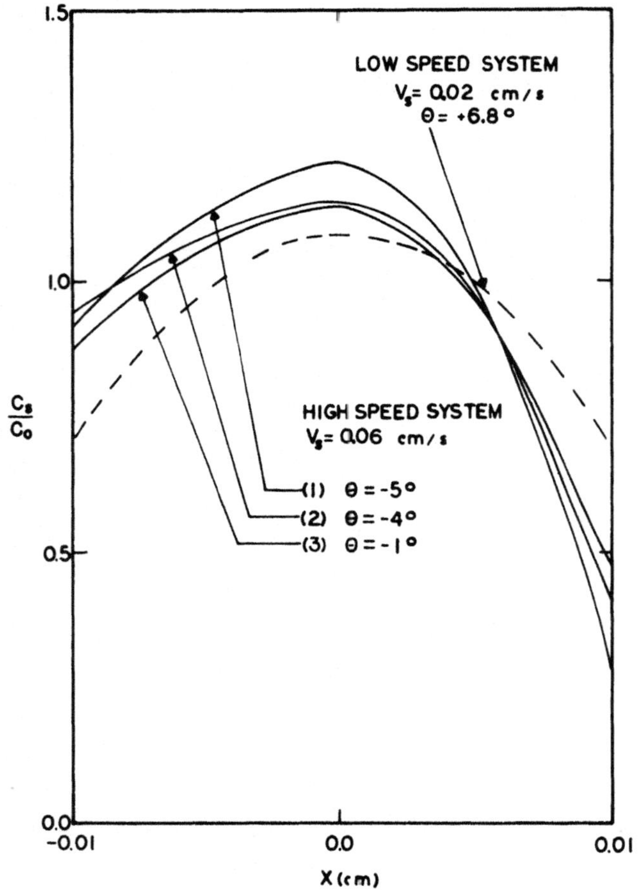

Fig. 24. Comparison of sheet interface impurity distributions for a low-speed system with $\theta = +6.8°$ (fig. 14) and a high-speed system with asymmetric die and several interface tilts.

even for thin ($\approx 300\,\mu m$) sheets. Calculated profiles all exhibit radial segregation, with sheet centerline concentrations greater than those at the surface. This is a consequence of interface curvatures which reflect solidification fronts that are concave toward the melt, and because k_0 for the impurity species modeled is less than unity.

The magnitude of the normalized centerline to surface segregation $\Delta C/C_0 = (C_s|_{x=0} - C_s|_{x=t/2})/C_0$ is greater in all cases than predicted solely on the basis of the calculated curvature magnitude (fig. 11). The simplest analytical expression for radial segregation gives (Coriell and Sekerka 1979)

$$\Delta C/C_0 = -(V_s/D)\Delta f(1-k_0).$$

More complicated segregation behavior is examined by Kalejs et al. (1983a). In the region of validity of this expression, the interface curvature contribution in the case of the growth conditions in fig. 12 gives $\Delta C/C_0 = 0.13$ for the interface with $2t = 0.01$ cm. This is considerably lower than the value $\Delta C/C_0 = 0.50$ reflected by the calculated results of fig. 12. Interface tilt, e.g. as in fig. 21, introduces further significant deviations in lateral segregation from behavior expected on the basis of interface curvature alone. These results illustrate the fact that interface convection is responsible for the radial segregation enhancement and likely dominates curvature effects over a wide range of growth parameters for an EFG silicon sheet.

The relationships between the influence of the interface tilt and radial segregation suggested by the results of fig. 21 can be explored further. This segregation behavior can be expressed in a more general form by utilizing a parameter analogous to a Peclet number for interface mass transfer $Pe_t = W_t V_t/D$ to describe regimes of radial segregation associated with interface tilt. For an interface tilt angle θ, $V_t \sim V_s \sin\theta$ is the transverse velocity typical of the interface radial convection field, and $W_t = 2t \sec\theta$. This shows that the extent of sheet surface concentration asymmetry varies as D^{-1} at constant tilt angle. The greater value of D for aluminum therefore constrains transport to a diffusion controlled regime ($Pe_t \lesssim 1$) more so than for carbon and reduces asymmetry in its radial segregation (fig. 21).

2.5.3. Carbon and oxygen

The study of the influence of carbon and oxygen on EFG sheet properties has been complicated because of lack of detailed knowledge of sources of these impurities, the possibility of formation of SiC and SiO_2 phases, and the discovery of an important source of these impurities during growth from reactive ambient gases in contact with the meniscus surface (Kalejs et al. 1981). The effect of these impurities on material properties is examined in section 4. The discussion here examines effects of ambient-related transport processes for oxygen and carbon on interface melt impurity distributions.

The calculations of the model of section 2.4.1 have been extended to model ambient gas–meniscus melt carbon and oxygen transport (Kalejs and Chin 1982). For this purpose, sources and sinks are included in the diffusion equation on the meniscus surface to represent reactions according to whether the impurity is added (e.g., oxygen) or removed (e.g., carbon) from the meniscus volume. Calculated oxygen sheet interface profiles through the sheet thickness obtained for the case where the only source is the meniscus surface (incoming melt concentration $C_0 = 0$) are given in fig. 25. Growth speed is the variable shown to most strongly influence the sheet impurity distribution with a meniscus source. The predicted transport

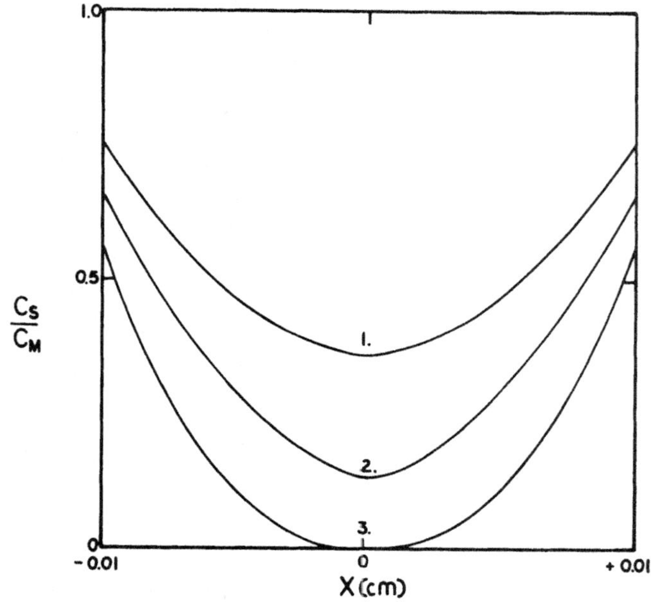

Fig. 25. Calculated impurity distributions through sheet thickness for uniform meniscus source concentration $C = C_M$ and impurity parameters $k_0 = 1.25$ and $D = 5 \times 10^{-5}\,\mathrm{cm^2/s}$. Results are for a constant sheet thickness of 0.02 cm and meniscus height of 0.026 cm, and growth speeds of 0.02 cm/s (case 1), 0.04 cm/s (case 2), and 0.06 cm/s (case 3).

under representative experimental conditions is able to account for levels of interstitial oxygen introduced in EFG ribbon from CO_2 gas deliberately added to the meniscus ambient during growth.

Calculations similar to these have been made to model processes that produce meniscus carbon depletion, such as through formation of SiC surface films. Evidence for carbon transport via meniscus–ambient gas-related processes is not as clear as for oxygen, and the study of carbon is further complicated because the melt is already supersaturated in carbon (see section 4.1).

3. Thermal stress

The origins of thermal stress are well known and have been discussed in detail for silicon sheet growth from the point of view of thermoelastic theory (Kalejs et al. 1980b, Gurtler 1980). Only recently has a model been developed by Lambropoulos et al. (1983) that allows incorporation of plastic deformation effects in sheet stress distribution calculations. The predictions of this model are examined here. Stress manifestations that need to be controlled in sheet growth are illustrated by application of the model to the case of 10 cm wide silicon ribbon EFG.

Thermal stresses build up rapidly as the sheet leaves the interface because nonuniformities in cooling lead to large second-order derivatives in the temperature profile (table 3 and fig. 9) in all practical growth systems that achieve high throughputs, e.g., widths greater than 2–3 cm combined with speeds of 2 cm/min and above. There are two types of stress manifestations. When thermal stress exceeds the critical yield stress, plastic deformation and stress relief occur through generation and motion of dislocations. In the other case of importance, stress build-up may be too rapid to be accommodated by creep. Then the sheet buckles at a critical stress which is a function of its width and thickness. The buckle produces nonuniformities in stress through the sheet thickness. If buckling occurs at temperatures where stress relief by creep is still possible, stress in the buckle will be relieved by dislocation generation also, and the buckle becomes a permanent deformation. Growth conditions are found for silicon sheet for which all of these manifestations of stress are observed.

Temperature distributions designed for high-speed (>2 cm/min) sheet growth lead to axial sheet profiles illustrated by the two general cases given in fig. 26. Profile 1 represents the 10 cm cartridge system discussed in section 2.3.1 as case C in table 2. Profile 2 is derived from profile 1 by eliminating the reheat region beyond about 0.4 cm from the growth interface and imposing a linear profile about 2 cm from the interface. These profiles have interface gradients of the order of 1000°C/cm in the sheet and high interface second derivatives of the temperature profile (see fig. 9 and table 2).

In principle, an 'ideal' profile exists for the sheet such that the stress at all points in the sheet is identically zero. This profile has the requirement $\partial(\alpha_T(\partial T/\partial y))/\partial y = 0$, or $\alpha_T(\partial T/\partial y) = \text{const.}$ α_T is the thermal expansion coefficient and is known to be temperature dependent. It is not possible to meet this requirement in the design of practical systems while maintaining high interface gradients. The focus of current research is thus to attempt to define system configurations which produce cooling profiles that minimize the plastic deformation and buckling. An approach developed by Lam-

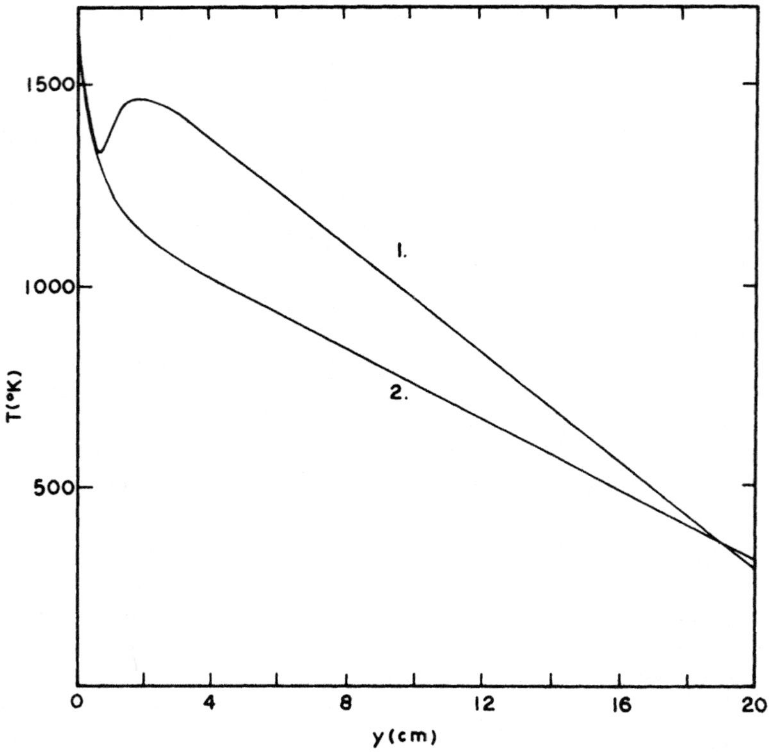

Fig. 26. Temperature profiles along growth direction used in modeling of sheet stresses: (1) EFG system, (2) idealized system.

bropoulos et al. (1983) that utilizes a numerical finite element scheme to calculate stresses with plastic deformation is the topic of the next section.

3.1. *Plastic deformation model*

The analysis treats a ribbon of negligible thickness and of width $2H$, extending from $x = +H$ to $x = -H$, and of length $L = 20$ cm. The interface is located at $y = 0$ (fig. 26). Steady-state growth at a speed V_s and plane stress conditions are other assumptions. Complicated creep response is reported for silicon at the high temperatures of interest ($\geqslant 1000°C$). Although primary creep is also under consideration, as is discussed below, the results presented here deal only with the effects of steady-state creep for illustrative purposes, with representative creep relations as given in table 4. The temperature profiles imposed on the ribbon are those shown in fig. 26.

Comparisons of stress distributions in an elastic sheet with those in a sheet capable of creep according to the 'low' and 'high' creep conditions defined in table 4 are given in figs. 27 and 28, where the axial (σ_{11}) and transverse (σ_{22}) stress components are plotted. Calculations have not been performed for the constitutive law in table 4 which gives the highest intensity creep. The high stress gradients and strain responses that arise near the interface with this intense creep require more refined computational meshes than are practical to incorporate into the numerical calculation scheme utilized. The necessity to account for such intense creep near the growth interface leads to consideration of an aspect of stress generation that is brought up for the first time with the present model. This is the question of the nature of the growth interface stress boundary condition that is demanded by the physics of the solidification process.

Table 4
High-temperature creep laws used in modeling. $\varepsilon_{ij} = C[\exp(-\beta / T) / T] (\sigma_e / \mu)^{n-1} S_{ij}$.

Case	C (K/GPa-s)	β (°K)	n	ε (s^{-1})*
Myshlyaev et al. (1969) 'Low-creep' condition	1.05×10^{29}	59760	5	1×10^{-6}
'High-creep' condition	1.05×10^{31}	59760	5	1×10^{-4}
Siethoff and Schröter (1983)	5.85×10^{22}	41800	3.6	41×10^{-4}

* Calculated strain rate for $\tau / \mu = 10^{-3}$ and $T = 1300$ K, with pure shear ($\tau = \sigma_{12}$). $\sigma_e = \sqrt{(3/2) S_{ij} S_{ij}}$, $S_{ij} = \sigma_{ij} - \frac{1}{3} \sigma_{kk} \delta_{ij}$, $\mu = 63.7$ GPa.

The steady-state model does not require specification of a particular interface stress distribution, nor does it provide information on the interface stress. The results presented here are for zero axial (σ_{11}) and transverse (σ_{22}) interface stress components. The solutions in figs. 27 and 28 even for the elastic cases thus differ from all previously reported thermoelastic model distributions (Gurtler 1980). The latter predict a large non-zero interface stress component σ_{22} (σ_{11} is zero there), and, consequently, produce different stress distributions elsewhere in the sheet.

The zero interface stress condition appears to be the most acceptable choice from an intuitive standpoint because the interface is bounded by the melt and cannot support a shear stress. Additional arguments in support of the choice of $\sigma_{22}(0) = 0$ have been made (Hutchinson 1983, Kalejs 1984) by considering the impact of non-zero σ_{22} on the present solutions in the presence of the creep conditions of table 4. These lead to the conclusion that the influence of any anticipated interface non-zero σ_{22} on the solutions in figs. 27 and 28 will be negligible at distances greater than 1 mm from the interface.

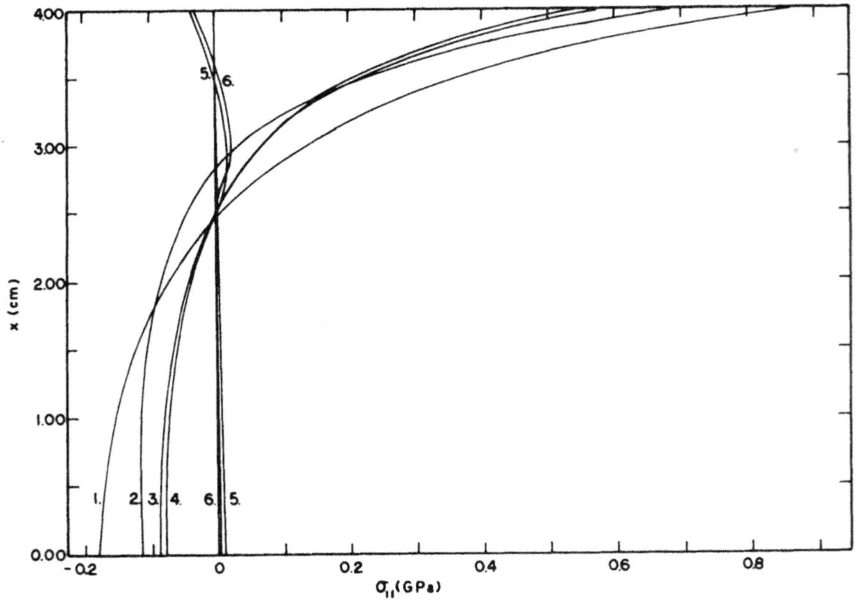

Fig. 27. σ_{11} variation across the sheet width at $y = 20$ cm (room temperature) for the high-creep case. Elastic solutions for idealized system (1) and EFG system (2) are compared to the solid with creep: idealized system at $V_s = 3$ cm/min (3), and at $V_s = 6$ cm/min (4); EFG system at $V_s = 3$ cm/min (5), and at $V_s = 6$ cm/min (6).

All these arguments for use of a model that includes plasticity are based on the assumption that the yield stresses are exceeded. Values of yield stress for silicon near the melting point as low as 0.6 MPa have been reported (Mil'vidskii and Bochkarev 1978). Yield stresses are very sensitive to temperature. Other data in the range of 1300–1400°C (Kalejs 1984) show that significant creep can be induced in four-point bending with stresses greater than 5 MPa. In order to have conditions that justify application of the present plasticity model at distances greater than 1 mm from the interface as described above, a flow stress of only 20 MPa near the melting point is required.

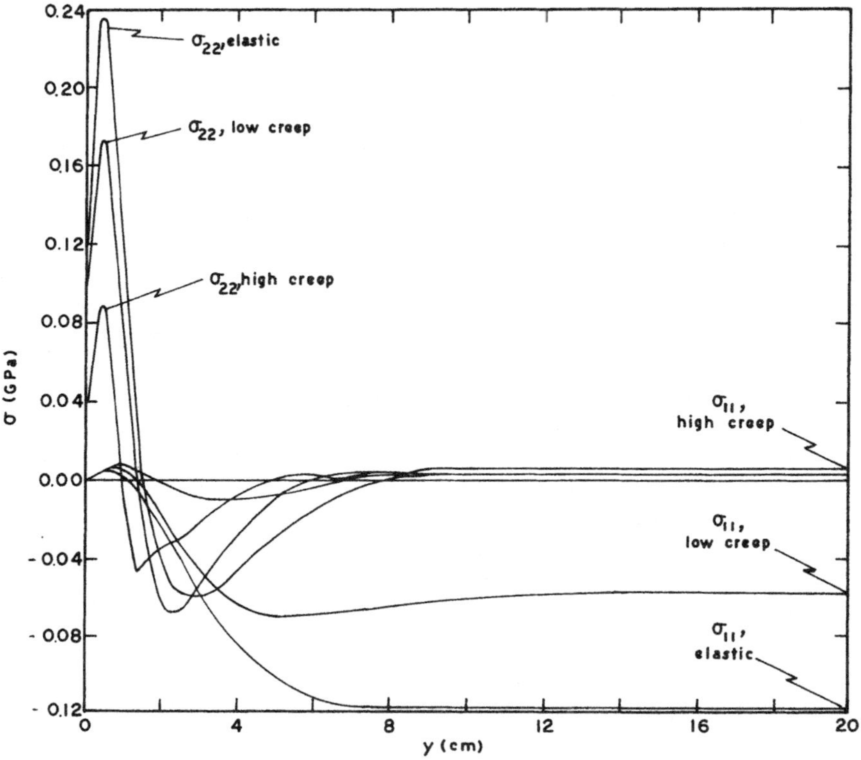

Fig. 28. σ_{11} and σ_{22} variations along growth direction near the sheet centerline ($x/H = 0.025$) for an EFG system at $V_s = 6$ cm/min.

3.2. Stress influence on sheet properties

The above analysis demonstrates that the growth regime for EFG silicon sheet production is one in which plastic deformation is a significant mechanism for stress relief. The extent to which stresses are reduced by creep from those in an elastic solid is particularly sensitive to details of the temperature profile in the sheet. Experimental data are not available to test the model at a quantitative level at this time. However, observations on defect configurations and buckling in 10 cm ribbon growth can be used to relate model predictions to growth conditions, to help guide development of low-stress sheet growth systems. These are discussed below.

3.2.1. Residual stress

The residual (room temperature) stress (σ_{11}) variation across the ribbon half width is shown in fig. 27. Ribbon edge elastic stresses are comparable for the two temperature profiles modeled. When 'high' creep conditions exist, an extremely large reduction in residual stress occurs for profile 1, but not for the idealized profile. Growth speed variations between 3 and 6 cm/min have a comparatively weak influence in comparison.

Figure 28 illustrates the variation of the principal stress components σ_{11} and σ_{22} along the ribbon centerline for the EFG system. In comparison to the large decrease in the residual stress σ_{11} with the high-creep conditions, variations produced in σ_{22} are not as extreme: σ_{22} remains much larger than σ_{11} and has its peak values within about 0.5 cm of the interface in all cases. σ_{22} contributes both to defect density and buckling, and thus influences σ_{11} also. Maximum strain rates of $3\text{–}6 \times 10^{-4} s^{-1}$ are predicted (Lambropoulos et al. 1983) which have their peak values in the region of the maximum in σ_{22} and are already very small beyond about 1 cm from the interface.

3.2.2. Defects

Dislocation densities in EFG sheets can range to values as high as $1 \times 10^7/cm^2$, and dislocations are very inhomogeneously distributed. The other dominant defects are twin boundaries. These usually occur in bundles and may be of first or higher order. Under some growth conditions subsurface structure is formed, as illustrated in fig. 23. Identification of macroscopic stress distributions responsible for a given defect configuration is hampered because of a number of complications. Yield stress and dislocation multiplication leading to stress relief are known to be functions of material parameters such as orientation, impurity content and initial defect configuration. An EFG sheet is not a single crystal, and contains the unique twin boundary structure, as well as high and inhomogeneously distributed concentrations of carbon. These factors make it unlikely that a given thermal stress, such as the high temperature σ_{22} component shown in fig. 28, is relaxed uniformly across the ribbon width. There is evidence that stress accommodation takes place over highly localized regions, for example, as shown in fig. 29. Very high density dislocation bands such as these are

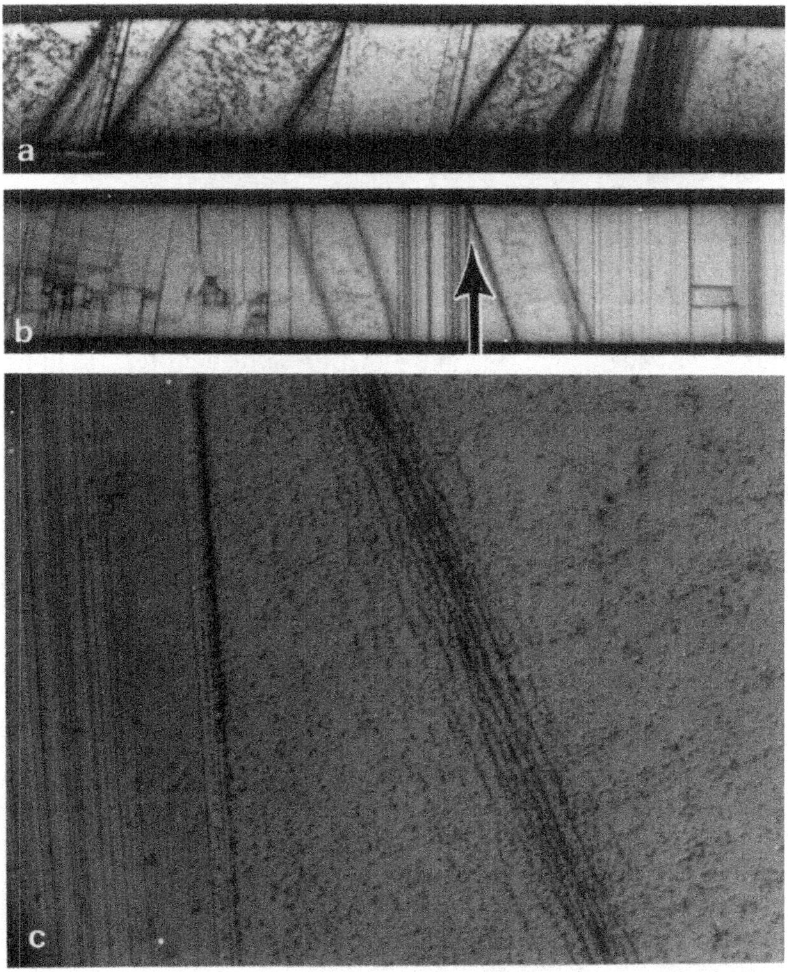

Fig. 29. Cross-sectional micrographs of an 10 cm wide EFG ribbon grown at 3.5 cm/min, showing regions of high dislocation density bands: (a) $2t = 0.15$ mm, magnification 70.4 ×, (b) $2t = 0.26$ mm, magnification 70.4 ×, (c) high magnification (654 ×) of sample (b) at location shown by arrow.

produced over regions of the order of the sheet thickness, but cover a relatively small fraction of the total cross-sectional area of the sheet. These regions appear to represent stress relief produced by development of Lüders instabilities through very rapid dislocation multiplication. The twin structure, such as shown in fig. 29, inhibits dislocation motion and tends to trap dislocations within the cross section. This will prevent uniform stress relief and lead to local residual stress concentrations in the sheet after cool down. The twins themselves appear to be growth-related and not stress-induced.

Although primary creep has not been modeled, it is likely to provide the chief mechanism for stress relaxation in the sheet at the highest temperatures. Primary creep has been studied by Kalejs (1984) in the temperature region between 1200–1400°C at stress levels approximating those reached in sheet growth. Stresses of the order of 10 MPa at strain rates of 10^{-4}–$10^{-3} s^{-1}$ are shown to induce dislocation densities between $1 \times 10^5 / cm^2$ to above $1 \times 10^7 / cm^2$ in times as short as 1–10 s. These dislocations appear along the major (111) slip planes and rapidly multiply and undergo glide to redistribute internal stresses. Thus, it is certain that the large high temperature σ_{22} stress predicted in the thermoelastic limit (e.g. as in fig. 28) is rapidly relaxed under typical growth conditions for growth speeds of the order of 3 cm/min. Significant primary creep is also found for times as short as 1 min at 1000°C. This will produce appreciable stress relaxation of σ_{11} stresses further away from the growth interface. The 'high' creep conditions of table 4 are even not as intense as the demonstrated primary creep magnitude. The model results thus provide the basis for concluding that accommodation of high thermoelastic stresses, which otherwise would produce sheet fracture and buckling, by creep is instrumental in allowing flat and low residual stress sheet to be grown in the EFG system considered.

3.2.3. Buckling

The thermoelastic residual stress (σ_{11}) magnitude predicted for the cartridge system in fig. 26 exceeds the critical buckling stress estimated for a 10 cm wide ribbon. For a sheet thickness of 300 μm, this is of the order of 25–50 MPa (Gurtler 1980). However, in the presence of the expected 'high' creep intensity, the characteristics of deformation and temperature region where the buckling occurs is now closely related to the sheet creep history and cannot be predicted using thermoelastic stress analysis alone.

Buckles in 10 cm wide ribbon become prominent as growth speed is increased in the range from 3 to 4 cm/min at thicknesses between 200–

300 μm. Average residual stresses stay low, however, of the order of 5–10 MPa; these buckles appear to be permanent, i.e., they are not elastic deformations at room temperature, and are 'frozen in' by high temperature stress relief.

The σ_{11} stresses given in figs. 27 and 28 are usually associated with the buckling stress. σ_{11} attains its steady-state value at distances of a few cm or more from the growth interface. However, the buildup of large σ_{22} stresses already within 1 cm of the interface cannot be ignored in considering buckling phenomena. Experiments show buckling modes that may be associated with both of these components can occur. A longer wavelength buckle, of the order of the ribbon width, is one manifestation reported by Kalejs et al. (1980b). This is likely produced by σ_{11}. However, relatively short wavelength buckles of the order of 1 cm are also observed, and these probably originate from the action of σ_{22} closer to the growth interface.

3.3. Stress control

The model and experimental observations discussed above show that creep limitations on sheet properties for 10 cm wide ribbons are imposed already at growth speeds of the order of 3 cm/min. In the case of the EFG system shown in fig. 8, it is the existence of large interface second-order derivatives, as shown in fig. 9, that is responsible for creep. It is evident from fig. 9, and also from table 2, that basic heat transfer conditions related to silicon material parameters restrict the sensitivity of the second-order derivative to variations in system component temperatures, as represented by T_A, T_B, and T_C. Consequently, the high interface value d^2T/dy^2 can be reduced only through a decrease in the interface gradient dT/dy. The latter defines the growth speed capability of a system. Conversely, if temperature gradients increase or higher sheet pulling rates are made viable through extension of the afterheater or raising its temperature, it will necessarily be at the expense of more severely defected material and/or buckling.

It is not possible to decide on the basis of the available data whether the temperature profile for the EFG system given in fig. 26 represents a minimum stress configuration, viz., with respect to already having the lowest possible defect density and highest buckle threshold speed, for the design speed capacity. At the interface gradients available (table 2), maximum growth speeds are estimated to be in the 4.5–5 cm/min range for a 200–300 μm thick 10 cm wide ribbon. Experimental work indicates that steady-state growth at 200 μm and 4.0–4.2 cm/min is sustainable. However, this ribbon already exhibits deviations from flatness, i.e., buckling, that are of the order of several times its thickness. This buckling, rather than residual stress or changes in defect density, has provided the most severe limitation on ribbon usefulness. Defect densities do not appear to vary measurably with speed over the range of speeds between 2.5–4.0 cm/min, and residual stress also remains essentially unchanged in the 5–10 MPa range of the threshold of measurement. More qualitative information on creep response appropriate to sheet stress relief and on interface temperature fields is required before the limitations of creep on the growth speed can be more precisely specified.

A strategy that can produce sheet with low residual stress but a high defect density while maintaining a high growth speed has been suggested to be workable with a profile of the general shape of profile 2 in fig. 26. This is accomplished if the nonlinear region of large curvature near the interface can be confined to temperatures where creep produces essentially complete stress relief. The present calculations indicate that, given 'high' creep conditions, it is not sufficient to impose a low-stress (linear here) region beyond 2 cm from the interface to produce a sheet with acceptable residual stress levels (e.g., as for the EFG curves 5 or 6 in fig. 27). As the growth speed requirements are raised above the speeds modeled here, it becomes necessary to compress the nonlinear region even more to reduce stress, a task which is difficult to accomplish because of weak radiation coupling between the sheet and surrounding system components of the type modeled in fig. 8.

Thermoelastic theory predicts that horizontal isotherm variations may compensate vertical profile curvature in minimizing stress (Morrison et al. 1976). Due to growth stability considerations, horizontal isotherm variations large enough to compensate for the interface nonlinearities cannot be imposed in practice. However, it has been shown from temperature field modeling that this strategy may be useful to reduce the impact of axial temperature profile nonlinearities in regions removed from the interface, such as shown between 1 and 5 cm in fig. 9 (Mataga et al. 1987).

4. Impurity and defect influence on material quality

The study of impurity limitations on material electronic quality in relation to its suitability for terrestrial solar cell fabrication generally has several purposes. One is to define growth-related factors that provide basic limitations on quality (Davis et al. 1980); the other is to assess the feasibility of using reduced purity feedstock (Bathey and Cretella 1982), where potential cost reductions are traded off against degradation of cell performance. Two categories of quality degradation mechanisms are relevant to defining the tolerance of as-grown material. In the one case, degradation arises from an increase in the level of an electrically active impurity in solution in the silicon matrix. Most often this is reflected in the bulk diffusion length which impacts on cell short-circuit current. At the other extreme, when reduced quality material is acceptable because of cost benefits, loss of material perfection or quality occurs more frequently as a result of the generation of impurity complexes and extended defects, e.g., grain boundaries and dislocations. As well as potential contributors to bulk lifetime degradation, these defects are particularly detrimental to cell voltage and fill factor in cases where they extend into the junction region. The latter often dominate cell performance in EFG sheet because of efforts to maintain high productivity in combination with high impurity levels.

Cell performance is determined by complex interactions among impurities, point defects and extended defects, which take place during post-solidification furnace cooling and in subsequent cell fabrication steps. The discussion of section 1 (see table 1) highlights growth factors that contribute to significant differences between Czochralski silicon and EFG sheet properties in the as-grown state. These differences further produce variations in material response to cell fabrication processes.

In conventional material, transition-metal impurities are considered to be the most deleterious defects in producing bulk lifetime and solar cell performance degradation (Davis et al. 1980). The mechanisms by which this occurs are not unambiguously identified, however, even in the highest lifetime dislocation-free silicon that can be produced (Graff and Pieper 1975). More recently, lifetime effects arising from complexes associated with oxygen in Czochralski silicon have been identified (Nauka et al. 1983). Introduction of extended defects, such as dislocations and grain boundaries, complicates study of bulk lifetime limiting parameters. It is the conventional view that dislocation electrical activity is produced predominantly by impurity decoration (Menniger et al. 1980, Queisser 1983).

There is no evidence to indicate that transition-metal impurities dominate cell performance of EFG materials to the same extent as in conventional Czochralski or float-zone silicon. Although total concentrations are several orders of magnitude greater in the former than in the latter two because of the $k_{eff} = 1$ segregation condition, yet observed cell performance is higher than expected on the basis of models for defect-free silicon. This increased tolerance of EFG sheets is usually attributed to the influence of defects which act to getter the electronically active impurities. The most common extended defects are twin boundaries and dislocations. First-order twin boundaries are not electrically active, but activity in dislocations associated with higher-order twins and twin bundles is observed (Cunningham et al. 1982). A detailed discussion of defect structure in EFG materials is given by Ast and Cunningham (1985).

Defects formed as a result of interface instabilities caused by high melt impurity levels, dislocations generated in plastic deformation, and complexes associated with carbon are all potential candidates to limit the lifetime of EFG materials. Experimental evidence on point defect participation in microdefect formation processes during cell fabrication has recently been obtained. This leads to consideration of models for interaction of carbon and oxygen with point defects in formation of impurity complexes during growth and processing which can influence recombination activity.

Carbon and oxygen influence on defect formation processes are examined in sections 4.1 and 4.2, respectively. The effects of impurity-related interface instability and stress that lead to formation of grain boundaries and dislocations are discussed in sections 4.3 and 4.4. Point defect models for defect formation during processing and heat treatment are considered in section 4.5.

4.1. Carbon

The graphite die provides an unavoidable source of the carbon found in EFG materials. In a departure from conventional silicon melt growth configurations, graphite as well as the usual silica crucibles are used to contain the bulk melt in EFG systems. Specific property changes that can be attributed to carbon, which arise additionally from the use of graphite in place of silica crucibles, have not been identified.

Partition of carbon among substitutional sites and SiC in EFG sheets has been well documented (Cretella 1983). However, melt processes that govern incorporation of carbon, SiC and formation of carbides other than SiC are not understood even on a qualitative level. Complications arise because factors that influence carbon precipitation such as temperature distributions in the die, nucleating agents and kinetic limitations cannot be specified at this time.

Measured carbon profiles through the ribbon thickness have been obtained using secondary ion mass spectroscopy (SIMS). A typical profile on an as-grown sample is shown in fig. 30. Redistribution of segregated carbon in the melt by interface shape and by convective flow field effects is predicted to be significant (sections 2.4 and 2.5). Clear evidence for aluminum redistribution is found (figs. 20 and 22). Absence of comparable carbon segregation under similar growth conditions supports arguments that carbon interface melt transport processes are influenced by second phase formation. Calculations that take into account second phase formation processes at the meniscus surface in modeling ambient effects have been done in order to estimate their impact on carbon levels (Kalejs and Chin 1982). Die interaction with the melt is not included in the modeling because kinetics governing SiC particle nucleation and growth are not understood.

Conditions of carbon supersaturation created in the meniscus at locations well removed from the interface are evidenced by formation of large ($\geq 10\,\mu m$) SiC particles at the die top. Substitutional carbon levels are typically in the range of $1–2 \times 10^{18} at/cm^3$ in the sheet. These are a factor of ten below the carbon levels in the die top melt estimated from known solubility data (Kalejs et al. 1981), and below measured total carbon levels (Cretella 1983). The large SiC particles entrained in the sheet surface from the die top cannot account for this discrepancy. Surprisingly, then, submi-

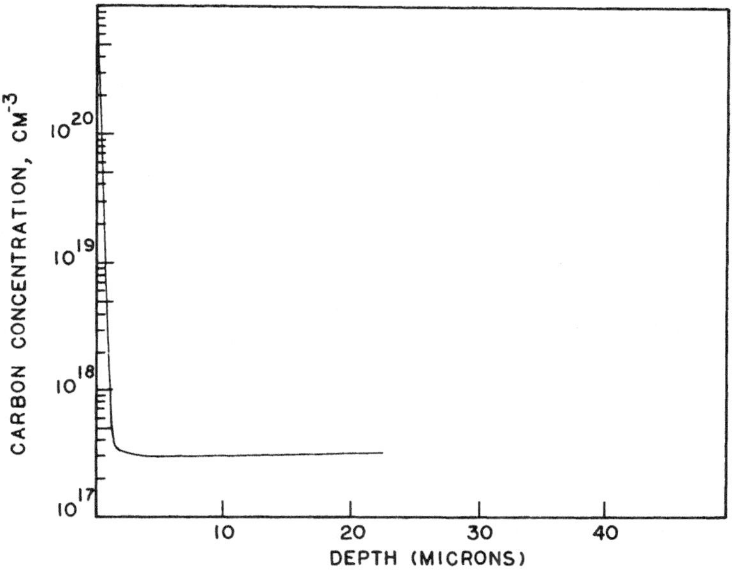

Fig. 30. Carbon concentration profiles through EFG sheet thickness as measured by secondary ion mass spectroscopy (SIMS) on as-grown material. Sample surface is at the left.

cron precipitates are not found in high-resolution transmission microscopy, with capabilities of detection down to 50–100 Å sizes (Strunk et al. 1980). Precipitates of micron size, which do not appear to originate from the die top, are occassionally found (Ast et al. 1982). Their distribution and frequency of appearance correlates with deliberate as well as accidental melt doping with metallic impurities such as aluminum and iron (Cretella 1983). These are thus thought to be metallic carbides or silicides which originate in the bulk or meniscus melts.

The absence of microprecipitates visible in TEM in as-grown EFG material is difficult to explain. However, etching studies show that microdefects are present in high densities. In addition to dislocation pits, these reveal a high density of shallow etch pits, not unlike those associated with swirls. Possible mechanisms for incorporation of high carbon concentrations under conditions that do not lead to identifiable precipitation are considered next.

Two-phase solidification of a silicon and SiC eutectic has been advanced to explain changes in material structure observed to be caused by variations in reactive ambient gas concentrations at the meniscus surface (Kalejs et al. 1981). This extends arguments proposed by Ravi (1977) for carbon incorporation in twin boundaries. The SiC phase can grow continuously as one component of a lamellar eutectic structure, the other component being silicon with carbon in solid solution, or it can form discontinuously as discrete crystals of SiC which are embedded in a continuous matrix of the silicon–carbon solid solution. It is proposed that these two structures correspond to the parallel twin boundary 'equilibrium structure' and 'large grain structure' produced under certain ambient conditions.

The eutectic growth mechanism leading to a dispersed SiC phase proposed above involves growth of SiC crystallites on the growth interface. It was estimated that precipitates could achieve diameters of 100 Å under typical sheet growth conditions before the advancing interface has cut off growth. The second possibility for formation of a carbon-rich dispersed phase is suggested by invoking point defect participation. If association of carbon and silicon takes place already near the interface prior to solidification, such clusters could form the nuclei for carbon-rich defects identified with the well-known B-swirls observed in float-zone and Czochralski silicon (Föll et al. 1977). Formation energies of agglomerates as compared to a SiC crystalline phase in the lattice may not differ greatly, and the defect configuration may depend only on supersaturation and kinetics. Undoubtedly, the high cooling rate of EFG sheets is a major factor in limiting growth of precipitates from such nuclei to a size that can be identified.

It has not been possible to image B-swirls by TEM. A search for carbon incorporated in twin planes of EFG sheet by TEM has been similarly unsuccessful (Ast and Cunningham 1982). It is thus not out of the question that none of the forms of discrete carbon-rich defects discussed can be resolved with present TEM methods for similar reasons related to insensitivity of TEM to carbon-based defects.

In summary, the available experimental data on carbon transport is interpreted to suggest that carbon does not undergo normal interface segregation in EFG sheet growth. Here the perspective is that the alternate processes by which carbon is incorporated involve partition of carbon between substitutional sites and one of several dispersed carbon-rich phases: lamellae, crystalline SiC microprecipitates or agglomerates of carbon and silicon self-interstitials, which cannot be detected by TEM.

4.2. Oxygen

Oxygen influence on EFG ribbon properties is clearly demonstrated and can be related to a number of growth variables. Material grown from quartz and graphite crucibles shows significant differences in response to solar cell processing (Ravi et al. 1981). Similar differences are produced in ribbons grown from graphite crucibles when oxygen is introduced by adding oxygen-containing gases, such as CO or CO_2, to an otherwise inert argon meniscus environment (Mackintosh et al. 1981). The analysis of section 2.5 illustrates the dependence of sheet oxygen levels on growth variables in the case when the sole source of oxygen is provided at the meniscus surface.

The most obvious difference in as-grown material produced by either of the above changes which make oxygen available during growth is an increase of interstitial oxygen concentrations. Interstitial oxygen is usually at the infrared spectrophotometry detection limit of about 5×10^{15} at/cm^3 in so-called 'oxygen-lean' ribbon grown from graphite crucibles under an inert argon ambient. Average levels of interstitial oxygen in 'oxygen-rich' ribbons grown with a quartz crucible, or with a graphite crucible combined with CO_2 concentrations of 2000–4000 ppm mixed with argon fed to the meniscus surface, are in the range $1\text{–}5 \times 10^{17}$ at/cc. Consistent changes in substitutional carbon do not accompany these interstitial oxygen increases.

Differences in as-grown material defect structure between ribbon grown from quartz and graphite crucibles are not found. The only effects on sheet structure related to oxygen are demonstrated for material grown from graphite crucibles. At intermediate CO concentrations, estimated at about

100–200 ppm in the neighborhood of the meniscus surface, parallel boundary 'equilibrium defect structure' is suppressed by an increase in the frequency of growth of large grains. These can be as wide as several cm and persist over distances of tens of cm along the growth direction. However, no indications of changes in microscopic defect structure as seen in TEM are found (Ast and Cunningham 1982), nor are there observed signifcant differences in material electronic properties between normal and large grain sheet, either before or after processing. These observations are not inconsistent with the models of carbon incorporation presented in section 4.1. The most obvious role of oxygen during growth in this situation is that it acts as a nucleating agent for carbon precipitates or agglomerates in the meniscus region. This role is also supported by the absence of SiO_2 precipitates even in oxygen-rich as-grown sheet. The absence of visible oxygen precipitates is not surprising. Oxygen precipitation will be inhibited by the low supersaturation, by the fast cooling rate and by high concentrations of carbon and dislocations. However, evidence is discussed below suggesting that oxygen influences carbon microdefect recombination properties during post-growth cooling and subsequent heat treatments even at the low oxygen/carbon ratios typical for EFG material.

The onset of large grain growth is not consistently observed and thus has been difficult to relate to growth parameters. It is found that a number of growth variables influencing interface shape also are of importance in promoting large grain formation. It is apparent, then, that the defect structure in EFG sheets is closely related to the oxygen–carbon interface melt balance. However, much uncertainty surrounds the mechanisms by which carbon and oxygen participate in defect formation even in conventional silicon, and it is not possible to construct detailed models at the atomic level for these processes.

4.3. *Constitutional supercooling*

Interface breakdown produced by constitutional supercooling because of excessive impurity accumulation at the growth interface generates structures detrimental to material quality. The variables which define the regime of interface stability against constitutional supercooling are the axial temperature and the impurity concentration gradients at the interface. These take on extreme values in the EFG process. Constitutional supercooling was previously examined for EFG (Swartz et al. 1975), but not for the die geometries under consideration here, for which k_{eff} is more strictly of the order of unity (see section 2.5). The liquidus slope, m, for most impurities is represented to a sufficient degree of approximation for present purposes using ideal solution behavior with m = 500°C at. fract. of the solute. Two typical contaminating impurities in EFG sheet growth, carbon and aluminum, have a similar potential for producing interface breakdown at considerably different melt concentrations because of more than an order of magnitude difference in k_0 (0.06 for carbon versus 0.002 for aluminum). For $k_{eff} = 1$, comparable interface concentrations, C_0/k_0, arise at bulk melt concentrations of $\sim 1 \times 10^{18}$at/cm^3 for carbon and $\sim 3 \times 10^{16}$at/cm^3 for aluminum.

The absence of evidence of carbon in boundaries in TEM and of redistribution at the interface argues against normal segregation models for carbon and in favor of the dispersed phase growth model discussed in section 4.1. However, it may still contribute to interface breakdown, as suggested by Rao et al. (1980). Evidence of constitutional supercooling due to aluminum has been observed in ribbons grown from melts doped deliberately with aluminum. Interface breakdown produced by aluminum is illustrated in the photograph in fig. 31a for the central region of a ribbon

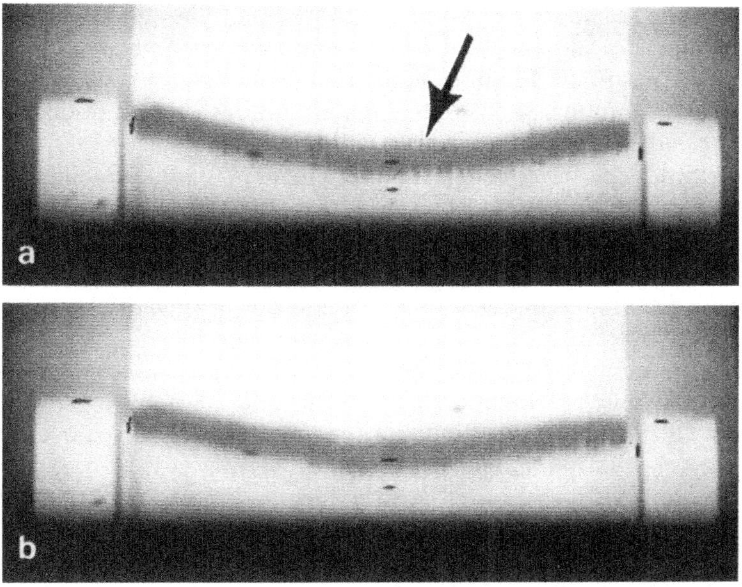

Fig. 31. Effect of constitutional supercooling produced by aluminum on interface shape: (a) interface perturbations in thick ribbon growth are identified by arrow, and (b) smooth interface after adjustment of growth conditions to produce thinner ribbon. The reference lines locate positions of die tip before seeding.

grown from a melt doped to 5×10^{15}at/cm^3. An increase in meniscus height, as in fig. 31b, is sufficient to eliminate interface irregularities. A possible explanation is that the meniscus height increase produces thinner ribbon, hence increases the axial temperature gradient (proportionally to $t^{-1/2}$) to above the threshold necessary to suppress constitutional supercooling. However, a straightforward application of conventional theory does not adequately explain these observations. For the given bulk melt level, this theory predicts that an axial temperature gradient of only 5°C/cm ($m\,\mathrm{grad}\,C_l$) is required for the ribbon of fig. 31a in order to suppress constitutional supercooling. For the known ribbon growth conditions, an interface temperature gradient of several hundred °C/cm is not unreasonable to expect, however. This example serves to illustrate that interface breakdown in EFG can occur already at lower bulk concentrations of contaminants than judged by application of constitutional supercooling relationships, and other growth-related factors need to be considered in the study of constitutional supercooling effects.

Solar cells have been made on 5 Ω cm (2.5×10^{15}at/cm^3) aluminum doped ribbon (Ho and Kalejs 1980). These show no differences in performance from comparable boron-doped ribbon. Onset of constitutional supercooling for aluminum at higher bulk melt concentrations precludes identification of bulk lifetime degradation with increasing concentrations. Similar constitutional supercooling manifestations have been documented in solar cells fabricated from cast silicon produced at increased growth rates which result in reductions in interface temperature gradients (Helmreich and Geissler 1984).

The behavior of impurities other than aluminum is also being studied (Cretella et al. 1985). Complications in understanding of segregation behavior of iron and copper arise with the onset of precipitation. Impurity synergistic effects are in evidence. This work indicates that both bulk and meniscus melt interactions that may involve carbon and oxygen are important in governing precipitation mechanisms of other impurities. Just as for the case of aluminum, precipitation and structural changes occur as the bulk melt concentration of any such impurity is raised. Bulk lifetime degradation leading to poorer cell performance as a result of the addition of transition metal impurities Fe and Mo, which is produced independently of structural changes, is not in evidence up to crucible melt contamination levels of the order of 1×10^{18}at/cm^{-3}.

4.4. *Dislocations*

Carbon is known to produce significant lattice strain because its effective atomic radius is smaller than that of the host silicon atom. The strain $\Delta a/a$ is related to carbon concentration C_s according to (Baker et al. 1968) $\Delta a/a = -6.5 \times 10^{-24} C_s$ (at/cm^3). For EFG sheets, $\Delta a/a \simeq 10^{-5}$. This strain may promote formation of complexes involving impurities and point defects, and influence residual stress relief leading to dislocations. Processed EFG sheets show evidence of stress relief through formation of both of these types of defects (Ast et al. 1982). Dislocation networks are observed, which are not present in as-grown sheet. Nodes often are decorated with very small precipitates of size 100 Å and smaller, which are too small to be identified according to composition. However, it appears they could be carbon-based, and form at sites provided by carbon-based nuclei, as discussed in section 4.1.

Oxygen influence on creep response has been noted in section 3. Oxygen, on account of its larger effective atomic radius, will compensate carbon lattice strain. Thus, the oxygen/carbon ratio influences dislocation generation both during the creep taking place in furnace cooling and in stress relief during processing. This relation has not been studied in detail up to now.

4.5. *Point defect models*

Recently, point-defect models have been successful in accounting for impurity diffusion and for defect formation processes in silicon (Gösele and Tan 1983). At high temperatures, silicon self-interstitials will be the dominant point defect present during post-solidification cooling under certain growth conditions (deKock 1977). Mechanisms have been proposed that involve these interstitials in formation of extended defects, such as cross-sectional structure in ribbon during growth (Rao et al. 1980) (see fig. 23 in section 2.5). Point defects may influence configurations of defects formed also in the PH_3 diffusion, since self-interstitial supersaturation is known to be promoted by high phosphorus surface concentrations. A model for point-defect participation in formation of microdefects is proposed here to account for not only post-solidification effects but also for EFG material lifetime responses during subsequent heat treatment and processing. This model is based on the concept presented in section 4.1, which assumes that the dominant microdefect in EFG materials arises at the site of a discrete carbon-rich nucleus.

Growth conditions in EFG favor formation of a carbon-based nucleus similar to the agglomerate of carbon and silicon self-interstitials proposed for the B-swirl by Föll et al. (1977). The ratio of the growth speed to the interface axial temperature gradient, $V_s/\mathrm{grad}\,T$, appears as the controlling parameter in defining the B-swirl formation regime (Voronkov 1982). Gösele (1983) has pointed out that this ratio has comparable values in the EFG systems discussed here and in conventional float-zone or Czochralski systems because both grad T and V_s scale proportionately (see table 1). Typically, the onset of B-swirls in FZ growth occurs for $V_s/\mathrm{grad}\,T \sim 3 \times 10^{-6}\,\mathrm{cm^2/°C\text{-}s}$ and extends over a growth speed range of about a factor of ten (deKock 1977). For EFG, the threshold may be attained at $V_s \sim 0.2$ cm/min with grad $T \sim 1000$°C/cm. Although conditions may be favorable for formation of the basic carbon self-interstitial nucleus, the configuration of the defect found at this site in the as-grown material is a function of growth speed, of cooling rate, of oxygen and point defect availability in each system.

It has been shown recently in several studies that silicon self-interstitials participate in development of defects in EFG materials during processing. TEM studies in processed material show that the formation of interstitial-type dislocation helices takes place during the 900°C 30-minutes PH_3 diffusion (Ast et al. 1982). Additional confirmation of self-interstitial involvement is obtained from study of carbon diffusion during PH_3 ambient anneals (Kalejs et al. 1984). Comparison of carbon out-diffusion profiles taken by SIMS in carbon-rich CZ and EFG silicon is given in fig. 32 for several anneal ambients and times. CZ silicon is a single-crystal material with a dislocation density of 10^4–$10^5/\mathrm{cm}^2$. The profiles are essentially identical for the two materials.

Several aspects of carbon and point defect interaction mechanisms are deduced from this data. First, the carbon out-diffusion with a PH_3 ambient anneal is observed to be anomalously large. This is attributed to enhancement of the carbon diffusion by the self-interstitial supersaturation produced by the phosphorus (Gösele and Tan 1983). The nitrogen ambient anneal shows the expected normal carbon diffusion rate. The oxygen ambient produces an intermediate diffusion rate. The second feature is carbon accumulation in the junction region leading to a carbon peak at about 0.2 μm from the surface observed after the 160-minute PH_3 ambient anneal. It is present both in the high oxygen content CZ and oxygen-lean ribbon, indicating that oxygen is not a major constituent of the defect formed there. The peak is not present after the oxygen ambient anneal. These results suggest that, in addition to carbon, phosphorus and silicon

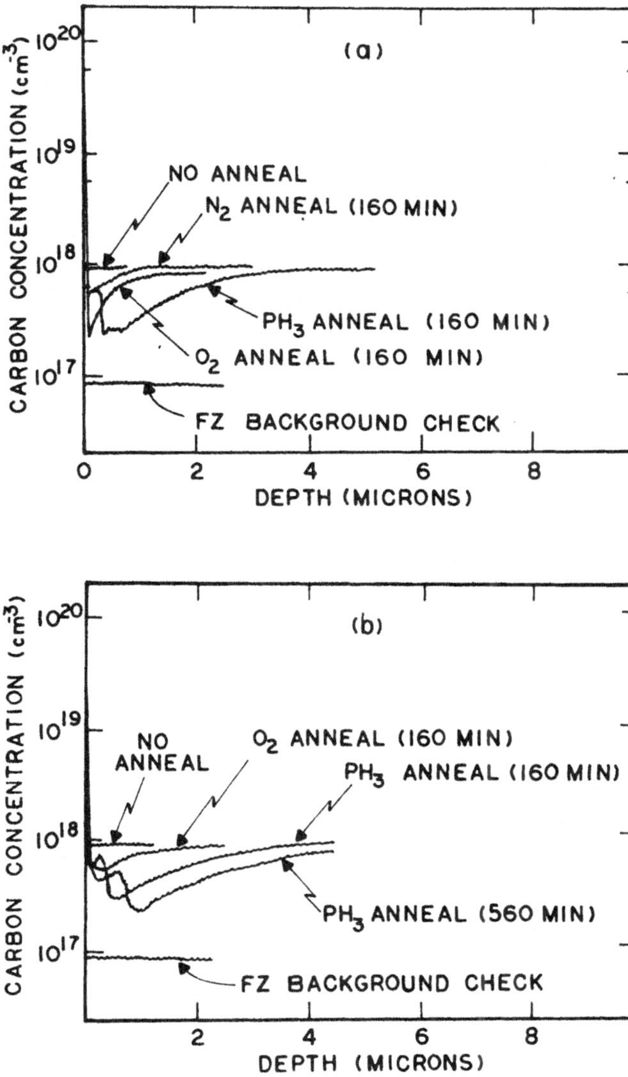

Fig. 32. Carbon concentration out-diffusion profiles for carbon-rich Czochralski silicon (a), and oxygen-lean EFG sheet (b), measured by secondary ion mass spectroscopy after anneals at 900°C with different ambients. Reference (no anneal) and background (FZ background check) carbon levels are also indicated. Sample surface is at the left.

self-interstitials are the most probable constituents of these defects. Microdefects of ~ 100 Å size have been observed in normally processed ribbon in high resolution TEM (Ast et al. 1982). They are too small to allow identification of their composition. The carbon diffusion data of fig. 32 give strong indication that such defects are likely to be carbon-based precipitates, at least in the near-junction region (see Feng et al. 1986).

The importance of the data of fig. 32 for the present arguments is that they show that self-interstitial supersaturation is maintained during diffusion of EFG material, even in the presence of its typically high dislocation density (10^6–$10^7/cm^2$), which can provide sinks for point defects. Evidence that the self-interstitial supersaturation produced by phosphorus in-diffusion promotes formation of carbon-based defects in the junction region further supports the model proposed in section 4.1 for dispersed carbon phase formation during growth via nuclei which contain self-interstitials as a dominant constituent.

These considerations can further help in interpreting data obtained in high-temperature annealing studies in ribbon as well as in carbon-rich Czochralski silicon. It is found in Czochralski silicon (Zuhlener 1983) that, while as-grown substitutional carbon is essentially stable against anneals of 50 hours at 1300°C for concentrations of less than 5×10^{17}at/cm^3, carbon removal from solution proceeds to decrease carbon concentrations to well below this level in material with as-grown concentrations greater than this. This behavior does not correspond to that found for oxygen during annealing at these temperatures, and strengthens arguments that carbon-based defect formation processes are governed by mechanisms different than those for oxygen. Experiments in carbon-rich float-zone silicon provide an additional perspective (Bean and Newman 1972). It was shown there that introduction of high dislocation densities inhibits carbon precipitation during high-temperature annealing. It is therefore possible to argue, in view of the evidence for self-interstitial participation in defect formation obtained from PH_3 diffusion studies, that carbon microdefect formation is controlled as much by the availability of point defects (viz., self-interstitials) as by supersaturation.

It is assumed, now, that carbon microdefect formation and growth in EFG sheet also will be controlled by silicon self-interstitial availability (supersaturation) both during post-solidification cooling and in subsequent heat treatments. Since the self-interstitial supersaturation is greatly enhanced during heat treatment in phosphorus ambients, which accompany cell junction formation, in particular, their effects on the configuration of carbon microdefects in as-grown sheets may be anticipated to be a factor in producing changes in recombination and in sheet structure observed in processing.

An important question that arises regards the contribution that carbon, self-interstitials and associated defects may make in influencing the recombination properties of EFG material. A study of response of cell performance to preprocessing anneals of up to one hour at 1200°C provides insight in this area (Kalejs and Ladd 1984). The results of this study are presented in table 5. A number of different silicon materials have been subjected to 10-minute and one-hour anneals prior to processing into solar cells with the regular 900°C PH_3 gas junction diffusion. This anneal benefits only the oxygen-lean EFG material.

Previous attempts to explain recombination behavior in EFG solar cells were based on models that assume that lifetime is dominated by gettering of deep level impurities by oxygen-based defects, either oxygen-vacancy complexes (Ravi et al. 1981), or extended defect sites which form by heterogeneous nucleation of oxygen clusters prior to precipitate formation (Ho and Wald 1981). Oxygen complexes which act as gettering centers in ribbon are energetically favored to dissociate during the 1200°C anneal because its interstitial oxygen concentration is considerably below the reported solubility limit of 7×10^{17}at/cm^3 (Craven 1981) at the anneal temperature. The consequences should be poorer cell performance in annealed than in unannealed ribbons.

The contrary response of ribbon cells with low interstitial oxygen to the 1200°C anneal is accounted for here by assuming that carbon microdefects dominate recombination behavior. Recombination activity may originate from the carbon–silicon self-interstitial nucleus itself, analogously to that suggested for oxygen-based microdefects (Nauka et al. 1983, Mao et al. 1984), or the microdefect may change bulk lifetime by acting as a gettering center for deep level impurities.

Ribbon lifetime responses to changes in oxygen availability during growth and to self-interstitial supersaturation during processing provide information on possible relationships between microdefect composition and recombination activity. When oxygen concentrations in ribbon are increased, either by replacing graphite by silica crucibles or by adding CO_2 to the growth ambient, cell bulk lifetime is improved. This is illustrated by comparison of the cell efficiencies of unannealed oxygen-lean (Type A) and oxygen-rich (Type C) EFG ribbon given in table 5. Reduction in self-interstitial supersaturation during high-temperature anneals is also shown to influence recombination properties. Cells made with a 1075°C CVD source diffusion show no light enhancement, and furthermore

Table 5. Material and cell processing data for 1200°C anneal experiments. The anneals were carried out in an argon ambient.

Type	Material Variables				Anneal time (min)	Cell Parameters				
	As-grown dark SPV L_n[a] (μm)	ϱ ϱ-type (Ω cm)	$[O]_i \times 10^{-17}$ (at/cm^3)	$[C]_s \times 10^{-17}$ (at/cm^3)		V_{oc} (V)	I_{sc} (mA/cm^2)	FF	η[b] (%)	η/η_0[c]
FZ	150	5	≲0.1	≲0.1	0	0.562	22.5	0.762	9.35	1.00
					10	0.538	21.0	0.756	8.25	0.88
					60	0.513	18.4	0.757	6.90	0.74
CZ	110	1	9.8	4.5	0	0.580	21.6	0.752	9.38	1.00
					10	0.570	21.0	0.750	8.95	0.95
					60	0.554	19.0	0.760	8.07	0.86
Silso	57	5	1–6	1–6	0	0.531	20.9	0.725	8.08	1.00
					10	0.517	21.8	0.677	7.65	0.95
					60	0.503	20.9	0.678	7.09	0.87
Ribbon										
Type A (Argon growth ambient)	41	4	≲0.1	9.0	0	0.472	13.7	0.644	4.18	1.00
					10	0.508	16.4	0.653	5.41	1.29
					60	0.521	17.4	0.691	6.36	1.52
Type B (Low CO_2 growth ambient)	37	4	≲0.1	21	0	0.490	15.8	0.691	5.30	1.00
					10	0.501	16.7	0.688	5.80	1.09
					60	0.504	16.9	0.699	6.02	1.14
Type C (High CO_2 growth ambient)	49	4	0.5	12	0	0.524	18.8	0.760	7.47	1.00
					10	0.511	18.9	0.716	6.92	0.93
					60	0.521	18.8	0.740	7.26	0.97

[a] See Bell and Freedman (1978) for details of measurement of diffusion length.
[b] No AR coating. Cell data are taken under AM1 conditions and 28°C.
[c] η_0 is reference efficiency without pre-processing anneal.

the deficiency in cell performance of oxygen-lean with respect to oxygen-rich ribbon is much reduced (Ravi et al. 1981). In addition, ribbon annealed at 900°C in an inert ambient, such as nitrogen or argon, exhibits no light enhancement regardless of oxygen content (Ladd 1982). This is explained with reference to the carbon microdefect model as the result of a reduction in microdefect electrical activity because of a decreased availability of silicon self-interstitials during the diffusion on account of lower phosphorus surface concentrations used in the diffusion. It is concluded from these observations that self-interstitial addition to or removal from the microdefect sites influences their recombination activity or increases gettering efficiency for impurities, while oxygen availability during growth tends to stabilize the defect nucleus against recombination activity changes during subsequent heat treatment and processing.

It was noted in section 4.4 that processing is accompanied by structural changes attributed to stress relief, and by microprecipitate formation. This makes it possible that observed recombination activity changes in EFG material do not arise solely from isolated microdefect sites, and that their interaction with dislocations is a factor that must be accounted for in a comprehensive model. The processes by which stress relief leading to formation of defects during the annealing and diffusion discussed above take place need to be studied, and some preliminary work in this area has been done to date (Gleichmann et al. 1985). The interpretation of the results is complicated by the necessity to account for effects of interactions among point defects, impurities, dislocations and the twin boundary structure common to EFG sheet. However, it is possible that microdefect electrical activity is governed only by availability of point defects and impurities, predominantly oxygen and carbon, even though the microdefects become incorporated into dislocations. It is this point of view that is taken here in presentation of a tentative model proposed by Gösele (1983) that accounts for recombination activity changes produced at the site of carbon self-interstitial agglomerates.

The microdefect configuration advanced is similar to that proposed for oxygen-based complexes, such as the 'new thermal donor' complex. There it is postulated that this agglomerate is the first stage of complex or precipitate formation which does not necessarily require carbon or self-interstitials but can also proceed only with vacancy participation. Electrical activity is attributed to interface states formed between the complex atoms and the silicon matrix.

This model is attractive as a candidate for EFG silicon also, except that carbon and self-interstitials replace oxygen and vacancies as the dominant constituents. The experimental data do not indicate whether an excess of self-interstitials is associated with the most detrimental configuration of a carbon-based nucleus insofar as degradation of lifetime is concerned. Oxygen, however, appears to act to stabilize the configuration against the high-temperature anneal. These constituents may further participate to transform the defect nucleus into a light enhancement center. The effect of a high-temperature anneal in the case of an oxygen-lean ribbon may be expected to lead to several changes in the defect configuration. Silicon self-interstitials may be expelled, or alternatively more carbon may be added, in either case producing a defect that is carbon-enriched. Simultaneous gettering of electrically active impurities during any such transformation of the defect configuration, either in heat treatment or processing, then can further produce bulk recombination activity changes. These are in evidence in the form of an unusual decrease in recombination activity of dislocations and grain boundaries observed in the oxygen-lean cells of table 5 which were annealed at 1200°C (Gleichmann et al. 1985).

It is possible to account for other unique EFG sheet electrical properties by invoking participation of carbon microdefects. One of these is an increased tolerance for deep-level impurities, which presumably attain much higher levels in EFG sheet than in CZ silicon because of unfavorable interface segregation conditions. Also, solar cell performance of EFG material is increased by hydrogen passivation in a very uniform way, and to a degree more striking than in other conventional material (Hanoka et al. 1983). Both of these results may be attributed to carbon microdefect domination of recombination sites which have a high affinity for hydrogen. One other solar cell material, ribbons grown by the ribbon-against-drop (RAD) technique shows as large a gain in cell performance as EFG material with hydrogen passivation (Hervo et al. 1984). This material has a comparably high carbon concentration and is produced with the same dynamic interface conditions favoring the self-interstitial supersaturation needed for formation of carbon microdefects in EFG sheet.

The effects of intermediate temperature annealing and of heat treatment associated with the phosphorus diffusion junction formation step on bulk lifetime and on cell performance have been examined by Ho et al. (1983, 1984). This work studied annealing effects in the temperature region below 900°C, and utilized oxygen-rich ribbon of varying oxygen concentrations. In the lower temperature anneals, defects similar in configuration to oxygen-based donor and new thermal donor complexes found in Czochralski silicon may increasingly become more influential in oxygen-rich EFG material and lead to recombination responses different from those observed after high temperature annealing. There is not sufficient data available to allow interpretation of this data in terms of behavior of carbon microdefects. Other studies have demonstrated EFG material to have nonuniform response to processing on a microscopic scale (Ravi et al. 1981, Bell and Hanoka 1982). This diversity of behavior likely has origins in complicated interactions among microdefects, principal extended defects such as dislocations and twins, and metallic impurities.

5. Summary

A review has been presented of current understanding of the relationships that arise between process variables and material properties of silicon sheet produced for photovoltaic applications by the EFG technique. The development of models to account for heat and mass transfer effects and for thermal stress generating mechanisms has aided the identification of the growth system characteristics that dominate EFG sheet properties in areas related to sheet dimensional uniformity, interface configurations, stress, and impurity effects that impact on solar cell performance.

Acknowledgements

I gratefully acknowledge the many contributions of my colleagues at Mobil Solar to the work described here. I also wish to thank Prof. R. Brown, Prof. J. Hutchinson, Prof. D. Ast, and Prof. U. Gösele for valuable collaboration in various areas of the material presented.

Acknowledgement is given for permission to reproduce figures and tables from the Electrochemical Society (fig. 25) and North-Holland Physics Publishing Amsterdam.

Appendices

A1. Parameters used for calculations

The following tables list the parameters used in the model described in the text for calculating domain dimensions (table A1), for silicon and graphite thermophysical constants (table A2), and for baseline case operating variables in dimensionless form (table A3).

Table A1
Dimensions of growth system.

Variable	Dimensional value (cm)
Thickness of die top (ℓ_1)	0.005
Thickness of die bottom (ℓ_2)	0.15
Half-width of gap in graphite capillary (ℓ_3)	0.0125
Length of die (ℓ_4)	0.25
Length of model system (ℓ_5)	5

Tabel A2
Material properties of silicon.

Property	Value
Thermal conductivity of melt k_ℓ	0.64 W / K cm
Thermal conductivity of solid k_M	0.22 W / K cm
Thermal conductivity of die k_d	0.43 W / K cm
Heat capacity of melt $c_{p\ell}$	1.0 J / gm K
Heat capacity of solid c_{ps}	1.0 J / gm K
Heat capacity of die c_{pd}	1.7 J / gm K
Density of melt ϱ_ℓ	2.52 gm / cm^3
Density of solid ϱ_s	2.30 gm / cm^3
Density of die ϱ_d	2.1 gm / cm^3
Melting temperature T_M	1683 K
Heat of fusion ΔH_f	1800 J / gm
Three-phase angle ϕ_0	11.0°
Surface tension γ	720 dyne / cm
Emissivity ε	0.64

Table A3
Operating conditions for heat transfer to uniform surrounding temperature.

Dimensionless Group	Value
Biot numbers, $\mathrm{Bi}_i \equiv hL^* / k_i$, $i = \ell$, s, d	0
Radiation number, $R_\ell \equiv \sigma\varepsilon_i L^* T^{*3} / k_i$	7.2 $\times 10^{-4}$
Melting temperature, $\tilde{T}_M \equiv T_M / T^*$	0.95
Stefan number, $S \equiv \Delta H_f / c_{ps} T^*$	1.0
Bond number, $\mathrm{Bo} \equiv gL^{*2}\Delta\varrho / \gamma$	2.1 $\times 10^{-3}$
Surrounding temperature, $\tilde{T}_\infty \equiv T_\infty / T^*$	0.2
Reference temperature, T* (K)	1783
Reference length, $L^* = 2\ell_3$ (cm)	0.025
Surrounding reference temperature, T_∞ (K)	357
ℓ, s, d – liquid, solid, die phases	

A2. List of symbols

a	Lattice constant
C	Solute concentration
C_I	Melt interface solute concentration
C_M	Meniscus surface solute concentration
C_0	Bulk melt solute concentration
C_s	Sheet interface solute concentration
D	Melt solute diffusion coefficient
f	Function representing interface configuration
h_{eff}	Static head
H	Sheet half-width
k_{eff}	Effective melt solute segregation coefficient
k_0	Equilibrium melt solute segregation coefficient
ℓ_1–ℓ_5	Calculation domain dimension length scales
L	Sheet length
m	Liquidus slope
Pe_t	Peclet number for interface melt
R	Radius of curvature of interface
R_s	Spreading resistance
s	Meniscus height
t	Sheet half-thickness
T	Temperature variable
T_1, T_2	Die tip temperatures
T_i	Die base melt inlet temperature
T_0	Die base temperature
$T_\infty^0(y)$	Axial environment temperature distribution
T_∞^0	Die base environment temperature
T_A, T_B, T_C	Prescribed cartridge system component temperatures
V_t	Sheet thickness (radial) velocity component
V_s	Sheet pull speed
w_c	Capillary width
w_d	Die width
W_t	Sheet thickness characteristic distance
x	Transverse (width) coordinate
y	Axial (growth direction) coordinate
α_T	Thermal expansion coefficient
θ	Interface tilt angle
θ_c	Apparent melt–die contact angle
ϕ_0	Melt–solid wetting angle

ϱ	Resistivity
σ_{11}	Axial sheet stress component
σ_{22}	Transverse sheet stress component

References

Ast, D.B., B. Cunningham and H.P. Strunk, 1982, EBIC/TEM Investigation of defects in solar cell silicon, in: Grain Boundaries in Semiconductors, Proc. Materials Res. Soc. Ann. Meeting, Boston, 1981, eds H.J. Leamy, G.E. Pike and C.H. Seager (North-Holland, Amsterdam) p. 167.

Ast, D.G., and B. Cunningham, 1982, Private communication.

Ast, D.G., and B. Cunningham, 1985, in: Silicon Processing for Photovoltaics I, eds C.P. Khattak and K.V. Ravi (North-Holland, Amsterdam) pp. 247–306.

Bathey, B.R., and M.C. Cretella, 1982, J. Mater. Sci. **17**, 3077.

Bean, A.R., and R.C. Newman, 1972, J. Phys. Chem. Solids **32**, 1211.

Bell, R.O., 1982, Private communication.

Bell, R.O., and G.M. Freedman, 1978, Minority carrier diffusion lenght from spectral response measurement, in: Proc. of 13th IEEE Photovoltaic Spec. Conf., Washington, 1978 (IEEE, New York) p. 89.

Bell, R.O., and J.I. Hanoka, 1982, J. Appl. Phys. **53**, 1741.

Chalmers, B., A.I. Mlavsky, T. Surek, J.C. Swartz, R.O. Bell, D.N. Jewett, D.A. Yates, K.V. Ravi and F.V. Wald, 1975, Continuous silicon solar cells, Final Technical Report NSF/RANN/SE/Gi03796X/FR/75/1, August 1975.

Ciszek, T.F., 1976, J. Appl. Phys. **47**, 440.

Coriell, S.R., and R.F. Sekerka, 1979, J. Cryst. Growth **46**, 479.

Craven, R.A., 1981, Oxygen precipitation in Czochralski silicon, in: Semiconductor Silicon 1981, Minneapolis, 1981, eds H.R. Huff, R.J. Kriegler and Y. Takeishi (Electrochemical Society, Pennington, NJ) p. 254.

Cretella, M.C., 1983, Private communication.

Cretella, M.C., B.R. Bathey and N.W. Marr, 1985, Proc. 6th EC Photovoltaic Solar Energy Conf., eds W. Palz and F.C. Treble (Reidel, Dordrecht) p. 929.

Cunningham, B., H.P. Strunk and D.G. Ast, 1982, Appl. Phys. Lett. **40**, 237.

Davis Jr, J.R., A. Rohatgi, R.H. Hopkins, P.D. Blair, P. Rai-Choudhury, J.R. McCormick and H.C. Mollenkopf, 1980, IEEE Trans. Electron Devices **ED-27**, 677.

DeKock, A.J.R., 1977, Point defect condensation in dislocation-free silicon crystals, in: Semiconductor Silicon 1977, Philadelphia, 1977, eds H.R. Huff and E. Sirtl (Electrochemical Society, Pennington, NJ) p. 508.

Ettouney, H.M., 1983, Heat, Mass and Momentum Transfer in Edge-defined Film-fed Crystal Growth, Ph.D. Thesis MIT, Massachusetts, June 1983.

Ettouney, H.M., and R.A. Brown, 1982, J. Cryst. Growth **58**, 313.

Ettouney, H.M., and R.A. Brown, 1984a, Mathematical modeling of high-speed ribbon systems: A case study of edge-defined film-fed growth, in: Proc. Flat-Plate Solar Array Project Research Forum on High Speed Growth and Characterization of Crystals for Solar Cells, Port St. Lucie, 1983, ed. K.A. Dumas (Jet Propulsion Laboratory, Pasadena) p. 311.

Ettouney, H.M., and R.A. Brown, 1984b, J. Appl. Phys. **55**, 4384.

Ettouney, H.M., R.A. Brown and J.P. Kalejs, 1983, J. Cryst. Growth **62**, 230.

Feng, S.Q., J.P. Kalejs and D.G. Ast, 1986, Mat. Res. Soc. Symp. Proc., Vol. 59, Oxygen, Carbon Hydrogen and Nitrogen in Crystalline Silicon, eds J.C. Mikkelsen Jr, S.J. Pearton, J.W. Corbett and S.J. Pennycook (MRS, Pittsburgh) p. 439.

Föll, H., U. Gösele and B.O. Kolbesen, 1977, J. Cryst. Growth **40**, 90.

Frost, H.J., and M.F. Ashby, 1981, Deformation mechanism maps for ceramics, Cambridge University Publication CUED/C/MATS/TR.79, May 1981.

Gleichmann, R., J.P. Kalejs and D.G. Ast, 1985, in: Mat. Res. Soc. Symp. Proc., Vol. 36, Impurity Diffusion and Gettering in Semiconductors, ed. R.B. Fair, C.W. Pearce and J. Washburn (MRS, Pittsburgh) p. 181.

Gösele, U., 1983, Private communication.

Gösele, U., and T.Y. Tan, 1983, The role of vacancies and self-interstitials in diffusion and agglomeration phenomena in silicon, in: Proc. ESSDERC '82, Agglomeration Phenomena of Point Defects in Silicon, Munich, 1982, eds E. Sirtl and J. Goorissen (Electrochemical Society, Pennington, NJ) p. 17.

Graff, K., and H. Pieper, 1975, J. Electron. Mater. **4**, 281.

Gurtler, R.W., 1980, J. Cryst. Growth **50**, 69.

Hanoka, J.I., C.H. Seager, D.J. Sharp and J.K.G. Panitz, 1983, Appl. Phys. Lett. **42**, 618.

Helmreich, D., and J. Geissler, 1984, Fast silicon ribbon growth methods, in: Proc. 5th EC Photovoltaic Solar Energy Conf., Athens, 1983, eds W. Palz and F. Fittipaldi (Reidel, Dordrecht) p. 955.

Hervo, C., M. Mautref, C. Belouet, C. Fages and J. Hervo, 1984, High-speed silicon ribbon growth by the RAD process, in: Proc. 5th EC Photovoltaic Solar Energy Conf., Athens, 1983, eds W. Palz and F. Fittipaldi (Reidel, Dordrecht) p. 971.

Ho, C.T., and J.P. Kalejs, 1980, Private communication.

Ho, C.T., and F.V. Wald, 1981, Phys. Status Solidi (a) **67**, 103.

Ho, C.T., G. Moeller and J.D. Mathias, 1983, Solid State Electron. **26**, 247.

Ho, C.T., G. Moeller, F.V. Wald and M. Spitzer, 1984, Sol. Cells **11**, 29.

Hutchinson, J.W., 1983, Private communication.

Kalejs, J.P., 1978, J. Cryst. Growth **44**, 329.

Kalejs, J.P., 1982, Large area silicon sheet by EFG, Final Report DOE/JPL 954355/81-21, September 1982.

Kalejs, J.P., 1983a, J. Cryst. Growth **61**, 485.

Kalejs, J.P., 1983b, Stress studies in EFG, Quarterly Report DOE/JPL 956312/05, December 1983.

Kalejs, J.P., 1984, Stress studies in EFG, Quartely Report DOE/JPL 956312/06, February 1984.

Kalejs, J.P., and R.O. Bell, 1984, Experimental aspects of the study of stress generating mechanisms in silicon sheet growth, in: Proc. Flat-Plate Solar Array Project Research Forum on High Speed Growth and Characterization of Crystals for Solar Cells, Port St. Lucie, 1983, ed. K.A. Dumas (Jet Propulsion Laboratory, Pasadena) p. 367.

Kalejs, J.P., and L.-Y. Chin, 1982, J. Electrochem. Soc. **129**, 1356.

Kalejs, J.P., and L.A. Ladd, 1984, Appl. Phys. Lett. **45**, 268.

Kalejs, J.P., G.M. Freedman and F.V. Wald, 1980a, J. Cryst. Growth **48**, 74.

Kalejs, J.P., B.H. Mackintosh and T. Surek, 1980b, J. Cryst. Growth **50**, 175.

Kalejs, J.P., M.C. Cretella, B. Chalmers and F.V. Wald, 1981, Effects of ambient gas changes during growth on the properties of EFG silicon ribbon, in: Proc. Symp. Electronic and Optical Properties of Polycrystalline or Impure Semiconductors and Novel Silicon Growth Methods, St. Louis, 1980, eds K.V. Ravi and B. O'Mara (Electrochemical Society, Pennington, NJ) p. 242.

Kalejs, J.P., L.-Y. Chin and F.M. Carlson, 1983a, J. Cryst. Growth **61**, 473.

Kalejs, J.P., H.M. Ettouney and R.A. Brown, 1983b, J. Cryst. Growth **65**, 316.

Kalejs, J.P., L.A. Ladd and U. Gösele, 1984, Appl. Phys. Lett. **45**, 540.

Ladd, L.A., 1982, Private communication.

Lambropoulos, J.C., J.W. Hutchinson, R.O. Bell, B. Chalmers and J.P. Kalejs, 1983, J. Cryst. Growth **65**, 324.

Mackintosh, B., J.P. Kalejs, C.T. Ho and F.V. Wald, 1981, Multiple EFG silicon ribbon technology as the basis for manufacturing low-cost terrestrial solar cells, in: Proc. 3rd EC Photovoltaic Solar Energy Conf., Cannes, 1980, ed. W. Palz (Reidel, Dordrecht) p. 553.

Mao, B.-Y., J. Lagowski and H.C. Gatos, 1984, Appl. Phys. Lett. **44**, 42.

Mataga, P.A., J.W. Hutchinson, B. Chalmers, R.O. Bell and J.P. Kalejs, 1987, J. Cryst. Growth, in press.

Menniger, H., H. Raidt and R. Gleichmann, 1980, Phys. Status Solidi (a) **58**, 173.

Mil'vidskii, M.G., and E.P. Bochkarev, 1978, J. Cryst. Growth **44**, 61.

Morrison, A.D., K.V. Ravi, C.V.H. Rao, T. Surek, D.F. Bliss, L.C. Garone and R.W. Hogencamp, 1976, Large area silicon sheet by EFG, Annual Progress Report, ERDA/JPL 954355/76-11, September 1976.

Myshlyaev, M.M., V.I. Nikitenko and V.I. Nesterenko, 1969, Phys. Status Solidi (a) **36**, 89.

Nauka, K., H.C. Gatos and J. Lagowski, 1983, Appl. Phys. Lett. **43**, 241.

Queisser, H.J., 1983, Electrical properties of dislocations and boundaries in semiconductors, in: Defects in Semiconductors II, Proc. Materials Res. Soc. Ann. Meeting, Boston, 1982, eds S. Mahajan and J.W. Corbett (North-Holland, Amsterdam) p. 323.

Rao, C.V.H.N., M.C. Cretella, F.V. Wald and K.V. Ravi, 1980, J. Cryst. Growth **50**, 311.

Ravi, K.V., 1977, J. Cryst. Growth **39**, 1.

Ravi, K.V., R.C. Gonsiorawski, A.R. Chaudhuri, C.V.H. Rao, C.T. Ho, J.I. Hanoka and B.R. Bathey, 1981, Ribbon growth-cell processing: Synergistic effects, in: Proc. 15th IEEE Photovoltaic Spec. Conf., Orlando, 1981 (IEEE, New York) p. 928.

Rudolph, P., 1982, Profilzüchtung von Einkristallen (Akademie-Verlag, Berlin).

Sachs, E.M., 1980, J. Cryst. Growth **50**, 102.

Sachs, E.M., and T. Surek, 1980, J. Cryst. Growth **50**, 114.

Siethoff, H., and W. Schröter, 1983, Scr. Metall. **17**, 393.

Strunk, H., B. Cunningham and D.G. Ast, 1980, JPL Technical Report No. 6, DOE/JPL 954852, December 1980.

Surek, T., and B. Chalmers, 1975, J. Cryst. Growth **29**, 1.

Surek, T., B. Chalmers and A.I. Mlavsky, 1977, J. Cryst. Growth **42**, 453.

Surek, T., S.R. Coriell and B. Chalmers, 1980, J. Cryst. Growth **50**, 21.

Swartz, J.C., T. Surek and B. Chalmers, 1975, J. Electron. Mater. **4**, 255.

Sze, S.M., 1969, Physics of Semiconductor Devices (Wiley-Interscience, New York) p. 25.

Tatarchenko, V.A., and E.A. Brener, 1980, J. Cryst. Growth **50**, 33.

Taylor, A.S., R.W. Stormont, C.C. Chao and E.J. Henderson, 1981, Large area silicon sheet via EFG tube growth, in: Proc. 15th IEEE Photovoltaic Spec. Conf., Orlando, 1981 (IEEE, New York) p. 589.

Ukhanov, Y.I., 1962, Sov. Phys.-Solid State **3**, 1529.

Ungar, L.H., and R.A. Brown, 1983, Private communication.

Voronkov, V.V., 1982, J. Cryst. Growth **59**, 625.

Wald, F.V., 1981, Crystal growth of silicon ribbons for terrestrial solar cells by the EFG method, in: Crystals: Growth, Properties and Applications, Vol. 5, ed J. Grabmaier (Springer, Berlin) p. 149.

Zulehner, W., 1983, Oxygen in silicon and its preparative aspects, in: Proc. ESSDERC '82, Agglomeration Phenomena of Point Defects in Silicon, Munich, 1982, eds E. Sirtl and J. Goorissen (Electrochemical Society, Pennington, NJ) p. 89.

CHAPTER 5

EPITAXIAL SOLAR CELLS ON RE-SOLIDIFIED METALLURGICAL-GRADE SILICON

R.V. D'AIELLO, P.H. ROBINSON and E.A. MILLER

RCA Laboratories
Princeton, NJ 08540, USA

Silicon Processing for Photovoltaics II, edited by C.P. Khattak and K.V. Ravi

Contents

1. Introduction

This chapter describes approaches towards the fabrication of low-cost solar cells which rely upon the epitaxial growth of high-quality layers on substrates derived from metallurgical-grade (MG) silicon. The basic strategy is to produce a very low-cost substrate by recrystallizing raw MG silicon, and then to grow a thin layer (20–30 μm) of sufficient purity and crystallographic perfection, containing a proper doping profile, in which high-efficiency cells can be fabricated. The conceptual steps in the fabrication are shown in fig. 1. It is important to note that the potential for high-efficiency cell performance need not be sacrified with such a structure because theoretical calculations of Hovel (1975) have shown that for suitable silicon structures which retain the back-surface-field (BSF) effect, the peak efficiency is relatively insensitive to the active layer thickness down to a thickness of ~25 μm. Thus, only about 1 mil (25 μm) of high-quality silicon need be grown on the MG substrate. In addition, referring to fig. 1, once the epitaxial layer is grown, the structure is quite similar to bulk cell-blanks, so the remaining process steps required to complete the solar cell may be chosen from a host of low-cost processes already developed.

The advantages of such an approach to low-cost cell fabrication are summarized below:

- Raw material for the substrate is MG silicon which is already produced in very large quantities (500 000 tons/yr) at low cost (~1.50 $/kg).
- The active film is grown epitaxially with precise control of thickness and doping profile to optimize solar-cell performance.
- Film thickness of ~25 μm is sufficient for high efficiency with the BSF automatically incorporated at the substrate/epi-layer interface.
- The resultant structure is entirely of silicon, eliminating potential thermal and electrical mismatch problems.
- Once the epitaxial layer is grown, the structure is compatible with already developed low-cost cell processes.
- Some simplifications in processing are possible because the substrate is not active in photovoltaic conversion but is highly conductive.

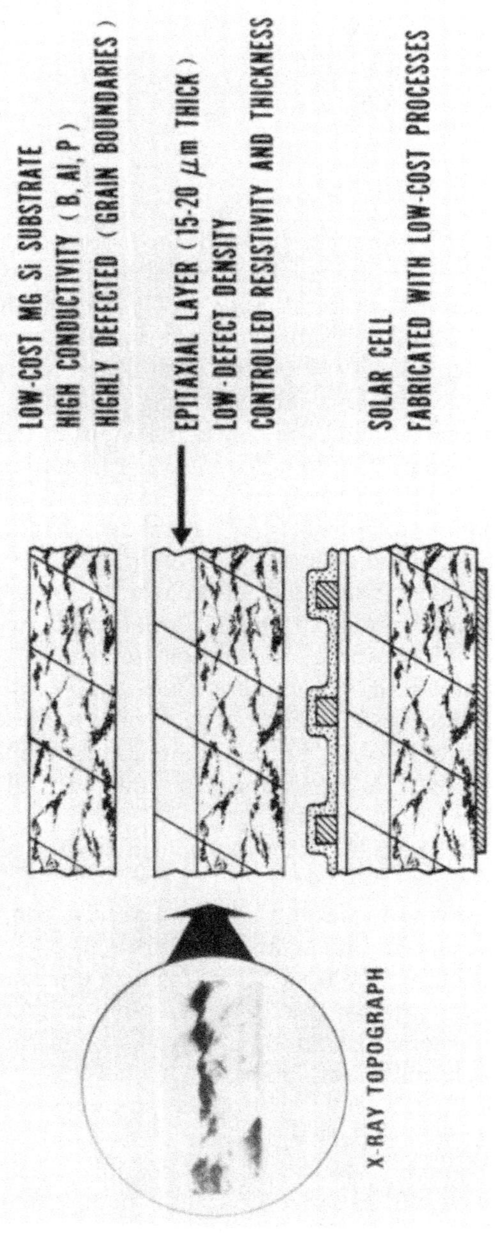
LOW-COST MG Si SUBSTRATE
HIGH CONDUCTIVITY (B, Al, P)
HIGHLY DEFECTED (GRAIN BOUNDARIES)
EPITAXIAL LAYER (15-20 μm THICK)
LOW-DEFECT DENSITY
CONTROLLED RESISTIVITY AND THICKNESS
SOLAR CELL
FABRICATED WITH LOW-COST PROCESSES
X-RAY TOPOGRAPH

– Reliability is expected to be as good as present cells made from bulk silicon.

The properties of MG silicon, reviewed in section 2, show that this material contains substantial quantities of impurities known to cause degradation of solar-cell performance. The first problem is then to select a recrystallization technique which will remove a large fraction of these impurities, yield a single crystal or large-grain polycrystal, while at the same time have a large-scale production potential at low cost. The approaches taken by several workers to achieve suitable silicon substrate material are also reviewed in section.

Although epitaxial growth of the active layers offers several technical advantages, the high growth temperatures (1000–1200°C) required, combined with the use of MG substrates, can lead to contamination of the grown layers either directly by diffusion from impurity sources in the substrates, or indirectly by contamination of the susceptors and subsequent transport into the growing films. In addition, substrates obtained from recrystallized MG silicon are likely to contain defects, and to be large-grain polycrystalline in structure, so that defect generation during epitaxial growth, and problems associated with grain boundaries need to be considered. A discussion of these problems along with the epitaxial methods and results obtained with various MG substrates are given in section 3.

The problem of the cost of epitaxial growth on a large scale is central to this approach since present commercial epitaxial reactors are generally quite costly and slow because they are built to satisfy the specialized needs of the semiconductor industry, where a premium is placed on precise control of the uniformity of thickness and resistivity. A description of the application of a novel epitaxial reactor as an approach taken to solve this problem along with results obtained are presented in section 4.

2. Metallurgical-grade silicon, ingot solidification and substrate formation

Metallurgical-grade silicon is produced primarily for use in the aluminum and steel industries, and has a current worldwide production rate of about a half million metric tons per year. It is extracted from quartzite by carbonaceous reduction at high temperature (1850°C) in very large arc furnaces. The MG silicon is taken from the furnace in liquid form about 2 tons at a time and quickly solidified in molds, after which it is broken up into small chunks for sale at about 1.50 $/kg.

These chunks are physically quite porous, having a silicon content of

Table 1
Spark source emission spectrographic analysis of MG silicon.

Element	ppmw	Element	ppmw	Element	ppmw	Element	ppmw
B	<10	Cu	10	Mo	<10	Ba	<1
Al	1300	Fe	1500	Ni	38	P	<50
Ca	14	Mg	<0.5	Ti	190	Sn	<10
Cr	11	Mn	170	V	<3	Sr	<3

97–99% with the remainder typically made up of the elements listed in table 1. About 1% of the MG silicon so produced is extensively purified, generally by the Siemens process, and sold to the semiconductor industry at about 80 $/kg.

The high impurity content, especially some of the heavy metals, represents a major problem in using MG silicon as a substrate to make solar cells since many of these elements can readily diffuse from the substrate into the growing epitaxial layer and cause a reduction in the minority carrier lifetime. Various methods have been applied in attempts to reduce the impurity concentrations while at the same time, or in a later process, produce a suitable material for substrates. Such multistep processes usually involve treatment of the raw MG silicon by acid leaching, and/or the passage of hot halogen gases followed by a form of unidirectional solidification. Chu et al. (1977, 1978) have demonstrated a reduction in Fe content by extended leaching in aqua regia, and a further reduction of Fe and other metallic impurities by heat treatment with phosphorus pentoxide, followed by a recrystallization by unidirectional solidification and zone-melting. Hunt et al. (1976) have developed purification techniques using reactive gases followed by directional solidification to produce a solar-grade material. McCormick (1980) modified this technique for application to fabricating epitaxial substrate material with the designation upgraded metallurgical grade (UMG) silicon. Kotval and Strock (1978) and Hanoka et al. (1978) used several steps involving:

(1) dissolution and recrystallization from an aluminum solution,
(2) leaching of the resultant crystallized silicon platelets with HCl, and
(3) purification and ingot formation by slagging and unidirectional solidification.

This material is called DS/RMS (directionally solidified/refined MG silicon). Secco D'Aragona and Liaw (1981) studied sequential steps of purification involving

- acid leaching
- phase separation
- reactive gas treatment
- liquid–liquid extraction, and
- impurity redistribution using Czochralski ingot pulling.

Their studies have indicated that ingot-pulling provided the most effective purification. The extensive work of Khattak and Schmid (Khattak and Schmid 1978, 1979, Khattak et al. 1982) using the heat exchanger method (HEM) to cast large ingots in crucibles by directional solidification has shown that this method rejects most impurities to the last material to freeze resulting in an overall purification of the cast crystal.

Because of the high cost of sawing ingots into wafers, several attempts were made to produce sheet material directly, or from purified MG silicon. In the work of Chu et al. (1978), the directional zone melting and solidification was conducted on graphite slabs to form substrates for the epitaxial growth. Ciszek et al (1981) produced a ribbon from melted commercial grade MG silicon by an edge-supported pulling (ESP) technique. These methods have shown promise, but have not yet been taken beyond the laboratory stage.

The purification and solidification methods described above generally reduce most impurity concentrations to the level of several to several hundred ppm, and result in a p-type silicon material of ~0.01–0.1 Ω cm resistivity with residual dopant impurities of B and Al in excess of P. Because of the low resistivity and the still relatively high concentration of impurities, these materials have been primarily used as substrates for epitaxial growth and subsequent solar-cell fabrication. A description of a typical epitaxial growth, and a review of solar-cell results obtained with representative samples of the substrate materials is given in section 3. The remainder of this section is devoted to a discussion of the results and some interesting problems associated with ingot solidification and substrate formation by the HEM process. This work was conducted in a joint program between RCA Laboratories and Crystal Systems Inc., the details of which have been published by Khattak et al. (1982) and D'Aiello et al. (1983). Here, only the salient features are discussed.

Rectangular ingots of 4–10 kg were directionally solidified by the HEM process using a selection of commercial grade MG feedstocks. Such ingots were examined physically and chemically to establish their crystallinity, defect density, and impurity content. Spark source emission spectrographic data taken on the feedstock and on samples taken from the center and close to the upper edge (last portion to freeze) of ingots solidified for 5 and 11 hrs are given in table 2. A review of this data shows that good segregation of impurities was achieved, as indicated by the high ratio of edge to center values and the levels of many impurities in the center of <1 ppm. These data, however, should be viewed with caution because only a small volume of material is examined, so that inhomogeneous distribution of impurities (clusters and complexes) can either indicate anomalously high values of some impurities or be missed entirely. Fabricating and carefully evaluating relatively large-area solar cells is perhaps a better test of material quality since the large area assures adequate sampling and, afterall, an understanding of the cell performance is the desired result.

In regard to inhomogenities, it was found that the amount and type of impurities present in the MG feedstock have a major effect on the quality and useful amount of silicon obtainable from an HEM ingot. Specifically,

Table 2
Spark source emission spectrographic analysis of typical meltstock and HEM solidified ingots. (All values in ppmw.)

Element		Solidification time: 5 h		Solidification time: 11 h	
	Sample	Sample		Sample	
	Meltstock	Middle	Edge	Middle	Edge
B	<10	<10	15	<10	<10
Al	1000	6	450	32	1070
Ca	100	<1	6	<0.2	10
Cr	1	<1	8	<1	16
Cu	42	11	51	3	30
Fe	1150	9	660	<1	1470
Mg	<1	<1	<1	<0.2	<0.2
Mn	85	<10	94	<2	18
Mo	17	<10	19	<5	<5
Ni	170	10	130	<0.5	190
Ti	170	<1	140	<0.5	190
V	270	<3	250	<0.5	<0.5
Zr				<5	54
Ba	7	<1	<1	<0.2	<0.2
C*	2500			40	200
P	<50	<50	<50	<50	<50
Sn	<3	<3	<3	<3	<3
Sr	<3	<3	<3	<5	<5

* Carbon values by fusion technique.

the high carbon content present in most MG feedstock is related to the formation of particulate inclusions (mostly SiC) during solidification. These particles were found to have a detrimental effect on solar-cell performance by causing shunting, thereby limiting the yield of useful substrate material (see, for example, Khattak et al. 1982). The particles are revealed when the substrate wafers are chemically etched. The material immediately surrounding the particle etches more rapidly than the bulk silicon, leaving a pit with the particle in the valley as seen in the inset of fig. 2. The size of the visibly detectable particles varies from 10 μm to as large as 500 μm in diameter, and their density within a given HEM ingot tends to be greater near the top and much lower near the seed or bottom end. The density and spatial distribution were found to vary somewhat with the source of MG feedstock. However, when pure semiconductor-grade feedstock was used in the same HEM apparatus, no particles were found, showing that particles are not equipment-related. An etched section show-

Fig. 2. SEM of typical particle as seen after etching; etched cross section of typical domestic MG singly solidified ingot showing particle distribution.

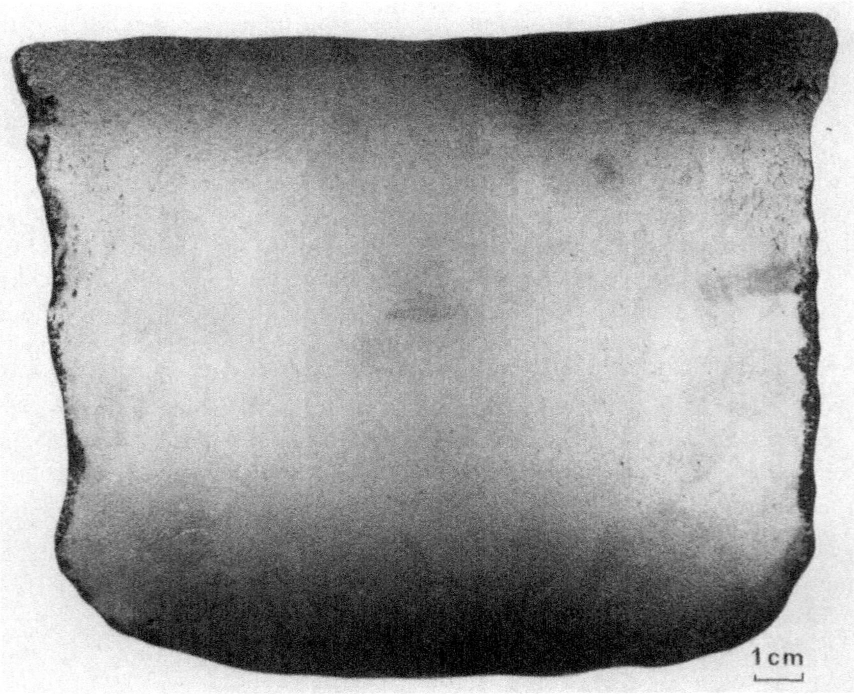

Fig. 3. Particle distribution of a South African sample after single directionally solidification.

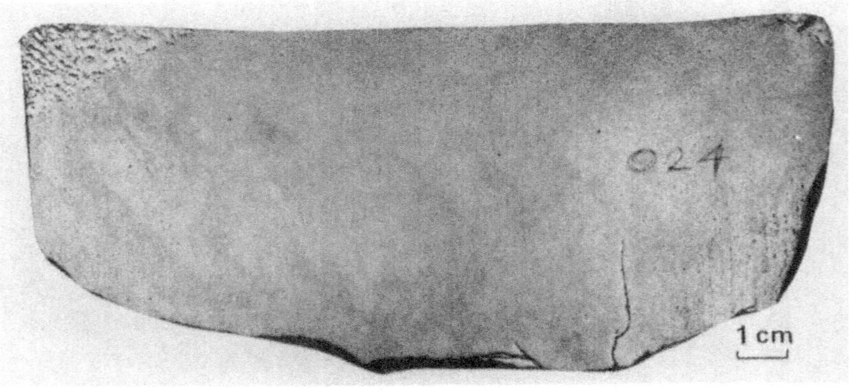

Fig. 4. Particle distribution of a South African sample after double directionally solidification.

ing the particle distribution in an ingot solidified from a typical MG feedstock is shown in fig. 2.

Physical and chemical analysis, powder X-ray diffraction, emission spectroscopy and Auger analysis were made, and have shown that the particles are primarily β-SiC, although Fe and Ca were also found to be associated with or in close proximity to the particles examined. Details of these analyses can be found in the work of Thomas et al. (1984).

MG feedstocks obtained from a variety of domestic and foreign manufacturers were evaluated; however, the MG silicon feedstock obtained from Silicon Smelters Ltd. (Pietersberg, South Africa) was found to be a much purer source than any of the domestic feedstocks. This supplier has one of the world's purest quartz deposits and uses a proprietary mixture of carbonaceous reductants. The arc furnace operation of Silicon Smelters Ltd. is different from others because they make silicon primarily for the silicone industry, which requires a purer grade of metallurgical silicon. This MG silicon was both singly and doubly directionally solidified by HEM, and the density of SiC particles observed upon etching was found to be considerably less than domestic sources of MG silicon, as shown in figs. 3 and 4.

Carbon analysis using the fusion technique was carried out on various MG feedstocks. The results on a number of samples from each source showed variation, which was to be expected due to the inhomogeneous nature of MG silicon; however, the Silicon Smelters Ltd. material showed a lower range of values, 20–1500 ppw, compared to 1000–3400 ppmw for the other sources.

The double solidification was performed by first removing about 5 mm of the outer surfaces of the singly solidified ingot and using the core of this ingot for the second solidification. Etched, a cross section of this ingot revealed a reduced particle count and a confinement of the particles to the top outer periphery of the ingot, as shown in fig. 4.

3. Epitaxial solar cells

3.1. Epitaxial cells grown on single-crystal semiconductor-grade silicon

It has been established (D'Aiello et al. 1976) that solar cells having excellent electrical characteristics and reasonably high (11–13.5%) conversion efficiency can be fabricated in thin (~25–50 μm) epi-layers grown on high-quality semiconductor-grade (SG) silicon substrates. These cell structures were generally grown in laboratory-scale horizontal reactors using dichlorosilane (SiH_2Cl_2), trichlorosilane ($SiHCl_3$) or silicon tetrachloride ($SiCl_4$) at growth temperature of 1100–1200°C. A typical resistivity profile of the grown p-layer of a n^+–p–p^+ cell structure used by D'Aiello et al. (1983) is shown in fig. 5. Starting from the substrate, the first material grown is referred to as the 'graded' or 'buffer' region, and in the case of structures grown on high-quality substrates, this region is made quite narrow (~2–5 μm) and serves to form the built-in electric field or back-surface-field (BSF) region. When MG substrates are used, the thickness of the low-resistivity portion of this region is often increased by 5–10 μm, and the graded region may be extended to the surface, there-by providing field-aided collection over the entire p-base of the cell. Solar cell structures are completed by forming an n^+–p junction either by conventional diffusion processes, or by epitaxial growth of an n^+-layer.

Data representative of the best, published, results achieved with epitaxial cells grown on high-quality silicon substrates are given in table 3. These data show that solar cells with excellent electrical characteristics can be obtained with epitaxially grown material. The solar-cell parameters, particularly J_{sc} and FF, should be compared with those given below for epi-cells made on MG substrates as these quantities are sensitive to life-time killing impurities and to the effects of grain boundaries.

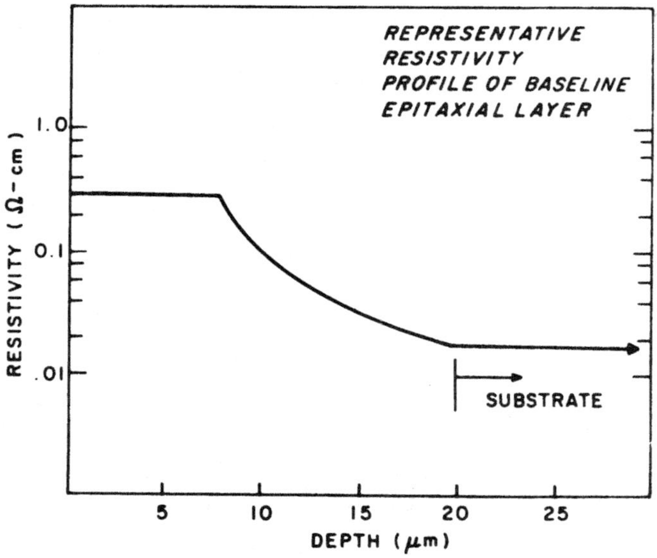

Fig. 5. Resistivity profile of epitaxial solar cell structure.

Table 3
Solar cell results achieved with epitaxial structures grown on SG substrates.

Epi-layer thickness (μm)	J_{sc} (mA/cm^2)	V_{0c} (V)	FF –	AM-1 η (%)	Cell area (cm^2)	Ref.	Comments
35	24.5	0.636	0.790	12.6	2.2	[1]	Epi-junction and graded-base epi-layer
40	25.0	0.603	0.800	12.3	2.2	[1]	Epi-junction, uniform-base epi-layer
15	28.2	0.605	0.810	13.2	4.5	[2]	Diffused junction
50	28.5	0.593	0.806	13.1	4.5	[2]	Diffused junction
50	26.6	0.559	0.740	11.0	97	[3]	Diffused junction

References:
[1] D'Aiello et al. (1976).
[2] Robinson et al. (1980).
[3] Secco D'Aragona et al. (1983).

3.2. *Epitaxial cells on MG derived substrates*

3.2.1. *Solar cell results*

The epitaxial method for cell fabrication has been applied to many of the MG-derived substrates described in section 2. Table 4 summarizes the published solar cell results achieved using epi-layers grown on particular MG substrates. Two key factors concerning the data in table 4 as they relate to the potential commercialization of this process should be mentioned. First, many entries show cell efficiency of over 10%, and such performance levels were achieved in commercially viable sizes of up to ~100 cm^2.

The data and referenced work in table 4 cover a wide range of substrate and solar-cell developments. Some comments and details related to the results obtained and which bare on future improvements follow.

3.2.2. *The effects of processing*

3.2.2.1. Epitaxial layer thickness. D'Aiello et al. (1978) and Robinson et al. (1980) have shown comparable results with 15 and 50 μm thick epita-

Table 4. Solar cell results achieved with epitaxial structures grown on various MG derived substrates.

Substrate	Ref.	Epi-layer thickness (μm)	J_{sc} (mA/cm^2)	V_{oc} (V)	FF	AM-1 η (%)	Cell area (cm^2)	Comments
UMG-IS	[1]	15	26.5	0.600	0.792	12.6	4.5	IS = Single-crystal, purest grade
UMG-IS	[2]	50	27.1	0.592	0.800	12.9	4.5	
UMG-IS	[2]	0	15.2	0.572	0.547	4.8	4.5	No epi-layer, diffused junction cell
UMG-2P	[2]	15	24.8	0.570	0.734	10.3	4.5	2P = Polycrystalline, low purity
UMG-2P	[2]	50	25.1	0.525	0.650	8.6	4.5	
RMS	[3]	17	25.0	0.555	0.745	10.6	4.5	
RMS	[3]	50	24.3	0.542	0.717	9.7	4.5	
CZ/RMS	[4]	0	20.3	0.560	0.722	8.5	4.6	No epi-layer, CZ pulled RMS substrate
CZ/MG	[5]	30–50	18.9	0.586	0.704	7.8	19.7	Twice pulled MG silicon
CZ/RMG	[5]	30–50	20.0	0.607	0.702	8.6	20.8	Pulled from 'refined MG silicion'
CZ/MG	[5]	0	9.5	0.510	0.600	2.9	4.2	No epi-layer
MG-Si	[6]	15	23.3	0.584	0.730	10.0	97	Substrates annealed prior to
MG-Si	[6]	50	26.3	0.590	0.750	11.6	97	epitaxial growth
HEM (SSMG)	[7]	20	21.9	0.579	0.754	9.7	10	SSMG = Singly solidified MG
HEM (DSMG)	[8]	20	23.7	0.580	0.777	10.7	10	DSMG = Doubly solidified MG
HEM (DSMG)		20	25.0	0.587	0.796	11.7	4.0	
MG/Graphite	[9]	25–30	23.5	0.560	0.746	9.8	9.0	
ESP-Ribbon (MG)	[10]	20	22.5	0.570	0.820	10.5	0.86	MG feedstock = DAR, see Hunt (1976)
ESP-Ribbon (MG)	[10]	20	20.5	0.530	0.750	8.3	0.73	Raw MG feedstock

References:
[1] McCormick (1980)
[2] Robinson et al. (1980).
[3] Robinson et al. (1978).
[4] Hanoka et al. (1978)
[5] Saitoh et al. (1978).
[6] Secco D'Aragona et al. (1981).
[7] D'Aiello et al. (1983).
[8] Khattak et al. (1982).
[9] Chu et al. (1979)
[10] Ciszek et al, (1981).

xial structures grown on RMS and UMG substrates. In their work, some indications of higher J_{sc} and η were seen for the class of substrate material which was further purified by double solidification. This is reasonable since thicker epi-layers would be beneficial only if accompanied by diffusion lengths longer than the thickness of the grown layer, a condition more attainable with purer substrates. A more pronounced effect is seen in the work of Secco d'Aragona et al. (1983), where gettering and annealing of the substrates were conducted prior to the epitaxial growth, and the cell results showed a monotonic increase in efficiency from 10% up to 12% for epi-thickness of 15 to 100 μm.

The choice of the optimium epi-layer thickness will depend upon the impurity content of the substrate, the cell process, and the cost per micron of growing the epitaxial layer.

3.2.2.2. Gettering and annealing. Various gettering and/or annealing heat treatments have been applied to the substrates prior to, or after epitaxial growth. Such treatments generally improve cell performance. These processes include; pregrowth heat treatments, HCl getter-etching in the epitaxial reactor, and in some cases thermal annealing after the epi-growth. Secco D'Aragona et al. (1983) in an extensive comparison of one- and two-step thermal annealing and phosphorus gettering of the substrates found that phosphorus gettering gives the same efficiency improvement as a two-step anneal but with a shorter annealing time. Chu et al. (1979/1980) have shown that solar cell structures subjected to heat treatment at 850°C after the epitaxial growth showed higher efficiencies than those without heat treatment.

3.2.3. Grain boundary effects

Although most of the substrates discussed here have quite large grain sizes (ranging from several millimeters to several centimeters), deletereous effects of the grain boundaries on cell performance have been detected. Some idea of the (order of) magnitude of these effects, and what can be done about them is described in work of Robinson and D'Aiello (1981). They used two different approaches to obtain higher efficiencies with epi-cells grown on MG substrates. The first relates to the geometric pattern of grain formation in directionally solidified ingots. When such ingots are used in the fabrication of solar cells, two attributes of the grains considered to be important for achieving high cell efficiency are large grain size and columnar structure.

Since MG substrates have a multigrained structure, the size, distribu-

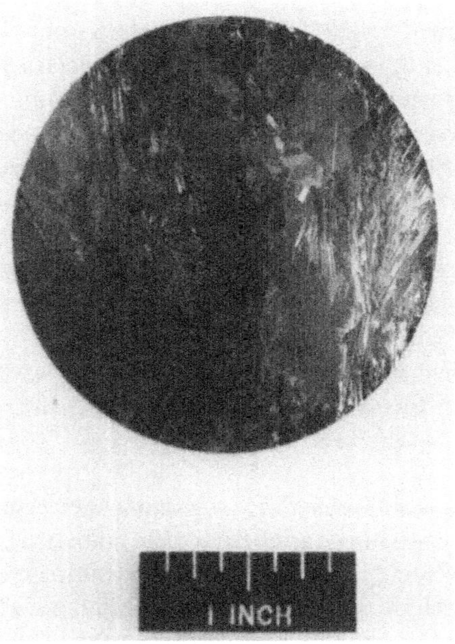

Fig. 6. Cross-sectional photomicrograph of a directionally solidified ingot cut perpendicular to the growth direction.

Fig. 7. Cross-sectional photomicrograph of a directionally solidified ingot cut parallel to the growth direction.

tion, and geometry of the grains can be important both in the epitaxial growth and in the final solar-cell performance. The size and geometry of the grains in wafers cut from this material are considerably different depending upon the direction of the cut relative to the direction of solidification of the ingot. Photographs of typical sections of unidirectionally solidified MG material cut both parallel to and perpendicular to the direction of solidification are shown in figs. 6 and 7. For the perpendicular cut, the grain sizes vary from about 1 mm to 5 or 6 mm in maximum length. In contrast, the parallel cut exposes grains of greater length (several cm) because of the columnar nature of the grain formation in the direction of solidification of the ingot. This can be utilized to advantage in the case of epitaxial cells since the growth of the active film will replicate the surface grain geometry resulting in columnar grains which are then transverse to the original direction, but of considerably larger area.

This approach was explored by the growth and fabrication of solar cells on both transverse and longitudinally cut sections. The pattern of solar cells used in these experiments and the typical geometry of the grains which resulted from parallel and perpendicular cuts are shown in figs. 8 and 9. Samples of 15 μm thick epitaxial structures were prepared and fabricated into solar cells for each of the two directional cuts. The average efficiency for all cells in each category shows that an almost one percent increase (10.5% versus 9.6%) is obtained when the parallel-cut substrates are used. Individual values of the cell parameters can vary greatly among cells on the same substrate, especially for the small-area cells, since their linear dimensions are of the order of the maximum grain size, so that some of these cells are located in areas containing only one or two grain boundaries. One such case, shown in fig. 9 upper right-hand cell, contains one grain boundary and has the highest set of solar-cell parameters of the group, contrasting with the left-hand cell which had the lowest in the group.

Seager and Ginley (1979) have provided considerable evidence that heat treatment of polycrystalline silicon in the presence of atomic hydrogen can cause a passivation of grain boundaries resulting in a reduction of barrier heights, electrical resistance and recombination. This passivation technique reduced detrimental effects of grain boundaries in 250 μm thick polycrystalline silicon solar cells. Although no direct solar-cell parameters were reported, the hydrogen passivation was shown to reduce both the diode reverse leakage current and the dark forward current. Such improvements in the *I–V* characteristics imply that even for solar cells made in relatively large-grain polysilicon, hydrogen passivation can result in increased values of open-circuit voltage (V_{oc}) and cell fill-factor (FF).

Fig. 8. Epitaxial solar cells made on an ingot cut perpendicular to the growth direction.

Fig. 9. Epitaxial solar cells made on an ingot cut parallel to the growth direction.

Robinson and D'Aiello (1981) applied this method and obtained direct experimental observation of increases in V_{oc}, FF, and cell efficiency by atomic hydrogen passivation applied to thin-film polycrystalline epitaxial silicon solar cells. The cells used in the experiment were n^+–p–p^+ structures formed by the epitaxial growth of 20 μm of silicon on polycrystalline upgraded metallurgical-grade (UMG) silicon substrates. The effect of the hydrogen treatment on the solar-cell performance is illustrated in fig. 10.

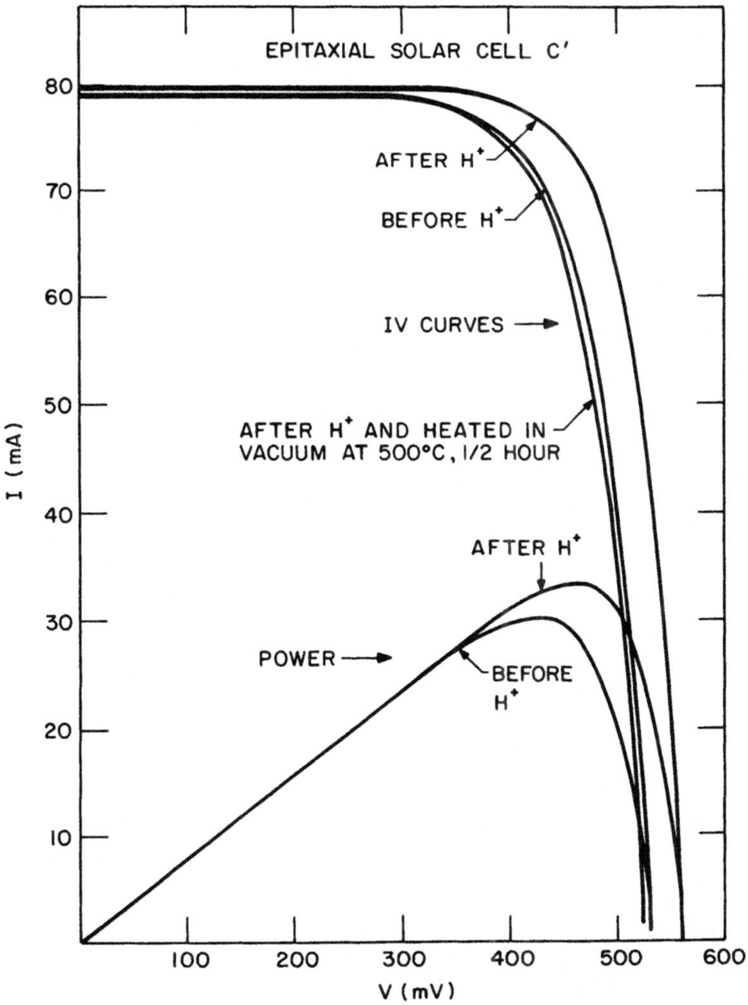

Fig. 10. Illuminated *I–V* curves before and after hydrogenation.

In these experiments, typical increases were V_{oc} (15–30 mV), FF (0.015–0.030), and the percentage increase in efficiency ($\Delta\eta/\eta$) ranged from 6.9–11%.

The above experimental data show that atomic hydrogen passivation can significantly improve the performance of polycrystalline silicon solar cells even in the case of relatively large-grain polysilicon. This may be especially true for thin-film epitaxial cells, since the improved cell performance was obtained with only a $3\frac{1}{2}$ h treatment. No attempt was been made to determine the minimum time/temperature cycle, but if penetration throughout the active cell volume is required, it should be considerably less for thin epitaxial cells than for the thick bulk cells reported previously.

3.2.4. *Impurity effects*

The residual impurities in the refined MG substrates are of sufficient diversity and concentration to prevent the use of such substrates for directly fabricating efficient solar cells. For example, the data in table 4 for the listed cases where no epitaxial layer was grown, generally show efficiencies of only 3–4%. The work of Davis et al. (1978) shows that the impurities such as V, Ti, Mn and Fe at the levels retained in the purified MG substrates (see, for example, table 2), if transferred to the epitaxial layer would likewise cause large reductions in cell efficiency for epitaxial cells. The good cell efficiencies obtained with epi-layers on such substrates implies that the impurity levels in the epi-layers are low. For some of the impurities (Ti, V, Mn), the diffusion rates are low and the epitaxial growth rate is sufficiently high so that even during growth at 1100–1200°C, no significant quantities of these impurities move into the grown layer. Some direct evidence of this was supplied in the work of D'Aiello et al. (1980) in which they compared solar-cell results for epitaxial cells and bulk cells made on substrates intentionally doped at $2\times10^{14}\,cm^{-3}$ with Ti. This concentration of Ti was found to cause a large (−45%) reduction of efficiency for the bulk cells compared to the control cells which had no added Ti, while the epitaxial cells showed efficiency reduction of only 12–15% from that of the controls.

The fast-diffusing impurities, such as Fe and Cu, would appear to have sufficient mobility as to contaminate the epitaxial layer. However, several factors tend to prevent, or greatly reduce this possibility. First, the MG substrates are polycrystalline, and highly strained due to point and line dislocations and defects. The grain boundaries and defects act as sinks, at whose sites impurities are trapped or form clusters of impurity complexes. Secondly, the standard procedure of HCl getter-etching the substrates just prior to epitaxial growth acts to deplete the surface layer of the substrates of impurities such as Fe and Cu by volatilizing them into the gas stream. The screening of impurities by the defects in the MG substrates, and the rapid growth rates (1–5 μm/min) possible by epitaxy, allow the grown layer to be relatively free of impurities contained in the MG material.

4. A low-cost approach to epitaxy

4.1. Reactor considerations

The method of producing solar cells described here relies on the epitaxial growth of a thin (20 μm) layer of homoepitaxial silicon, such as that shown in fig. 4. For such a solar cell to be cost effective, certain requirements must be met by the epitaxial reactor chosen. For example, large-area epitaxial solar cells are desired with a square or rectangular shape preferred for high packing density when they are assembled in the module. In addition, the reactor must have the capability of a high throughput of epitaxial wafers. The high throughput capability becomes a significant factor when one considers that the reactor must deal with large area substrates.

For the solar cell devices to be economically acceptable the wafer substrates would likely consist of low-cost metallurgical-grade silicon having a square shape with up to 10 cm per slide, $\approx 100\,cm^2$ area. To meet these requirements one must consider the technical and practical cost features of operating the epitaxial reactor when using large-area wafers of low-purity silicon.

The epitaxial reactors commercially available are primarily designed to deal with the tight specifications for integrated circuits and solid-state power devices. These reactor types fall into three categories: (a) horizontal reactor, (b) barrel reactor, (c) pancake reactor. A fourth type of reactor known as continuous chemical vapor deposition (CCVD) reactor was explained by David Bellavance and Roger Anderson (1983). At the time of writing not enough data on this reactor were available for comparison with the others. For comparison, several practical reactor designs are illustrated in fig. 11. Such reactors are usually heated with an RF energy coil located outside a quartz bell jar heating the graphite susceptor containing the silicon substrates. In the case of the barrel reactor an alternative radiantly heated reactor is available which uses lamps surrounding the quartz bell jar instead of the RF coil. For the case of the pancake reactor, the RF coil is located inside the quartz bell jar but shielded from corrosive reaction chemicals by a hydrogen gas curtain below the susceptor.

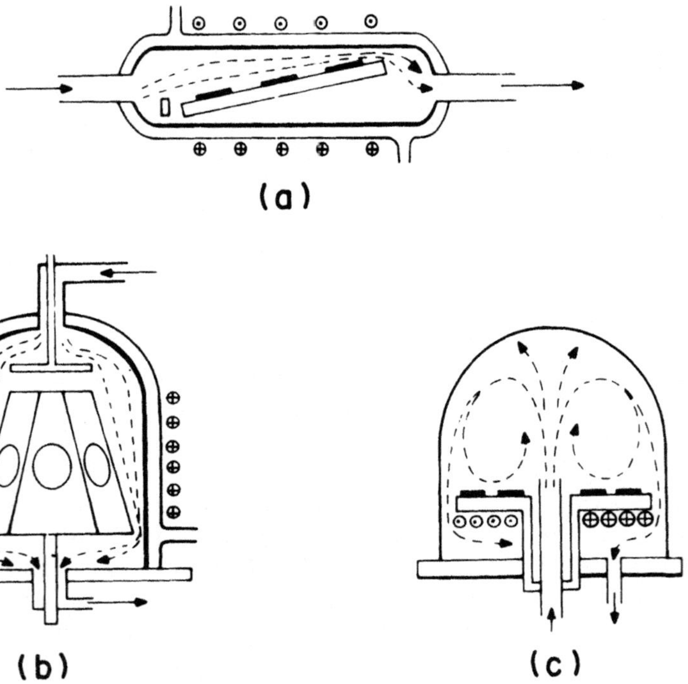

Fig. 11. Schematic representation of three conventional reactor designs.

Silicon epitaxial reactors use either silane or one of the silicon halide gases as source material for the silicon epitaxy. Where silane (SiH_4) and silicon dichloride ($SiCl_2H_2$) are used, the quartz bell jar wall must be kept cool to prevent the deposit of silicon on the walls. When silicon trichloride ($SiHCl_3$) and silicon tetrachloride ($SiCl_4$) are used, the bell-jar must be kept cool to prevent the deposit of silicon on the walls. When silicon trichloride ($SiHCl_3$) and silicon tetrachloride ($SiCl_4$) are used, a hotter wall temperature can be tolerated before wall deposition occurs. The choice of which silicon-source gas is used is made on the basis of growth rate, deposition temperature and economic considerations.

The type of reactors described here all use a single unitized graphite susceptor to hold the substrates. Because of their unitized construction, the ability to expand size to meet the requirements of a high throughput of large area wafers is limited by geometric and material constraints.

Economical operation of an epitaxial reactor is related to factors such as power usage and chemical consumption. These factors are strongly affected by throughput, i.e., number of wafers per run cycle. Thus, the choice of a reactor with economical power and chemical usage, required a novel reactor which could be expanded both in wafer-area size and throughput. Such a reactor was evaluated for the purpose of establishing its capabilities for large throughput of epitaxial solar-cell type wafers at reasonable economical cost.

A reactor design which seems suited to meet the solar-cell requirements is a high-throughput reactor which is best described as a rotary-disc type. The structure and operational characteristics of this reactor have been described by Ban (1978), Ban and Miller (1979) and Miller (1984). As shown in figs. 12 and 13, the susceptor is a vertical stack of individual composite discs each separated by spacers, and each capable of holding two wafers, one on top and one on the bottom. Gases are introduced by means of a distributor tube placed along the length of the stack. The susceptor stack rotates and is heated by a RF load coil placed around a quartz bell jar which surrounds the susceptor stack and distributor tube.

One of the main advantages of the rotary-disc reactor over other reactor types is a higher wafer stacking density for a given reactor volume. This efficient stacking density also gives the rotary-disc reactor a higher chemical efficiency and lower power consumption per wafer.

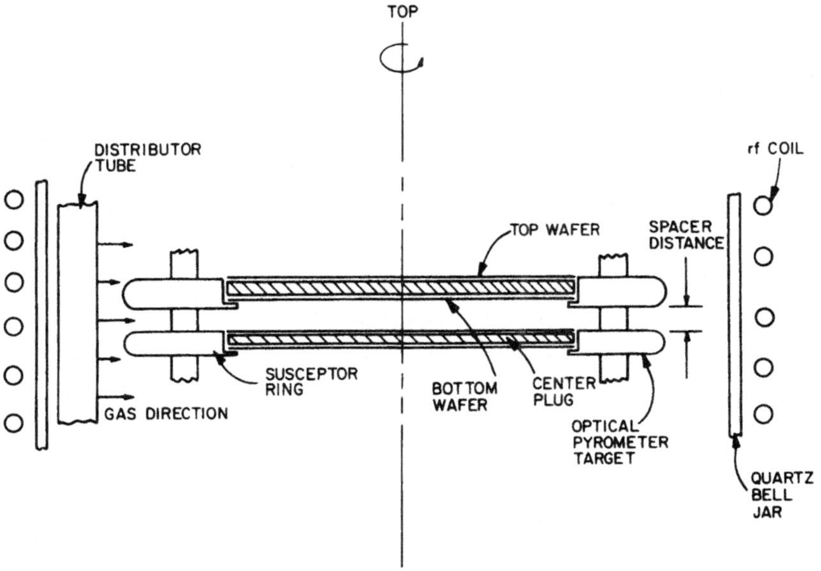

Fig. 12. Schematic of one section of a rotary-disc reactor.

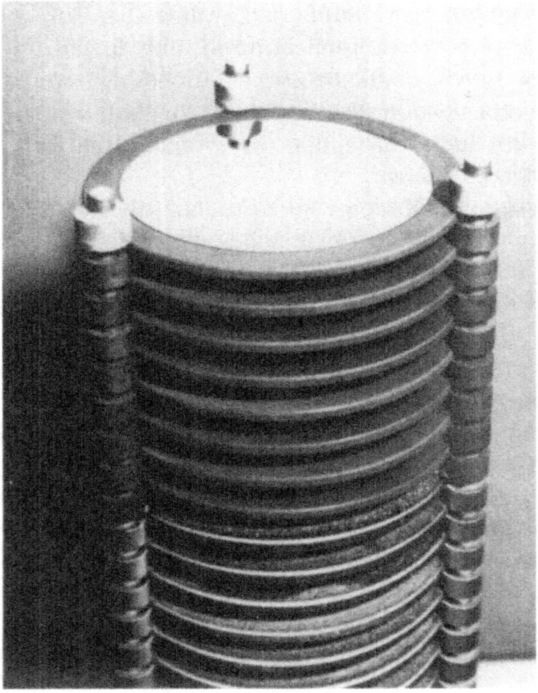

Fig. 13. Rotary-disc susceptor stack.

Table 5
Typical operational parameters for various practical epitaxial reactors.

Reactor type	Batch capacity	Hydrogen consumption (l/min/water)	$SiCl_4$ (gm / min per wafer)	Power	Growth (μm / min)
Horizontal	18	20	1.0	450 kHz / 150 KW / 325 KVA	1.5
Barrel	24	8.3	0.8	450 kHz / 150 KW / 325 KVA	1.5
Pancake	22	8.0	0.40	450 kHz / 150 KW / 325 KVA	1.5
Rotary disc	40	5.0	0.45	450 kHz / 150 KW / 325 KVA	1.5

Table 5 lists the key cost drivers per wafer in each of the reactor categories shown below. The values given in table 5 for the horizontal, barrel and pancake types are for state-of-the-art reactors designed for 100 mm diameter wafer capacity, while the values for the rotary disc are those reported for a prototype model. The 40 wafer (100 mm diameter) capacity of the rotary disc can be expanded simply by adding more susceptor rings to the stack length. Expansion to larger diameter wafers is also accomplished at no cost to wafer loading capacity. Thickness and resistivity uniformity can be controlled in the rotary-disc reactor by adjusting each region of the stack array. This feature, along with the ability to replace discrete parts of the susceptor array, gives this reactor a cost advantage over other types of reactors that use a single unified susceptor for a multitude of substrates.

4.2. Operational growth characteristics

The rotary-disc reactor has several variables some of which are not found in other reactor types. Apart from temperature, silicon-source concentration and substrate diameter, one must consider spacer distance between susceptor rings, ring diameter, distribution-tube slot width, gas velocity and positional direction of the gas stream at the time of growth. These variables were explored to optimize the uniform growth of 20 μm thick layers over large-area substrates.

4.2.1. Temperature

The temperature of each susceptor location is controlled by adjusting both the exterior RF load coil turns and the spacing between susceptor rings. Since the temperature of the substrate during growth is largely responsible for the growth rate, temperature uniformity is one of the parameters necessary for thickness uniformity wafer to wafer.

4.2.2. Spacer distance

As in the case of temperature, the spacer distance between wafers is important for growth-rate uniformity. This is a parameter unique for the rotary-disc reactor and, therefore, gives an additional degree of freedom in establishing growth uniformity, along with adjustment of conditions for the temperature profile. For example, a wider spacing allows more source gas to enter between the susceptor rings and, therefore, increases growth rate. This spacing option allows the adjustment of growth rate locally for a small temperature variation in the stack to achieve overall growth-rate uniformity on all wafers.

4.2.3. *Susceptor ring width*

The width of the susceptor ring is useful in controlling gas stream conditions as the gas enters the susceptor stack. A wider ring allows for more gas heating and the establishment of proper boundary conditions before the source gas encounters the substrate. Ability to control this unique rotary-disc parameter has significantly contributed to achieving thickness uniformity from center to edge on the wafer.

4.2.4. *Distributor-tube slot width and gas velocity*

The width of the distributor tube slot is used to choose the spread angle of the gas exiting the slot and the volume of gas delivered to the stack. These conditions are controlled to yield the desired gas velocity component for proper growth rate magnitude and uniformity.

4.2.5. *Gas directional control*

During epitaxial growth, the directional position of the gas stream is controlled by either sweeping the slot side to side or holding the stream direction pointed at a particular part of the rotating wafer for a period of time. This permits a control of the gas direction over the wafer which results in a more uniform growth rate. This degree of freedom allows for optimizing these operational growth characteristics for maximum growth rate, growth-rate uniformity from wafer to wafer and from center to edge of the wafer, and extension of batch size to yield greater throughput of wafers.

4.3. *Results using the rotary-disc reactor*

Runs made with a loaded stack of 40 three-inch-diameter single-crystal CZ wafers gave resistivity uniformity ±10%. Center-to-edge thickness uniformity for the wafers was ±6% and wafer-to-wafer thickness uniformity was ±10%. These results are well within the requirements for epitaxial solar cells. Selected solar cells were fabricated from the epitaxial wafers made in one run by a diffusion step to form the junction and then evaluated for cell characteristics (fig. 14). These data show good uniformity of cell efficiency in the 9.5–10.7% range.

In addition, the reactor was used to evaluate 10 cm × 10 cm square substrates. The equipment size restricted the stack capacity of ≈10 wafers, but enabled the evaluation of the conditions for uniform growth. A novel distributor tube was developed, and the susceptor design was modified to optimize the growth parameters. Results showed that the growth on 100 cm^2 area wafers gave thickness variations of ±5% on all points of the wafer.

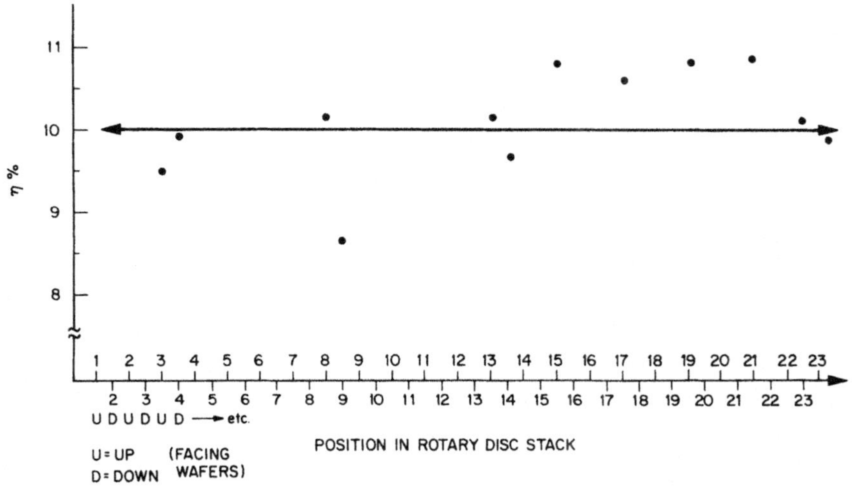

Fig. 14. Solar-cell efficiency as a function of position in an HTR rotary-disc reactor.

Experimental runs were also made in which 7.6 cm diameter control CZ wafers were mixed with MG/HEM substrates. The averaged results for five such runs are given in table 6. About 60% of the cells made on HEM substrates from ingot 010 showed shunting effects that were traced to the high particle content of this ingot, which accounts for the low values of open-circuit voltage (V_{oc}) and fill-factor (FF). In contrast, only 20% of the cells from ingot 018 exhibited obvious particle-related effects. Comparison of the three best 42 cm^2 area cells with the 2 cm^2 cells also from ingot 018 shows that only the fill-factor is lower for the large-area cells. The solar-cell values obtained for the control CZ wafers are similar to those obtained in

Table 6
Average cell parameters for five HTR runs.

Substrate	J_{sc} (mA/cm^2)	V_{oc} (mV)	FF	η_{ar} (%)	Cell area (cm^2)
Control (CZ)	24.01	579	0.75	10.5	42
Ingot 010 (Foreign source MG)	22.15	538	0.52	6.2	42
Ingot 018 (South African MG)	23.13	570	0.62	8.2	42
Ingot 018 (Best 3, no particles)	23.2	576	0.65	8.7	42
Ingot 018 (2 cm^2 cells)	23.4	581	0.79	10.7	2

runs in which no MG/HEM material was included, indicating that cross-contamination from the MG material during epitaxial growth is not a problem.

5. Summary

From the work presented in this chapter, the objective of providing a lower cost alternative to conventional cell fabrication by using epitaxial layers grown on MG-derived substrates has been demonstrated. Several methods of refining raw MG silicon to form these substrates were reviewed. Some of these provide quite pure material, but at high cost, while others provide a moderate degree of purification but represent processes more suitable for commercialization. In all cases, a grown epitaxial layer is required in order to achieve high-efficiency solar cells.

Various refined MG materials were used as substrates for epitaxial growth and solar-cell fabrication. These have resulted in reasonably high cell efficiencies. The high costs associated with epitaxial growth was addressed by the use and demonstration of a rotarty-disc reactor. This reactor was shown to have high throughput potential and can be used to grow on both round and square wafers, and is compatible with the use of MG derived substrates. No significant differences in thickness, electrical, or solar-cell parameters were observed in comparing this reactor to commercial or laboratory scale reactors. It was found that wide variations in epi-layer thickness and profile shape could be tolerated and still provide good cell performance alleviating the need for tight control of these parameters during epitaxial growth and thus allowing greater flexibility in future reactor designs.

The work described here has resulted in considerable progress in the development of the several technologies needed for the production of efficient solar cells using a process starting with raw MG silicon. With additional development, this process holds great promise for commercial application.

References

Ban, V.S., 1978, J. Electrochem. Soc. **125**, 317.

Ban, V.S., and E.A. Miller, 1979, Chemical Vapor Deposition 7th Int. Conf., eds T.O. Sedgwick and H. Lydtin, The Electrochemical Society Proceedings **79-3**, 102.

Bellavance, D., and R.N. Anderson, 1983, Electrochem. Soc., Spring Meeting, Extended Abstracts **83-1**, 653.

Chu, T.L., S.S. Chu, K.Y. Duh and H.I. You, 1977, IEEE Trans. Electron Devices **ED-24**, 438.

Chu, T.L., S.S. Chu, E.D. Stokes, C.L. Lin and R. Abderrassoul, 1978, Proc. 13th IEEE Photovoltaic Spec. Conf., Washington DC (IEEE, New York) 1106.

Chu, T.L., S.S. Chu and E.D. Stokes, 1979/1980, Sol. Energy Mater. **2**, 265.

Ciszek, T.F., M. Schietzelt, L.L. Kazmerski, J.L. Hurd and B. Fernelius, 1981, Proc. 15th IEEE Photovoltaic Spec. Conf., Kissimee FL (IEEE, New York) p. 581.

D'Aiello, R.V., P.H. Robinson and H. Kressel, 1976, Appl. Phys. Lett. **28**, No. 4, 231.

D'Aiello, R.V., P.H. Robinson and D. Richman, 1980, J. Electron. Mater. **9**, No. 2, 435.

D'Aiello, R.V., P.H. Robinson and E.A. Miller, 1983, RCA Review **44**, 30.

Davis, J.R., A. Rohatgi, P. Rai-Choudhury, P. Blais, R.H. Hopkins and J.R. McCormick, 1978, Proc. 13th IEEE Photovoltaic Spec. Conf., Washington DC (IEEE, New York) p. 490.

Hanoka, J.I., H.B. Strock and P.S. Kotval, 1978, Proc. 13th IEEE Spec. Conf., Washington DC (IEEE, New York) p. 485.

Hovel, H.J., 1975, Semiconductors and Semimetals, Vol. II, Solar Cells (Academic Press, New York) ch. 5.

Hunt, L.P., V.D. Dosaj, J.R. McCormick and L.D. Crossman, 1976, Proc. 12th IEEE Photovoltaic Spec. Conf., Baton Rouge LA (IEEE, New York) p. 125.

Khattak, C.P., and F. Schmid, 1978, Proc. 13th IEEE Photovoltaic Spec. Conf., Washington DC (IEEE, New York) p. 137.

Khattak, C.P., and F. Schmid, 1979, Proc. 2nd EC Photovoltaics Solar Energy Conf., Berlin (Reidel, Dordrecht) p. 106.

Khattak, C.P., F. Schmid, P.H. Robinson and R.V. D'Aiello, 1982, Proc. 16th IEEE Photovoltaic Spec. Conf., San Diego, CA (IEEE, New York) p. 128.

Kotval, P.S., and H.B. Strock, 1978, US Patent 4124410.

McCormick, J.R., 1980, Proc. 14th IEEE Photovoltaic Spec. Conf., San Diego, CA (IEEE, New York) p. 298.

Miller, E.A., 1984, Development and Operating Charateristics of a Epitaxial High-Throughput Reactor for Large Area Solar Cells (Private Communication).

Robinson, P.H., and R.V. D'Aiello, 1981, Appl. Phys. Lett. **37**, No.1, 63.

Robinson, P.H., and R.V. D'Aiello, 1981, Appl. Phys. Lett. **37**, 63.

Robinson, P.H., R.V. D'Aiello and D. Richman, 1980, Proc. 14th IEEE Photovoltaic Spec. Conf., San Diego, CA (IEEE, New York) p. 54.

Saitoh, T., N. Nakamura, H. Itoh, S. Matsubara, T. Warabisako and T. Tokuyama, 1978, Proc. 13th IEEE Photovoltaic Spec. Conf., Washington, DC (IEEE, New York) p. 479.

Seager, C.H., and D.S. Ginley, 1979, Appl. Phys. Lett. **34**, 337.

Secco d'Aragona, F., and H.M. Liaw, 1981, Proc. Symp. Materials and New Processing Technologies for Photovoltaics, eds J.A. Amick, R. Rai-Chaudhury, E. Sirtl and J.P. Dismukes (Electrochemical Society, Pennington, NJ) p. 119.

Secco d'Aragona, F., and H.M. Liaw, 1983, Sol. Cells (to be published).

Thomas III, J.H., R.V. D'Aiello and P.H. Robinson, 1984, J. Electrochem. Soc. (to be published).

CHAPTER 6

METALLURGICAL WAYS OF SILICON MELTSTOCK PROCESSING

JOSEF DIETL

Heliotronic GmbH
D-8263 Burghausen, FRG

Silicon Processing for Photovoltaics II, edited by C.P. Khattak and K.V. Ravi

Contents

1. Introduction

The photovoltaic conversion of solar energy into electricity appears to many as the most attractive long-range alternative to conventional energy sources. Being the predominant base material in the electronics industry, silicon also is the most widely used material in terrestrial photovoltaics. A basic requirement of large-scale solar cell fabrication thus is to provide solar-grade silicon at costs of lowest possible levels.

To initiate the necessary development on a broad basis, photovoltaic programs have been started around 1975 in different parts of the world [see ERDA Report (1976) for the USA, Hamakawa (1981) for Japan, and Van Overstraeten (1981) for Europe]. The best known are the ERDA/DOE program of the USA, conceptualized for the period 1975–1985, and the program of West-Germany (Wacker, AEG and Siemens). The essential part of the ERDA/DOE program is concentrated in the low-cost silicon solar arrays (LSSA) project of the US DOE Program (1978). The basic idea behind is demonstrated in fig. 1 by the targets in price development for solar cell modules per peak Watt ($/W). The time table is characterized by two different functional systems: In the earlier years, accomplishments can be expected only by semiconventional approaches on the basis of electronic-grade silicon, whereas unconventional techniques are planned for mass production of solar cell-based modules towards the end of the present decade. Similar ideas are the basis of the Wacker solar material program as commented by Sirtl (1981). It is attempted here to point out the necessity of a balanced cost reduction for all three technical disciplines of material preparation: refining, crystallization and slicing, as is illustrated in fig. 2 where columns B and C1/C2 stand for feasibility marks from the viewpoint of 1978.

In the late 1970s an abundance of different approaches existed in the field of semiconventional medium-cost and unconventional low-cost solar-grade meltstock processing as well. Much less diversification can be observed today, certainly caused by an increased knowledge about technical facts and chances, and their correlation with economic factors. The present chapter follows the steps of previous reviews (given, e.g., by Hunt 1974,

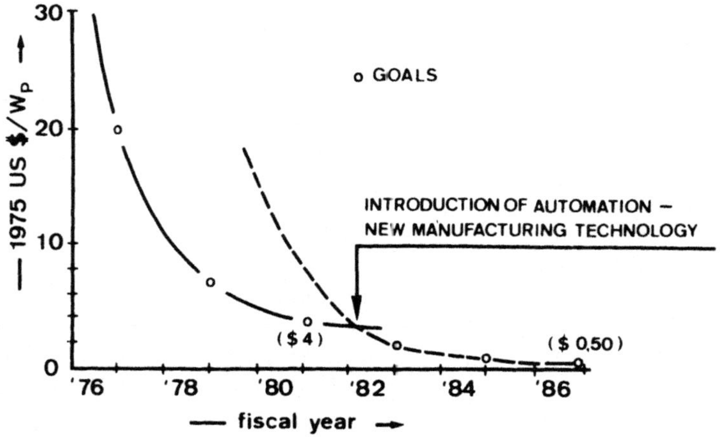

Fig. 1. Original time table of the low-cost silicon solar arrays (LSSA) project on generator costs – as a part of the US DOE Photovoltaic Program (1978); dashed curve: feasibility stage.

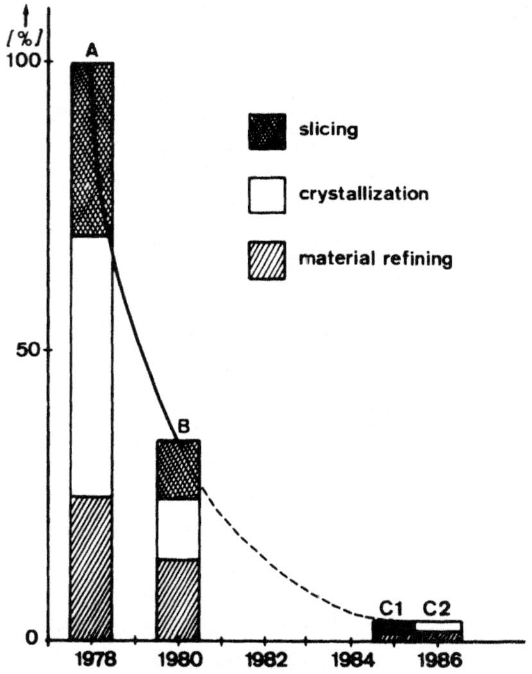

Fig. 2. Planned development of relative costs in the Wacker program on solar-grade silicon. A, Conventional production; B, Semiconventional technology mix; C, All-solar-grade technology; C1, sliced ingots; C2, foils (after Dietl et al. 1981); dashed curve: feasibility stage.

Dietl 1981a, Pizzini 1982, Sanjurjo 1982), concentrating only on the status of technical development of metallurgical ways of silicon meltstock processing. These processing routes via metallurgical-grade silicon of various purity ranges seem to have the highest potential to produce bulk-quality solar-grade silicon at real low cost.

The solar cells for terrestrial application, up to now, are produced almost exclusively from semiconductor-grade silicon, a material too expensive for competitive generation of electricity on a large scale. The successful evaluation of a metallurgically upgraded silicon meltstock for solar cell application (as reported elsewhere, e.g., McCormick 1980, Storti 1981, Khattak et al. 1981, Hanoka et al. 1981) could open up the way for expanding the solar market by changing its industrial and commercial feature.

2. Meltstock for terrestrial solar cells

2.1. Key role of silicon

Crystalline silicon is unchallenged as a semiconductor base material in the present electronics industry. The technology for semiconductor silicon has evolved over the past 35 years. Thus, an extensive knowledge is available on materials preparation and handling, materials properties and compatibility, crystal growth, defect characteristics and device fabrication. The unique properties of the native oxide layer have been an important part of the breathtaking evolution of this element as one of today most important and promising materials.

Silicon has a number of qualities which make it also one of the most desirable materials for terrestrial photovoltaic application. Besides the attractive combination of chemical, physical and electrical properties, the majority of reasons is correlated with a mature technology warranting economy and reproducibility of production on an already large industrial scale. An important role, however, in the different long-term projects on the terrestrial application of solar cells plays the availability of the base material. Silicon is obtained from silicon dioxide, the most plentiful mineral in the earth crust, as can be seen in table 1. This table shows the relative abundance of silicon and other materials which could be relevant as components of active photovoltaic materials. The practically inexhaustive resources of silicon form a sound basis even for the most pretentious programs on power generation by direct conversion of sunlight into electricity.

The world-wide industrial production capacity of silicon (Eastern Block

Table 1
Abundance of 'photovoltaic' materials in the earth crust (after Ciszek 1983).

Element	Weight %	Element	Weight %
Si	27.7	Ge	0.0007
Al	8.1	As	0.0005
Mg	2.1	Sb	0.0001
P	0.1	Cd	0.00002
S	0.05	In	0.00001
Zn	0.01	Se	0.000009
Cu	0.007	Te	0.0000002
Ga	0.002		

countries excluded) exceeds 2 million tons per year. Most of it is used as deoxidizer in the steel industry and as an alloy component in the aluminum industry, respectively. A quarter of the total silicon quantity consists of 'silicon metal', i.e. metallurgical-grade silicon (MG-Si) of 98% purity, as an average quality (Roskill Information Services 1976). 20% of the MG-Si is used for the preparation of organosilanes (silicone polymers) whereas only about 4% – i.e., 1% of the total production of silicon – is converted to high-purity 'electronic-grade' silicon.

Estimates on world's electrical power generation in the year 2000 point towards a value of approximately 20 TW. Assuming that 5% may then be replaced by photovoltaic generation of electricity would mean about 20 million tons of MG-Si – based on solar cells with 10% efficiency and a 1 kW/20 kg Si relationship. This is equivalent with an accumulative annual production of 2 million tons between the years 1990 and 2000. Figures like these are thought to illustrate clearly enough the orders of magnitude one has to consider. Any process study on this basis, therefore, has to be put in a sound economical framework from its very beginning. On the other hand, it is gratifying to learn that even this type of prognosis would mean only doubling of an already existing industrial capacity in terms of raw material.

2.2. *Semiconventional CVD processing*

The final goal of the solar program can be reached only by applying unconventional methods in every particular subtechnology. Semiconventional approaches – in material production, crystallization and slicing techniques as well – may help to ease the solar market situation. In currently produced solar cells the starting material is still conventional vapor-deposited silicon from the Siemens process. As predicted by Freiesleben (1980), this situation will presumably hold for several more years to come.

Attempts to basically improve the conventional trichlorosilane ($SiHCl_3$) chemistry into a truly low-cost process have little chance. Of the numerous versions examined to modify chemical vapor-deposition processing only three ways have survived: Common to all is the conversion of metallurgical-grade silicon into trichlorosilane, purification of this compound by distillation and subsequent reduction to silicon by methods less costly than the conventional Siemens process. This calls for processes with considerably lower energy consumption, increased yields and continuous production of silicon.

As stated by Sirtl (1983) the Wacker version remains in the framework of the conventional $SiHCl_3$ chemistry but makes use of a deposition reactor set-up, where the well-known thermal decomposition of $SiHCl_3$ on silicon rods is modified into a system that offers a much larger surface area.

As reported by McCormick (1982) in the Dow Corning version $SiHCl_3$ is catalytically converted into SiH_2Cl_2 (and $SiCl_4$) that can be decomposed at higher yield and lower temperatures in a Siemens-type reactor. Based on a 1000 t/year production capacity, the price is estimated at around 25 \$/kg, which is considerably less than the present 50–100 \$/kg.

The Union Carbide version is reported to convert $SiHCl_3$ by two catalytic disproportion processes via SiH_2Cl_2 into silane (SiH_4) (Breneman et al. 1978). SiH_4 is then pyrolytically decomposed at 100% yield in a fluidized bed reactor. This process was considered the most promising of more than ten different processes investigated at JPL (Lutwack 1981) to have the best potential for producing solar-grade silicon on a semiconventional basis.

As a result of basic costs, increasing complexity and need for safety precautions, no chance can be seen for any of the silicon products from the above cited processes to achieve a price level well below the 20–30 \$/kg range. As schematically indicated in fig. 3, all these semiconventional processes starting from $SiHCl_3$ provide the same amount of silicon from a generalized aspect. This also applies to the byproduct $SiCl_4$. Hence it follows, the more the production volume exceeds the salability limit of this byproduct, the more the recycling of $SiCl_4$ to $SiHCl_3$ becomes a necessity. Up to now, however, the various processes are in different stages of development. None of the corresponding silicon products – semiconventional in terms of cost and quality – are presently available in larger quantities.

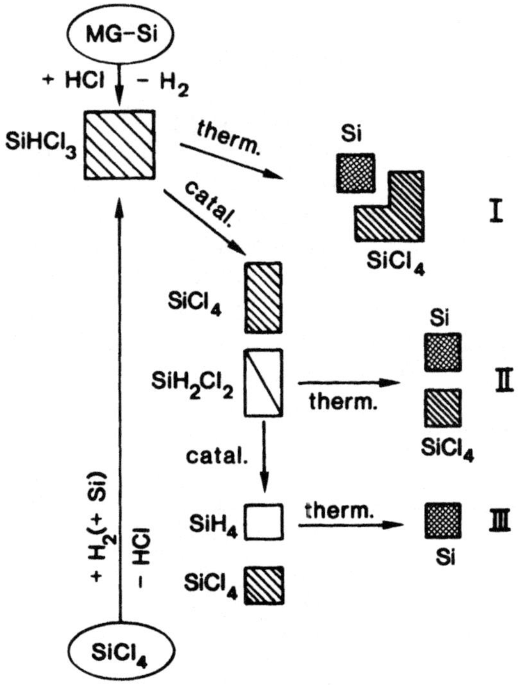

Fig. 3. Schematic diagram of theoretical yields of conventional and semiconventional CVD processes starting from $SiHCl_3$ (after Sirtl 1983).

3. Solar-grade specification

The term 'solar-grade' characterizes the material quality of a feedstock silicon which though less pure and lower in cost than 'electronic-grade' silicon, is, nevertheless, suitable for the production of solar cells with a conversion efficiency under AM1 (air mass 1) conditions of at least 10%. As pointed out by Sirtl (1979) with regard to real low-cost silicon production, we rather should make use of the term 'terrestrial solar-grade' (TSG). This expression has to stand for the particular impurity situation of the different processing sequences and the necessary simplification of any production technology.

In the meantime, the term 'solar-grade' has become a 'trade mark' for different products and often did not sufficiently distinguish between specialities in a chemical, crystallographic, or device-related sense. Primar-

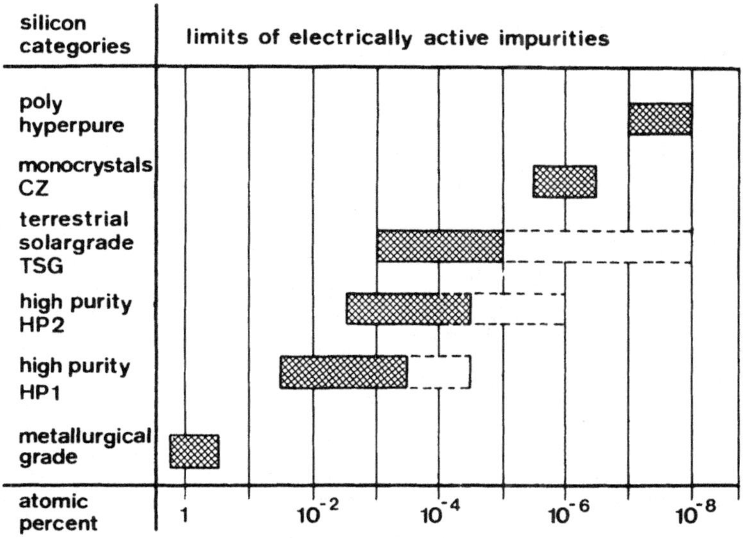

Fig. 4. Impurity ranges of different silicon qualities: HP1, basic stage; HP2, advanced stage of refining; TSG, terrestrial solar-grade. (After Dietl 1983a.)

ily, we would like to see this expression being confined to the chemical properties of the feedstock silicon.

In fig. 4 impurity ranges of presently defined silicon qualities are shown. This compilation covers more than eight orders of magnitude from metallurgical-grade silicon to hyperpure silicon including the more or less defined stages of refining, high-purity 1 (HP1) and high-purity 2 (HP2), prior to the TSG quality range. The special character of a solar cell as an electronic device allows for reduced purity in general, but demands hyperpure quality in terms of recombination-active elements as indicated by the blank areas in fig. 4.

3.1. Effects of impurities

Impurities in silicon may be expected to affect the solar cell performance and efficiency in a variety of ways. At high impurity concentrations, crystal growth can be perturbed by inclusions, precipitates (Helmreich and Ast 1980) or defects that may finally cause a structural breakdown (Hopkins et al. 1977). Still there is only limited information available on the influence of impurity-induced lattice defects like dislocations, stacking faults and twin boundaries, which are known to have harmful effects on solar cell efficiency as well.

Semiconductor properties may be influenced by electrically active impurity centers which reduce the minority carrier lifetime either by increased recombination or by scattering-induced mobility loss. Additional impurity-related effects include interface degradation, changes in series or shunt resistance (Wolf and Rauschenbach 1976) and increased junction excess current due to precipitates in the depletion region.

In order to understand the specific behavior and interaction of certain elements, typical impurities have to be investigated on their way through the different refining stages. Tremendous efforts have to be made to correlate the concentration of impurities with microstructural defects, electrical properties and solar cell characteristics. Such correlations are necessary for establishing the concentrations that can be tolerated without degrading the solar cell performance. The objective of various solar-grade definition projects (Davis et al. 1980, 1982, Hopkins et al. 1982) was to find out purity requirements and, later on, practicable specifications for solar-grade

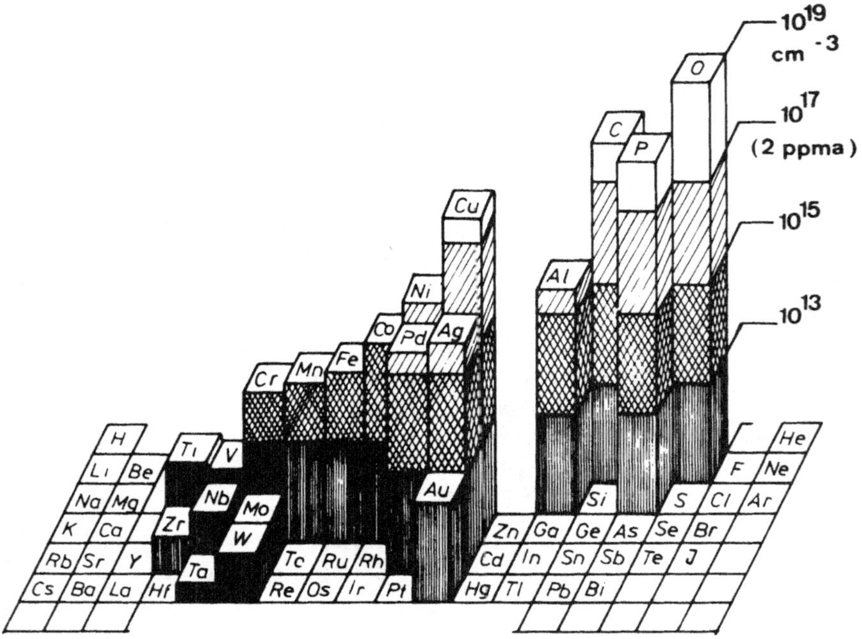

Fig. 5. Threshold values for solar cell degradation of different impurities in p-type Si (after Hopkins et al. 1982).

silicon. Most efforts have been concentrated on impurity effects in single crystals grown from some modifications of semiconductor-grade silicon within the conventional technology.

From these results, threshold values were determined for a lot of impurity elements. In fig. 5 the detrimental influence of the transitions metals in particular can be seen. These deep-level impurities, acting as carrier recombination centers, are present in metallurgical-grade silicon in the hundred and even thousand ppm range. To gain solar-grade quality they have to be removed to extremely low levels especially in p-type material, where impurities like titanium and vanadium seem to be even more detrimental. Based on the observation that impurity-induced performance loss is primarily due to a reduction of the base diffusion length, Davis et al. (1980) developed a model which predicts cell performance as a function of deep-level impurities. The effects of impurities in n-base and p-base devices differ in degree but submit to the same modeling analysis. A comparison of calculated and measured performance for simultaneously present impurities indicates that they interact in a number of ways: Copper appears to improve titanium-doped cells (Salama 1978), the presence of donors may increase the solubility of acceptors, and reciprocally. In other cases, pairs of impurities or more complex defects have been found, e.g., iron–oxygen complexes by Rozgony and Pearce (1978) or iron–boron complexes by Wünstel and Wagner (1982).

Additional efforts have been concentrated on other major impurities in metallurgical-grade silicon. For instance, main group metals like magnesium and calcium, of less significant interest in the field of conventional silicon research up to now, have been analyzed with regard to solubility and segregation coefficient data by Sigmund et al. (1980). Very limited information exists on the influence of the alkaline earth metals on solar cell performance except for magnesium which was investigated by Hopkins et al. (1977) in the 3×10^{12}–3×10^{14} at/cm^3 range without observing any degradation effect. The influence of shallow level impurities like boron, phosphorus and aluminum is substantially the same in electronic-grade and in solar-grade silicon but the tolerable concentrations have shown to be higher in the case of solar cell application. Informations available from investigations by Gutsche and Hill (1976) and Pizzini et al. (1980) referring to Czochralski- and float-zone crystals deal with these doping elements as primary impurities alone as well as in more or less complex matrices. The results show that boron and phosphorus can be allowed at concentrations up to 10^{17} at/cm^3 without significant degradation of cell performance. In the case of aluminum, where only 10% of the total amount is electrically active, this limit drops to 10^{16} at/cm^3.

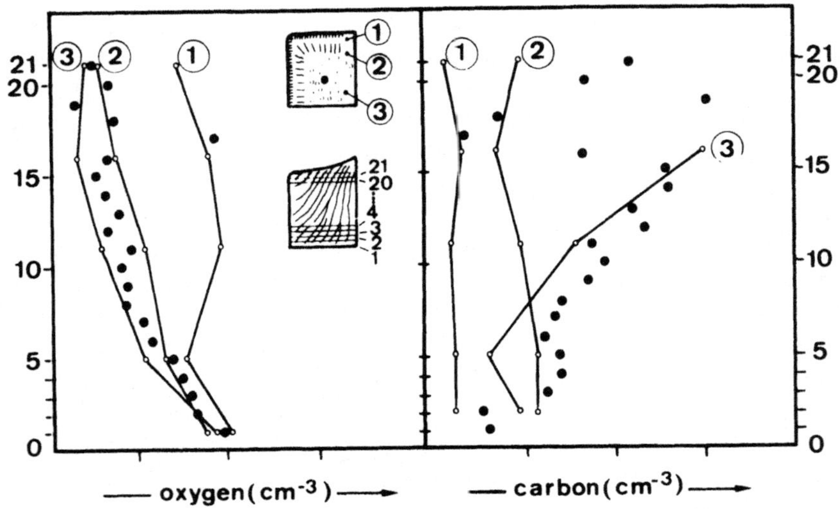

Fig. 6. Segregation of oxygen and carbon in a SILSO ingot. Different regions in horizontal cross-section (position ①, ②, ③) and vertical sequence of slices 1, ..., 21 are given in the insert (after Helmreich and Ast 1980).

Additional attention is focussed on non-metals like the chalcogens sulfur, selenium and tellurium (Holm and Sirtl 1981, 1982) or carbon and oxygen. A marked segregation close to theoretical values was found by Helmreich and Seiter (1979) for the latter slow-diffusing light elements as shown in fig. 6. The behavior of oxygen in solar cells as well as in other semiconductor devices relies on the electrical activity of dissolved oxygen, this activity depends on suitable thermal treatments and on the gettering of transition metal impurities (Rozgony and Pearce 1978) especially in polycrystalline material after heating at 600°C (Redfield 1981). Carbon is a predominant impurity in metallurgical-grade silicon resulting from the carbothermic reduction process in the arc furnace. The influence of carbon on the electrical properties of silicon is complex and still not well understood. A rapid increase of the leak current as well as the formation of stacking faults have been shown to be directly related to substitutional carbon (Abe et al. 1981). At higher carbon concentrations, silicon carbide precipitates are formed which cause shorting of the p–n junction (Khattak et al. 1981). Moreover, very complicated relationships between the nucleation rate of oxygen precipitates and carbon content have been observed by Craven (1981), although there is only limited experience on this carbon–oxygen interaction.

3.2. Dependence on subsequent processes

The results on the different effects of single or multiple impurities in silicon for solar cells should, however, be seen as a rough guideline rather than as the indication of precise threshold values. As most of the above-cited investigations refer to Czochralski or float-zone single crystals, the tolerable impurity concentrations given there must not necessarily apply to multicrystalline silicon.

Accumulation of impurities at grain boundaries combined with enhanced gettering can drastically change the situation. This does not mean that in every case the purity requirements are relaxed in comparison with single crystals but that the behavior of impurities may be different in polycrystalline material. Besides the mere geometrical effect on the minority carrier properties up to a grain size of 1 mm (as shown by Zehaf et al. 1982) the influence of increased diffusivity at the grain boundaries on the deactivation of impurities is evident.

The influence of iron, titanium and zirconium – some of the major contaminants of solar silicon feedstock produced via the carbothermic reduction of quartz – on electrical and photovoltaic properties of polycrystalline silicon has been studied by Pizzini et al. (1983), taking into account possible synergistic effects of oxygen and carbon. Only in the case of titanium an extended recombination area close to the grain boundaries was observed. This is assumed to be caused by an enhanced grain boundary segregation driven by a higher affinity to oxygen (and carbon?) for which the grain boundaries act as preferential sinks.

An additional key factor that will demand or can tolerate a certain quality of the feedstock material is the decision whether the subsequent crystallization is done by rapid sheet-like solidification with minor impurity segregation or by moderate ingot growth with strong segregation. The segregation behavior and, consequently, the electrical effects of impurities depend on the type of grain boundary (Davis et al. 1982). Extreme cases are given in fig. 7, where special grain boundary situations in highly contaminated multicrystalline silicon are shown.

Any subsequent annealing procedure intended, for instance, to avoid internal strain in the bulk material or to reduce the brittleness of ribbons or foils will change the impurity distribution. This can lead to levelling of accumulated impurities by back-diffusion to the inner part of the grains or, on the contrary, to migration of special impurities to gettering centers. At higher impurity concentrations, the formation of precipitates during the cooling-down cycle due to retrograde solid solubility can be predicted.

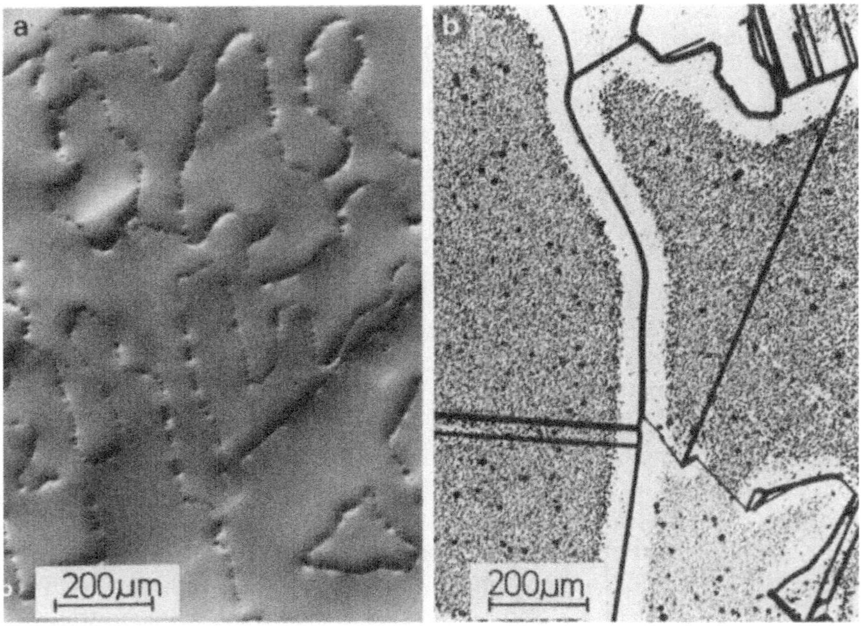

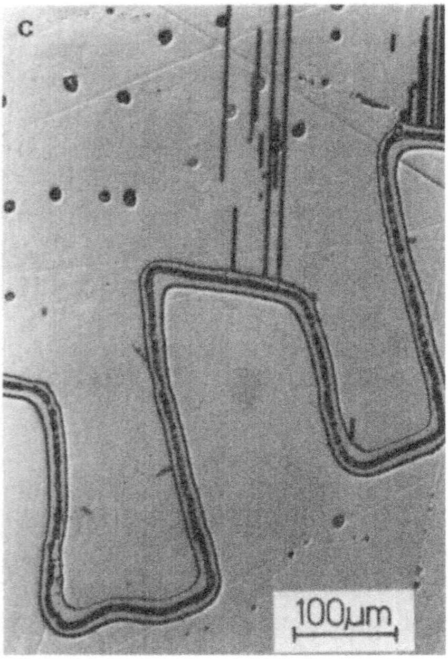

Fig. 7. Different metallographic features of highly contaminated multicrystalline silicon: (a) strongly carbon contaminated (SiC precipitates in low-energy grain boundaries); (b) carbon plus oxygen contaminated (oxygen migrates into low-energy grain boundaries); (c) high doping with electrically active elements (close to compensation?) (p–n junctions at low-energy grain boundaries). (After Sirtl 1983.)

However, the situation is even more complex: In addition to different crystal growth techniques various cell manufacture processes are applied. They also can be quite different in terms of their sensitivity to impurities. Particularly for their sensitivity to fast-diffusing impurities, which most of the transition metals belong to, close contacts with device processing engineers will be of mutual benefit when it comes to utilizing the gettering potential of certain processing steps. Therefore, good knowledge about the type of cell that terrestrial solar-grade silicon will be used for must be of great value to the material supplier. On account of the different crystallization techniques, annealing procedures and types of solar devices and their modification over the years, the final specification of terrestrial solar-grade silicon certainly will comprise a broad range of quality.

3.3. *Impurity analysis*

The analytical characterization of starting materials, intermediate qualities and end products has to be seen as an indispensable task in the framework of solar-grade silicon development. The upper concentration limits for electrically active elements in terrestrial solar-grade silicon lie in the range of 10^{-5}–10^{-7} (see fig. 4). A survey given by Millet (1980) of atomic absorption spectroscopy (AAS), atomic emission spectroscopy (AES), X-ray fluorescence, neutron activation and spark source mass spectrometry shows that this concentration range is still detectable by modern instrumental methods of chemical analysis.

Both AAS and AES require samples prepared as solutions. As the signals obtained from those instruments will not be related to element concentrations in a directly calculable manner, the measurements have to be calibrated by means of artificial standards. The signals may be influenced by interactions with the major components of the matrix or even other impurities. The advantage of the solution techniques is based on their relative independence from such interferences. A limitation of all solution techniques are the background signals caused by contamination. The contaminants may be impurities of reagents, dust in the air or derive from surface contamination of both sample and apparatus (Tschöpel 1980).

Spark source mass spectrometry (SSMS) seems to be one of the most useful methods of analyzing different silicon qualities. The principle advantage of this method is the broad range of impurity elements detected simultaneously with a sensitivity of down to 10^{-8} for directly sparked solid samples. As reported by Hurrle and Dietl (1981), silicon powder is mixed with gallium to form solid rods. The detection limits are in the range of

0.02–1 ppma with the exception of a few elements contained in the gallium or introduced during sample preparation. By using self-prepared standards, relative sensitivity factors for 35 impurity elements have been determined to enhance the accuracy of the results. The concentration values for most of the elements are proven to be accurate within a factor of two.

Methods of compositional analysis of the most important impurities in solar-grade silicon are discussed by Hunt (1983) from the viewpoint of a user who needs, for instance in the R&D field, rapid analytical feed-back so that decisions can be made to guide the R&D program. The doping elements boron and phosphorus can be determined with adequate accuracy and precision by resistivity measurements and by the Hall-effect, respectively. Another analytical method which has proven to be adequate for determining the boron content is given by the inductively coupled plasma emission spectroscopy (ICP-ES). Among the analytical techniques readily available, only SSMS and ICP-ES are sensitive enough to monitor metal impurities in the silicon end product. To measure carbon and oxygen concentrations, infrared spectroscopy can be used. However, if the boron level exceeds 3–4 ppma, silicon becomes opaque to infrared radiation. The ideal solid solubility of carbon in silicon is less than 10 ppma, but more than an additional 10 ppma can be dissolved under conditions of supersaturation. Carbon concentrations in excess of this lead to the formation of silicon carbide particles. For measuring the concentration of second-phase silicon carbide laborious methods including microscopic counting are required.

The composition and chemistry of both intergrain and intragrain regions in polycrystalline silicon have been investigated by Kazmerski (1983) using high-resolution surface analysis techniques. Segregated impurities localized at grain boundaries can be mapped by scanning Auger electron spectroscopy (AES), electron-energy loss spectroscopy (EELS), secondary ion mass spectroscopy (SIMS), X-ray photoelectron spectroscopy (XPS) and ultraviolet photoelectron spectroscopy (UPS). Direct correlation between impurity content and minority-carrier lifetime – evaluated by electron-beam induced current (EBIC) – is reported. Oxygen segregation to grain boundaries at temperatures above 650°C could be shown as well as the interrelationships between segregated oxygen and hydrogen passivation treatments. Camanzi et al. (1983) revealed by scanning transmission electron microscopy (STEM) elements like zinc, potassium and iron to precipitate within and at grain boundaries. Elements like aluminum, molybdenum and chromium, at the contrary, are not generally observed in such regions.

Impurities acting as carrier recombination centers were examined by deep-level transient spectroscopy (DLTS) (Miller et al. 1977). Deep-level analysis provides a means of measuring the energy density and capture rates of impurity-induced active centers and from these an evaluation of the concentration of electrically active contaminants present in silicon at levels well below the limits accessible by standard chemical methods can be done. As shown by Weber and Wagner (1980) DLTS can be used to investigate both silicon ingots and solar cells, as well as to observe changes in impurity activity due to thermal processing.

4. Silicon production

Economically useful deposits of silicon dioxide – as the primary silicon source – can be found all over the world. SiO_2 is composed of the two most abundant elements in the earth crust. In nearly all geological periods, SiO_2 has been concentrated in vast quartz deposits. The different SiO_2 qualities are compiled in fig. 8. The quality covers impurity ranges from <0.03% in quartz veins or rock crystals to about 1% in quartzite rock. Medium qualities are found in quartz sands and in chemically converted special silicates, either genuine or purified by chemo-mechanical treatments. The

mining product	quartzite rock	special silicates	quartz sand	purif.sands pegmatites	quartz veins rock crystal
up-grading	standard (tech.grade)	chemical conversion	standard (tech.grade)	chem.mech. treatment	standard
nature of particles	chunks	powder	grains	grains	nuggets
impurities % wt.	~ 1	< 0.3	>0.2	< 0.05	< 0.03

SiO_2 Qualities

Fig. 8. Different types of SiO_2 sources (after Sirtl 1979).

preference or necessity to use a certain quality depends largely on the type of reduction process or the sequence of subsequent refining steps. Prices of SiO_2 products reach from 0.02 $/kg for quartz sands up to above 10 $ / kg for rock crystals.

4.1. *Standard metallurgical-grade silicon*

The reduction of silicon dioxide with carbon in an electric arc furnace is a large-scale process which is technically well established since the turn of the century. In fig. 9 a technical set-up of a submerged arc furnace is shown. Commercial furnaces use about 10–30 MW of electrical energy to power carbon electrodes of about one meter in diameter. A medium-sized furnace represents a production capacity of the order of 20 000 tons of metallurgical-grade silicon (MG-Si) per year.

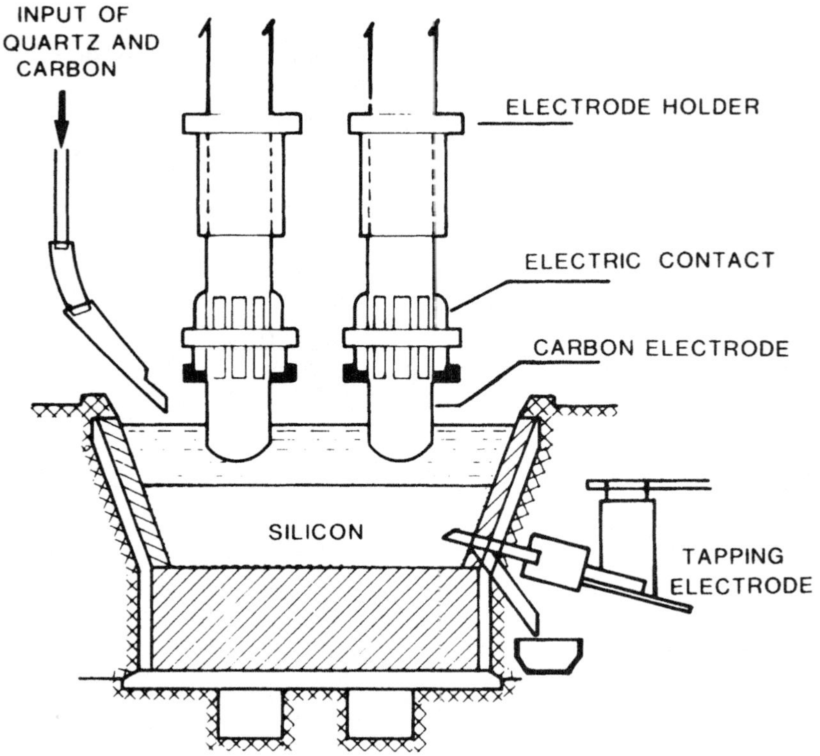

Fig. 9. Technical set-up of a submerged arc furnace for silicon production (after Sirtl 1979).

The nominal capacity of all existing furnaces in the Western World is known to be between 600 000 and 650 000 tons. As pointed out by Jasper (1983) approximately 500 000 tons can be seen as a readily available capacity. In 1982, consumption has been estimated to have arrived at about 420 000 tons, representing 84% of the effective capacity. Forecasts until 1990 talk of a growth rate for metallurgical-grade silicon of about 6% in the average of the typical four-to-five year cycles. The use of metallurgical-grade silicon in the photovoltaic field, if seen realistically, will have no impact on the next cycle, and probably not much on the next one after that.

To understand the carbothermic reduction of silica a number of chemical aspects have to be considered but also the sequence of the reactions, the kinetics and the transport of matter and heat in a working submerged arc furnace (as reviewed, e.g., by Motzfeld 1982). In the industrial electric furnace the mixture of quartz and carbonaceous reductants is charged from the top. At a temperature of about 1800°C the reductants will be largely converted to silicon carbide according to the reaction

$$2SiO_2 + 6C \rightarrow 2SiC + 4CO(g).$$

The charge, now essentially consisting of SiO_2 and SiC, descends into the hottest zone of the furnace where two concurrent reactions,

$$SiO_2 + 2SiC \rightarrow 3Si + 2CO(g), \quad \text{and}$$

$$2SiO_2 + SiC \rightarrow 3SiO(g) + CO(g),$$

take place with a high proportion of SiO in the reaction gas. On its way upwards it is cooled by its passage through the descending charge where it may condense by several reactions. SiO is thus retained in the charge and brought back to the reaction zone. The action is analogous to that of a reflux condenser with liquid silicon accumulating at the bottom of the furnace.

For the production of standard metallurgical-grade silicon, quartzite rock is the most widely used raw material. Its inexpensive mining requires little energy and has negligible environmental aspects. Coke, coal, wood chips and other forms of carbon are used as reductants. The kinetics of the reduction process described above depends on the structure and quality of the charge materials, particularly that of the reductants. A perfect carbon balance is of utmost importance to the arc furnace process. An undercarboned mixture can create a lot of slag in the furnace as the quartz can only be melted. It is practically impossible to reduce this low viscosity quartz slag even with additional carbon. An undercarboned mix can also

Table 2
Impurity concentrations (ppmw) in metallurgical-grade silicon from different suppliers (after Dietl 1981b).

Impurity	Concentration (ppmw) in MG-Si		
	A	B	C
Mn	260	500	50
Cr	25	20	50
Cu	25	50	20
Ni	110	30	10
Fe	3800	3500	5000
Al	1600	2400	2500
Ca	2700	2200	500
Mg	60	50	70
Ti	150	250	150
B	10	20	15
P	40	30	20

cause SiC-deposits when unreacted carbon gets wrapped by the molten quartz slag.

In the reduction process, impurities in SiO_2 are also reduced and, together with those from carbon mixtures, electrodes and other auxiliary materials contaminate the crude silicon to an amount of around two percents. In table 2 typical impurity levels in metallurgical-grade silicon from different suppliers are compiled. The main impurities are iron, aluminum and calcium in the range of some 1000 ppmw, manganese and titanium in the range of some 100 ppmw, and among the impurities below 100 ppmw, boron and phosphorus as the main doping elements. Another unavoidable impurity which is found in MG-Si in excess of saturation limits is carbon in the range from 100–1000 ppmw. Most of the impurity levels are far above those allowed for solar-grade silicon. Nevertheless, the worldwide industrial capacity and the low cost of the metallurgical-grade product make the upgrading to solar-grade quality increasingly attractive.

4.2. *Improved carbothermic process*

Since impurity levels in standard metallurgical-grade silicon are largely determined by the impurities present in the raw material, studies have been initiated to create in an arc furnace system a much higher degree of purity from where it is hoped to reach solar-grade quality in a last purification step, e.g., by controlled solidification.

4.2.1. Dow Corning approach

This approach has been investigated for the first time at Dow Corning (Hunt and Dosaj 1979). In addition to raw materials and refractories of convenient purity this process demands a specially designed arc furnace, the direct arc reactor (DAR). Certain quartz deposits have been identified as appropriate sources of silica with respect to purity and quantity of supply. Purifying carbon by high-temperature treatments in fluorocarbon environments has been successful, but resulted in materials of low reactivity. The best carbon reductant has been found as carbon black powder formed into pellets using sucrose as a binder. In a 200 kVA single-electrode arc furnace, which was constructed in a way as to minimize impurities released from the liner material, silicon was prepared at about 3 kg/h.

Due to the high operation temperature of the arc furnace the recovery of certain impurity elements is less than one and is inversely related to their vapor pressure, as is shown in fig. 10. Analytical measurements re-

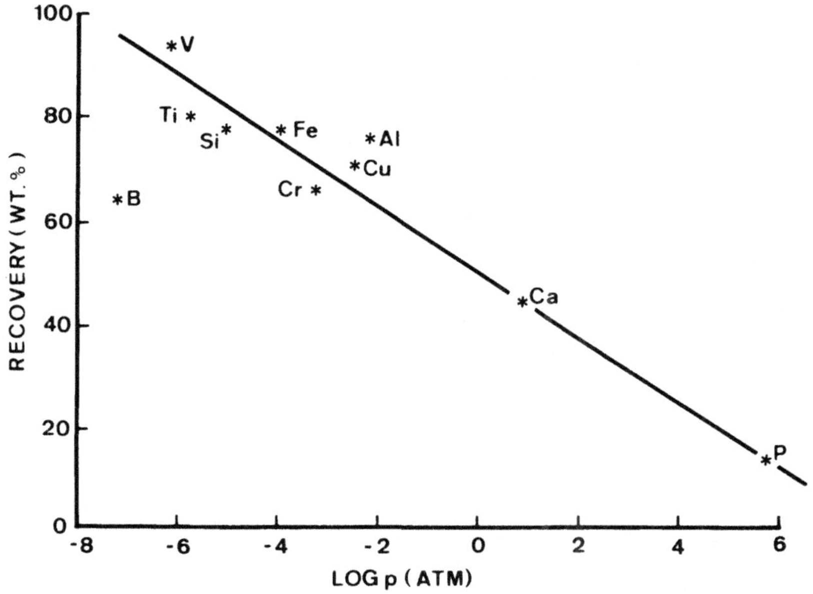

Fig. 10. Recovery of impurity elements in the arc furnace as a function of elemental vapor pressure at 1650°C (after Hunt et al. 1978).

vealed aluminum and iron concentrations of less than 100 ppmw each with levels of B, P and other impurities less than 10 ppmw, thus raising the purity of DAR-silicon to 99.99%. It ought to be mentioned, however, that the carbon content in silicon from carbothermic reduction in every experimental run was above saturation limit, irrespective of the way the improved furnace process was conducted.

This material, although of quite respectable purity as compared with standard MG-Si, still is not usable for solar cell fabrication and requires further purification, e.g., by crystallization under conditions of unidirectional solidification. It was demonstrated at Dow Corning that two Czochralski pulls were necessary to convert the DAR material into single-crystal ingots, of which the seed-end wafers were used to fabricate solar cells with efficiency values of up to 14% at an average of 9.3%.

Succeeding the early activities at Dow Corning at least five companies, namely Elkem (with Exxon), Heliosil (with Italian MG-Si manufacturers, as reported by Rustioni et al. 1982), Pechiney, Siemens and Solarex are or have been working on the attempt of utilizing different improved versions of the carbothermic process for the production of solar-grade silicon.

4.2.2. *Exxon/Elkem approach*

The motivation of a joint R&D agreement between Exxon Corporation and Elkem a/s was to explore the feasibility of practical production technology for low-cost solar-grade silicon. The sequence of manufacturing

Table 3
Impurity analysis of SiO_2 sources evaluated at Exxon/Elkem for improved arc furnace tests (after Amick et al. 1983a).

Impurity	Concentration (ppmw)		
	Lump quartz (1–2 cm)		Granular quartz (0.01 cm)
	Mt. Rose [1]	Malvern [2]	Type A
Al	30	60	20
Fe	20	30	10
Ti	1	<1	5
B	<1	1	0.5
P	<1	1	1

[1] Mount Rose Mining Co. Ltd., Armstrong, BC, Canada.
[2] Malvern Minerals Co., Hot Springs, Arkansas, USA.

steps selected for evaluation consisted of preparation of pure carbon and silica feedstocks and reduction of the feedstock to pure silicon in an improved arc furnace. Analytical data of impurities in the feedstock materials are listed in tables 3 and 4 as reported by Amick et al. (1983a). The methodology of feedstock selection and the methods for pelletizing the granular quartz and carbon black powder with water–sucrose solutions were taken over from the Dow Corning process.

The best silicon produced from a 100 kW arc furnace, whose construction was modified to establish the most suitable conditions for pure operation, was analyzed to have about 50–100 ppmw each of aluminum and iron, and about 10 ppmw titanium (table 5). All other impurities were found to be less than 10 ppmw, especially in the case of boron, where levels down to about 2 ppma could be reached.

After smelting, the silicon was subjected to chemical upgrading by acid leaching to remove most of the metal impurities. As shown in table 6, the major impurities after a two step etch (aqueous HF and HNO_3–HF–CH_3COOH acid mixture) are iron and aluminum at levels of a few ppma. Boron and phosphorus concentrations are nearly unchanged as compared with table 5.

After Bridgman recrystallization as additional refining step the phos-

Table 4
Impurity analysis of carbon sources evaluated at Exxon/Elkem for improved arc furnace tests (after Amick et al. 1983a).

Impurity	Concentration (ppmw)		
	Carbon blacks		Carbon electrodes [3]
	Type I [1]	Type II [2]	Pure quality
Al	2	20	<1
Fe	10	80	1
Ti	<1	10	<1
V	<1	<1	<1
Ca	<1	80	5
Mg	2	50	<2
B	1	<1	<0.5
P	<1	<1	<1

[1] Selected 'thermal black' from natural gas.
[2] Selected 'furnace black' from residual oil.
[3] Union Carbide Corporation, Sweden.

phorus content was lowered, and the recrystallized silicon was invariably p-type because of the residual boron level. Experimental results of this crystallization procedure are discussed by Amick et al. (1983b): While metallic impurities could be reduced to less than the detection limits, the principal residual impurities were oxygen and carbon. Silicon carbide particles could be observed with an infrared transmission microscope. After two Bridgman recrystallizations the carbon level was estimated as below about 5 ppma throughout most of the ingot. Two Bridgman recrystallizations substantially improved solar cell performance yielding efficiency values equivalent to those obtained from Bridgman recrystallized semiconductor-grade silicon.

Rice hulls as an alternative raw material for producing silicon have been

Table 5
Impurity analysis of Exxon/Elkem silicon as a function of raw material feedstocks (after Amick et al. 1983a).

Run	Raw material feedstocks		Impurity analysis of Si				
	C[a] / SiO_2[b] Pellets	SiO_2[b]	B (ppma)	P (ppma)	Al (ppmw)	Fe (ppmw)	Ti (ppmw)
17	Type I / Type A	Mt. Rose	3.6	2.4	190	160	16
19	Type I / Type A	Mt. Rose	4.1	1.4	290	190	22
20	Type I / Type A	Type A	1.8	2.1	70	90	10
24	Type I / Type A	Malvern	2.8	1.9	130	160	13

[a] See table 3.
[b] See table 4.

Table 6
Impurity content in Exxon/Elkem silicon after acid leaching (after Amick et al. 1983a).

Run	B (ppma)	P (ppma)	Fe (ppma)	Al (ppma)	Ti (ppma)
20	3.1	1.7	<4	<10	<2
29	2.5	4.5	11	101	<2
33–1	1.5	0.9	5	<10	<2
34–1	2.0	1.3	<4	<10	<2
34–2	3.1	1.3	8	10	<2

set about most recently under investigation in another joint R&D agreement between Exxon Corporation and Elkem a/s. Rice hulls are a high-volume, low-cost commodity which contains both silica and carbon. As reported by Hunt et al. (1983), impurity analyses have indicated that rice hulls have low concentrations (10–20 ppmw) of aluminum and iron, the major impurities in conventional raw materials used for silicon production. The doping impurities boron and phosphorus were found to be 1 and 40 ppmw, respectively.

Extruded pellets of coked rice hulls, with sucrose as a binder, were shown to be very reactive raw materials for the submerged arc furnace technology. Preliminary runs using these pellets resulted in the lowest energy consumption ever measured in the experimental furnace. An estimate of impurity concentrations in the reduction product was made from the fraction of the impurities in the raw material that ends up in the resulting silicon showing the potential of producing a quality with impurity levels one to two orders of magnitude lower than those in standard metallurgical-grade silicon. As the major impurities in rice hulls, Ca, K, Mg and Mn are reported to be drastically reduced to around 20 ppmw by hot hydrochloric acid leaching, this material should be valued as an interesting alternative for the production of solar-grade silicon.

4.2.3. Siemens approach

Another promising method to obtain silicon from an improved arc furnace process is followed at Siemens using starting materials of even higher purity. As discussed above, pure natural quartz – e.g. from deposits in the USA, Canada, Brazil or South Africa – has an impurity concentration which requires further purification of the silicon produced by carbothermic reduction. To overcome this problem, Aulich et al. (1982a) developed a process in which abundant and inexpensive quartz sand is purified using a combined glass melting/fiber leaching (FL) process.

The basic idea behind the FL process is to first provide a solvent for the different impurities using glass forming oxides and then to remove the solvent from the SiO_2-network by acid leaching treatment. After getting quartz sand into the vitreous phase by adding metal oxides such as Na_2O, CaO or Al_2O_3, thin fibers are drawn from the glass melt. By annealing the fibers a phase separation is effected into a phase which is rich in SiO_2 and a phase which contains the additives together with most of the impurities. The glass fibers are then treated with hot hydrochloric acid solution, and as a result the metal oxides and impurities present in the glass matrix are leached out by an ion-exchange mechanism: Metal ions are exchanged for hydrogen ions from the acid resulting in a porous layer of silica, so that the underlying glass layer is continuously attacked until most of the impurities are removed. To demonstrate the feasibility of the FL process in preparing high-purity SiO_2, sea sand with a high impurity level was used as starting material. Table 7 shows impurity analyses before and after applying the FL procedure. The low concentrations of boron and of the transition metals in particular, qualify the silica as an excellent starting material for high-purity arc furnace operation.

The production of solar-grade silicon requires the use of carbonaceous reactants comparable in purity to FL silica. In addition, the carbon must be highly reactive for the reduction process to succeed. As carbon black was shown to be of sufficient chemical reactivity, it was purified by leaching with hot hydrochloric acid for several hours in order to remove harmful impurities. In table 8 significant improvement in the purity of carbon black by HCl treatment is demonstrated.

For testing purposes, a research reactor (120 kVA connect load) made of high-purity graphite and a larger reduction furnace (550 kVA connect load) were used. As reported by Dietze et al. (1982), silicon produced initially from pure natural quartz and charcoal was turned into single-crystal ingots after two Czochralski pulls. Solar cells with more than 10% efficiency have been produced from this material. When using high-purity

Table 7
Impurity concentrations in silica purified with the fiber leaching (FL) process (after Aulich et al. 1982b).

Impurity	Concentration (ppmw)	
	Starting material sea sand	SiO_2 after FL-process
Na	10^4–10^5	<5
Mg	10^3–10^4	0.05–1
Ca	10^3–10^4	<5
Al	10^4–10^5	0.5 –1
B	1 –10	<1
Fe	10^3–10^4	<0.5
Ni	1 –10	<0.05
Mn	10 –10^2	<0.05
Cu	1 –10	<0.05
Ti	10^2–10^3	0.05–0.1

Table 8
Purification of carbon black with hot hydrochloric acid (after Aulich et al. 1982b).

Impurity	Concentration (ppmw)	
	Starting material	After purification
Al	<0.062	<1
Cr	<0.62	<10
Cu	0.25	<0.1
Fe	62	<1
Mn	0.62	<0.1
Mg	6.2	0.1–1
Mo	0.062	<1
Ni	3.1	<0.1
Ti	<0.124	<0.1
Ca	100–1000	<10
B	<1	<1

FL-silica and acid-purified carbon (formed into pellets with a high-purity phenol resin) the silicon obtained was of considerable purity revealing boron below 0.5 ppmw, phosphorus below 0.2 ppmw, and aluminum as well as transition metals below 10 ppmw (Aulich 1983).

4.3. *Metallothermic processes*

Thermodynamic calculations – as compiled by Hunt (1974) – on the chemical preparation of elemental silicon from common silicon compounds indicate that nearly 150 out of 200 tested reactions are thermodynamically feasible. In about 20 cases practicable routes to solar-grade silicon could be demonstrated. Besides the carbothermic route two of the feasible metallothermic processes represent high-potential candidates within the small number of low-cost processes.

4.3.1. *Aluminothermic reduction of quartz*

In fig. 11 the chemical potential, expressed by the temperature function of the standard free-enthalpy value, is depicted for both the carbothermic and the aluminothermic reduction according to the summary reactions

$$SiO_2 + 2C = Si + 2CO,$$

$$3SiO_2 + 4Al = 3Si + 2Al_2O_3.$$

Whereas the carbothermic reduction of SiO_2 is inefficient below 1700°C, the reduction with aluminum is a pronounced exothermic reaction over the whole temperature range. The reduction of quartz with aluminum at comparatively low temperatures is a promising way to produce high-purity silicon on a large scale. A suitable technical process on this basis, investigated at Wacker (Dietl et al. 1982a), requires liquid phases of both metallic and oxidic nature acting as solvents for the reaction products. The two liquid and immiscible systems are represented by an Al–Si alloy and an Al_2S_3–Al_2O_3–SiO_2 slag system.

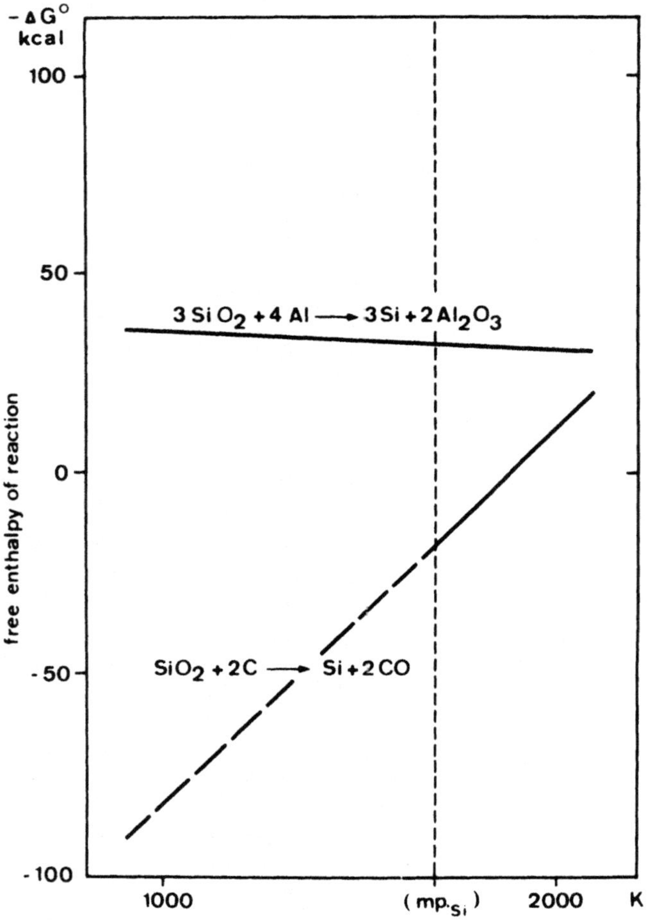

Fig. 11. Free-enthalpy values of carbothermic and aluminothermic reduction of SiO_2 (after Sirtl 1979).

The reduction process must not utilize chunks of quartz rock, as is required in the standard carbothermic arc furnce, but is based on particulate matter, like powder or sand, the latter being available in large quantities and, mainly, higher purity. To initiate the process, schematically drawn in fig. 12, quartz sand is introduced to the reaction vessel while stirring the liquefied slag to dissolve it there and mix it with aluminum. Droplets of aluminum being dispersed in the slag reduce SiO_2. The final products, silicon and aluminum oxide, are kept in the liquid state by dissolving in excess aluminum and aluminum sulfide, respectively. This reaction step is followed by a phase in which the Al droplets containing dissolved silicon separate from the slag as a liquid metal pool. The reaction takes place in a temperature range of 1100–1200°C where the solubility of Al_2O_3 in the slag is of the order of 25% by weight. To keep the slag for multiple use, most of the Al_2O_3 is crystallized by lowering the temperature of the slag, thus enabling its removal. Recycling of aluminum by reducing the by-product Al_2O_3 by electrolysis is a rather important part of the process.

At the temperature of the controlled reduction process, the aluminum melt is able to dissolve up to 65% by weight of silicon. When the slag phase and the Si/Al phase have separated, the Si/Al phase is slowly cooled causing the silicon to recrystallize in form of intergrown platelets. The platelets are separated from the liquid phase at a temperature slightly

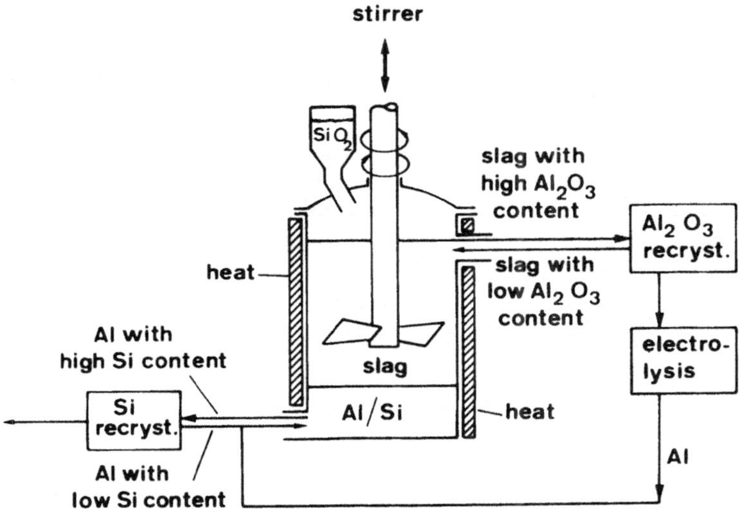

Fig. 12. Processing sequence of the aluminothermic reduction of quartz sand.

Table 9
Impurity concentrations (ppma) in Si from aluminothermically reduced quartz sand (after Holm 1982).

Impurity	Concentration (ppma)		
	Quartz sand	Si platelets	Milled and leached
Mn	<2	≤2	<1
Cr	<2	≤2	<1
Cu	<2	≤2	<1
Ni	<2	≤2	<1
Fe	90	1–3	<1
Al	700	2000–4000	~300
Ca	40	≤2	<1
Mg	25	≤2	<1
Ti	100	≤2	<0.1
B	5	≤3	<3
P	10	3–5	<0.5
C	–	<50	<50
S	–	~1	≤1

above the solidification of the Al/Si eutectic. Finally, the aluminum melt containing residual silicon is returned to the reaction vessel.

As shown by Dietl et al. (1982b), micrographs of intergrown silicon platelets reveal inclusions of Al/Si eutectic encapsulating different impurity precipitates. After removing adherent aluminum the platelets therefore are comminuted and leached with diluted hydrochloric acid. In table 9 impurity concentrations in silicon from aluminothermically reduced quartz sand are shown. In the milled and leached product most of the impurities are below 1 ppma. The remaining aluminum level is determined by its temperature-dependent solid solubility in silicon when recrystallizing from aluminum solution. To remove this residual aluminum, additional pyrometallurgical purification is required.

4.3.2. *Sodium reduction of SiF_4*

Silicon tetrafluoride (SiF_4) represents a secondary silicon source which is of economical interest in the field of solar-grade silicon production. SiF_4 is formed as an inexpensive by-product in the production of superphosphate fertilizers from phosphate ore. The ore always contains CaF_2, which is converted to HF when treated with H_2SO_4. HF then reacts with the SiO_2 present in the phosphate ore to form SiF_4. An amount of about 75 000 ton/year is recovered in the United States as fluorosilicic acid, H_2SiF_6. The method to obtain silicon by sodium reduction of SiF_4, starting from fluorosilicic acid, is known as the SRI (Stanford research institute) process.

As reported by Sanjurjo et al. (1981) the overall process consists of three major steps:

- Precipitation of Na_2SiF_6 followed by SiF_4 generation;
- Na reduction of SiF_4;
- Separation of Si from the Si-NaF mixture.

As shown in the flow-chart in fig. 13, NaF is added to H_2SiF_6 to form Na_2SiF_6 in the precipitation step:

$$H_2SiF_6(aq) + 2NaF(s) \rightarrow\rightarrow Na_2SiF_6(s) + 2HF(aq).$$

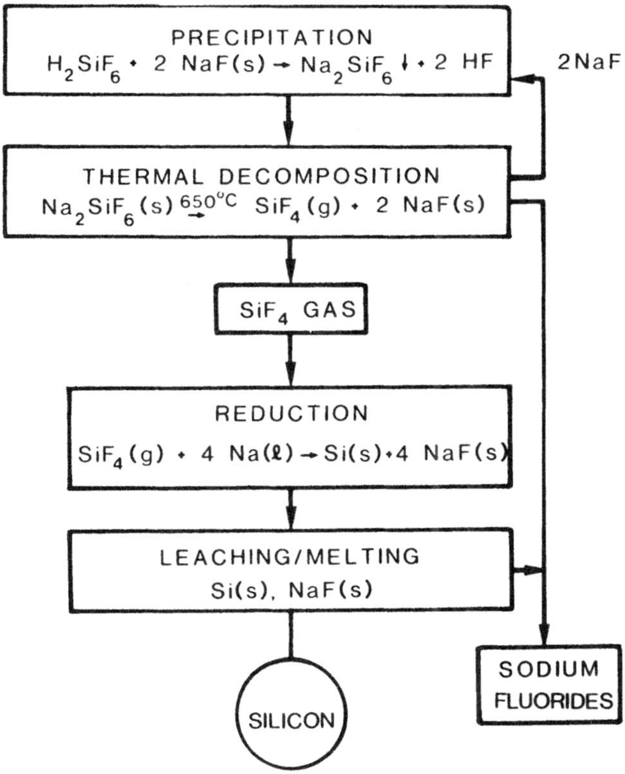

Fig. 13. Processing sequence of the SRI approach (after Kapur et al. 1979).

After filtration and drying, Na_2SiF_6 is thermally decomposed at 650°C to form $SiF_4(g)$ and NaF(s). From analytical measurements it is clear that most of the purification is accomplished as a result of the thermal decomposition. Most transition metal fluorides are stable condensed phases at the temperature of decomposition and, therefore, stay in the solid. SiF_4 is then reduced with sodium to produce silicon:

$$SiF_4(g) + 4Na(\ell) \rightarrow Si(powder) + 4NaF(s).$$

The exothermic character of this reaction provides sufficient heat to sustain the process at a temperature of 500°C. To separate silicon from NaF, the Si/NaF mixture is acid-leached or melted, forming two immiscible phases where liquid–liquid extraction provides additional purification. In table 10 impurity concentrations in the starting H_2SiF_6 and in the final silicon product are listed as revealed by plasma emission spectroscopy analysis. Of special interest for solar cell applications are the low concentrations of boron and phosphorus. The silicon produced by the SiF_4–Na reaction – if purified by a final process, e.g., by directional solidification as discussed by Kapur et al. (1979) – should be considered a potent material for photovoltaic application. However, as the reduction process yields four times as

Table 10
Representative impurity content in H_2SiF_6 and silicon obtained from SRI process (after Sanjurjo et al. 1981).

Impurity	Concentration (ppmw)	
	Starting material H_2SiF_6	Silicon product
B	1	0.1
P	33	0.2
Al	8	0.8
Ti	–	2.0
Mo	11	0.3
Fe	13	<7
Co	0.5	0.5
Ni	1.2	2
Zn	1.4	0.01
Na	460	1
K	9	0.01
Ca	110	1
Mg	55	0.1

much NaF as Si, low-cost silicon seems to be attainable by adequate by-product credit only.

4.4. Electrowinning and refining

Electro-deposition of silicon has a long history, dating back well over 100 years. As reviewed by Elwell and Feigelson (1982), electro-deposition may offer high cost effectiveness as a means of producing elemental silicon from an inexpensive source. Purification of the silicon product occurs intrinsically due to the difference in deposition potential between silicon and most of the impurity elements. Those impurities with higher deposition potential than the silicon will remain in solution during deposition. Those with lower deposition potential will co-deposit with the silicon but may be removed by a sacrificial predeposition procedure.

The use of electro-deposition to produce solar silicon can be classified as either electrowinning of silicon from some compound or electro-refining of metallurgical-grade silicon to obtain purer material. Recent studies on silicon electrowinning have concentrated on two systems. One of them uses K_2SiF_6 dissolved in KF/LiF or KF/LiF/NaF eutectic solvents at low temperature (750°C), the other one is based on a $SiO_2/BaO/BaF_2$ melt at temperatures above the melting point of silicon. K_2SiF_6, the silicon source in the low-temperature process, derives from fluorosilicic acid, the waste by-product of the fertilizer industry discussed above. Deposition of silicon from K_2SiF_6/fluoride melts onto graphite could be shown by Rao et al. (1980, 1981) to yield substantially inclusion-free layers of silicon. The best deposits contained only 8–10 ppmw of total impurities with Cr, Cu, Fe and Ni at about 2 ppmw each. The electrical resistivity reached values as high as 3 Ω cm, normally n-type, probably due to the presence of phosphorus at levels below the limit of detectability of emission spectroscopic analysis.

One major disadvantage of the low-temperature process is that the deposition rate is inevitably slow. Most systems used for electro-deposition of solid layers do not allow stable deposition at rates exceeding a few tens of micrometers per hour. This limitation does not apply to electrowinning of liquid deposits, as it is commercially used in the Hall–Heroult process for the production of aluminum.

The most suitable source material for electro-deposition of silicon above its melting point is SiO_2. In view of its high melting point, a ternary system has been used by reacting selected microcrystalline silica of high purity with an alkaline earth metal oxide adding the corresponding fluoride as a fluxing agent. Best results were obtained by De Mattei et al. (1981) using the SiO_2/BaO eutectic, with BaF_2 added to facilitate melting and to lower the viscosity of the ternary melt. The purity of silicon produced from this system at a temperature of 1450°C was 99.98% by weight, the major impurities being 20 ppmw Al, 20 ppmw Fe, 10 ppmw Ti, and Cr, Cu, Mn and Zr below 5 ppmw.

An alternative to the high-temperature process is to use an alloy cathode. Olson and Kibbler (1981) recently utilized a liquid tin cathode which is preferable to aluminum or copper since it does not introduce shallow levels or deep-level recombination centers for photovoltaic application. The silicon product contained about 10 ppmw of transition metals. Single crystals grown from this material were n-type with resistivities of 0.05–0.1 Ω cm probably due to the presence of residual phosphorus. Although the use of a liquid alloy cathode in the system $SiO_2/Na_3AlF_6/LiF$ permits electro-deposition at convenient temperatures around 850°C, the deposition rate is slow compared with processes where silicon is deposited as a pure melt.

Electrorefining as another electrometallurgical treatment has been shown to be an efficient, potentially low-cost method for purifying metallurgical-grade silicon. The basic components of such a process consist of an electrolyte, an impure silicon anode, and a cathode formed as deposition substrate. Recent progress was made by Olson and Carleton (1981) by alloying the MG-Si anode material with copper to a composite of Si and Cu_3Si. As electrolyte a molten mixture of K_2SiF_6 and LiF/KF eutectic was used at a temperature of 750–790°C. The silicon was deposited on thin sheets of graphite or silicon.

Electropurification is expectedly effective since the impurities which are less electropositive than silicon (e.g., the doping elements boron and phosphorus) remain in the anode at the low potential difference ($<100\,mV$) which is necessary to transfer silicon to the cathode. The major success of the process, however, is attributed to the extraordinary mass transport properties of Cu_3Si. As the impurities diffuse through the Cu_3Si phase at much slower rate than silicon, Cu_3Si has the properties of a silicon filter, capable of efficiently separating silicon from the impurities. This difference in diffusion kinetics is a more powerful purification mechanism than the electrolytic effect that can be realized with an unalloyed silicon cathode. One disadvantage of this process is the requirement that the copper has to be purified at intervals when the impurity levels build up to an extent that preferential diffusion of silicon no longer occurs.

Refining of metallurgical-grade silicon in a multiple cell, machined from

Table 11
Impurity concentrations in electrometallurgically refined Si and in starting MG-Si (after Olson 1983).

Impurity	Concentration (ppma)	
	MG-Si	Refined Si
Al	3400	1
B	17	0.7
Ba	9	<0.2
Ca	290	<0.07
Cu	–	0.2
Cr	40	0.2
Fe	>2500	0.1
Mg	85	0.9
Mn	550	0.03
Ni	39	<0.1
P	14.5	3
S	7	0.4
Ti	290	<0.1
V	250	<0.05
Zr	13	<0.03

a block of graphite, yielded silicon of promising quality, as demonstrated in table 11. After stripping the refined silicon from the graphite cathodes and boiling it in concentrated HCl to remove adhering electrolyte salt, single crystals were grown using the Czochralski technique. The resistivity of 0.1 Ω cm agrees well with the chemical analysis data in table 11, where the dominant donor impurity is phosphorus. Wafers from the seed-end of the boule were converted into solar cells with efficiency values around 12% indicating that this electrorefined silicon is of solar-grade quality.

Any attempt to calculate the cost of electro-deposited silicon requires, at present, extrapolation of a type that is subject to great uncertainty. From about 16 kWh of electricity as assumed by Elwell and Feigelson (1982) to be required for 1 kg of Si at a deposition potential of 2 V an energy payback time of 300 days for the device was calculated. Although these numbers may be optimistic, there seems to be a considerable possibility to achieve an interesting base material for photovoltaic application.

5. Metallurgical refining

The problem of upgrading typical products made in standard arc furnace systems has many facets. It starts with geographic and geological aspects of the plant in question, continues with efforts in modifying the reduction furnace process and ends with appropriate refining procedures including their most effective sequence. Optimization of the refining route is inseparably combined with economic aspects and with the conditions of subsequent crystallization and device processing.

Proceeding from the different conceptions of utilizing the reduction of quartz, three economically feasible routes for preparing solar-grade silicon meltstock can be seen. In fig. 14 a schematic flow-diagram indicates the

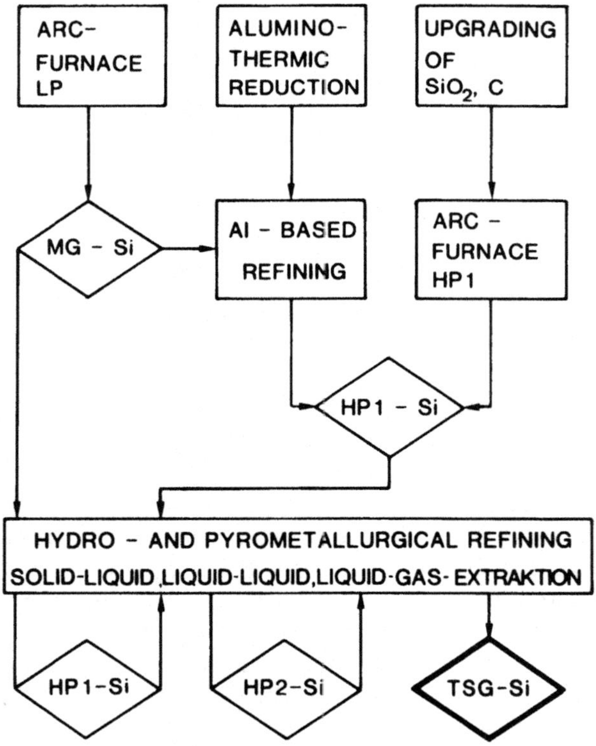

Fig. 14. Different approaches to large-scale production of silicon for terrestrial solar cells: HP1, basic stage; HP2, advanced stage of refining; TSG, terrestrial solar-grade. (After Dietl 1983a.)

processing sequences leading from different quartz qualities via intermediate purity stages to solar-grade silicon for terrestrial application. The route starting from the low-purity (LP) arc furnace requires more severe hydro- and pyrometallurgical refining to reach via the more or less defined stages of upgrading high-purity 1 (HP1) and high-purity 2 (HP2) terrestrial solar-grade (TSG) quality. The aluminothermic reduction process yields silicon which, like the product obtained from an improved carbothermic process in a HP1 arc furnace, shows HP1 quality (although different in character) that needs at least one further purification step.

Practice-oriented approaches have been started at several places to reveal the effect of different metallurgical refining procedures on certain elements and impurity groups. Basic hydrometallurgical and pyrometallurgical upgrading techniques with special purification effects according to solar-grade requirements are:

(a) Solid–liquid extraction (acid leaching);
(b) Liquid–liquid extraction (slagging);
(c) Liquid–gas extraction (gas blowing and evaporation);
(d) Aluminum-based refining (recrystallization from Al solution);
(e) Directional solidification.

Major efforts had to be made to install and improve a number of analytical methods to trace impurity elements, in some cases from the percent to the sub-ppm range. Analytical work for permanent control of most different variations of material and refining procedures can amount to a significant share of the total costs in this field.

5.1. Hydrometallurgical purification

In the field of upgrading crude silicon to solar-grade quality, solid–liquid extraction – known as acid leaching – has a long history as a very efficient purification step (see, e.g., Tucker 1927, Ohl 1946, Aas and Kolflaath 1974). Impurity precipitation in polycrystalline metallurgical-grade silicon in form of silicides, oxides, silicates or metal eutectics is a precondition to hydrometallurgical refining. The formation of precipitates is enhanced by both low segregation coefficient and low solid solubility of the impurity element in question. As shown in fig. 15, the striking correlation between segregation coefficient and degradation threshold value for solar cell efficiency seems to be helpful in photovoltaics: The lower the permissable impurity level the lower the segregation coefficient and, thus, the better the purification effect.

In fig. 16 the well-known solid solubility characteristics of impurity elements in silicon are shown. In most cases there is a maximum solubility around 1300°C, and retrograde behavior towards both lower temperature and melting temperature. The correlation between solid solubility data and segregation coefficient of different impurities is given in fig. 17. Besides the maximum solubility at lower temperatures (full line), as compiled by Fischler (1962), solubility data near the melting point are extrapolated from the solubility curves in fig. 16. Both lines in fig. 17 again demonstrate the attractivity of hydrometallurgical processing: The lower both solid solubility and segregation coefficient, the completer the accumu-

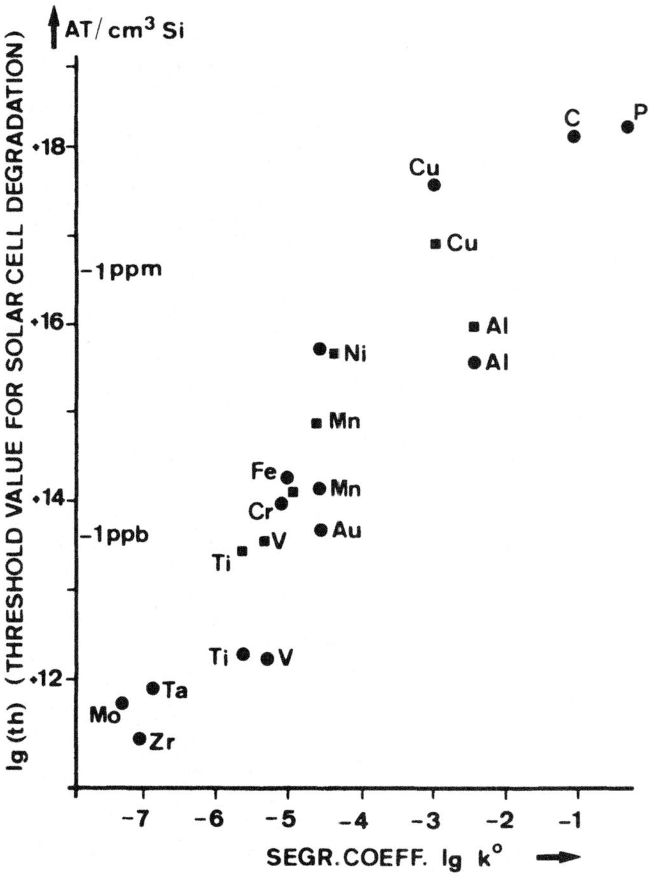

Fig. 15. Correlation of solar cell degradation and segregation coefficient for different impurities in Si: ● p-type; ■ n-type (after Hopkins et al. 1980).

lation of the given impurity element in precipitates and, thus, the better the chance for effective purification by acid leaching.

Solid solubility represents the limitation of hydrometallurgical refining for impurities like iron, aluminum, calcium, manganese and titanium which are present in metallurgical-grade silicon in concentrations exceeding solid solubility. Whether the higher or the lower solubility data have to be considered depends on the diffusion coefficient of the impurity in question. Slowly diffusing elements such as aluminum and calcium essentially remain segregated in precipitates, whereas elements with high diffusion coefficient (e.g., iron and manganese) tend to saturate the silicon crystal to maximum solubility by back-diffusion during the cooling cycle.

During solidification of metallurgical-grade silicon from the carbothermic process, impurities with low segregation coefficients and concentrations exceeding the solid solubility concentrate mainly along the grain boundaries of the resulting polycrystalline material. Figure 18 shows a micrograph of metallurgical-grade silicon with a grain size of around 100 μm. Along the grain boundaries, impurity inclusions are discernible; the dark spots inside of a grain have been identified as SiC particles

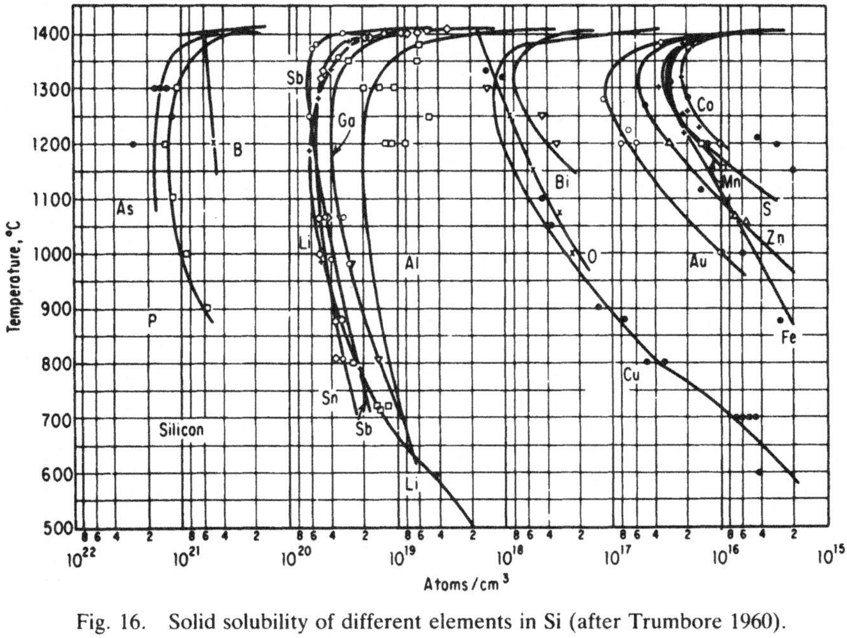

Fig. 16. Solid solubility of different elements in Si (after Trumbore 1960).

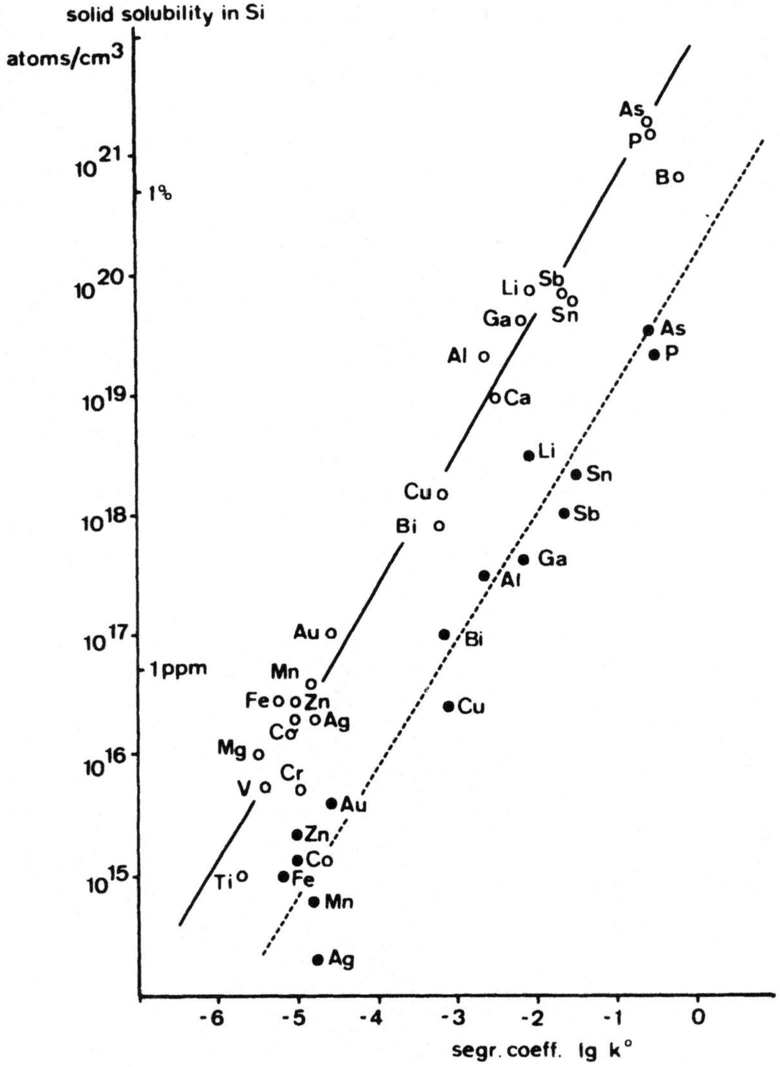

Fig. 17. Correlation of solid solubility and segregation coefficient for different impurities in silicon. ○ maximum solubility (after Fischler 1962); ● solubility near 1410°C (after Dietl 1983b).

originating from the carbothermic process. In fig. 19 the elemental composition of a typical impurity precipitate is given as investigated using an electron microprobe. Picture (a) represents a secondary electron image of the microscopic structure, the following pictures reveal the distribution of the impurity elements Fe, Al, Ca, Ti and Mn, respectively. With the exception of Fe and Mn, all other elements are localized in different areas of the precipitate.

For effective refining, the metallurgical-grade silicon has to be crushed and the precipitates exposed to the action of an acid. The leaching mixture must be able to attack the impurities but should not dissolve the silicon itself. In fig. 20 a flow-chart of the hydrometallurgic refining process is given. Lumps of MG-Si first have to be crushed by a jaw and a roll crusher to a particle size below 2 mm. To avoid dangerous powder explosions, the subsequent pulverization is achieved by milling under wet conditions. The addition of water allows a liquid cyclone to be used to classify the milled product and to recycle the coarse-grained portion. The leaching equipment consists mainly of a vessel in which the pumpable silicon–acid slurry is agitated by an impeller to enhance the leaching effect. The silicon powder

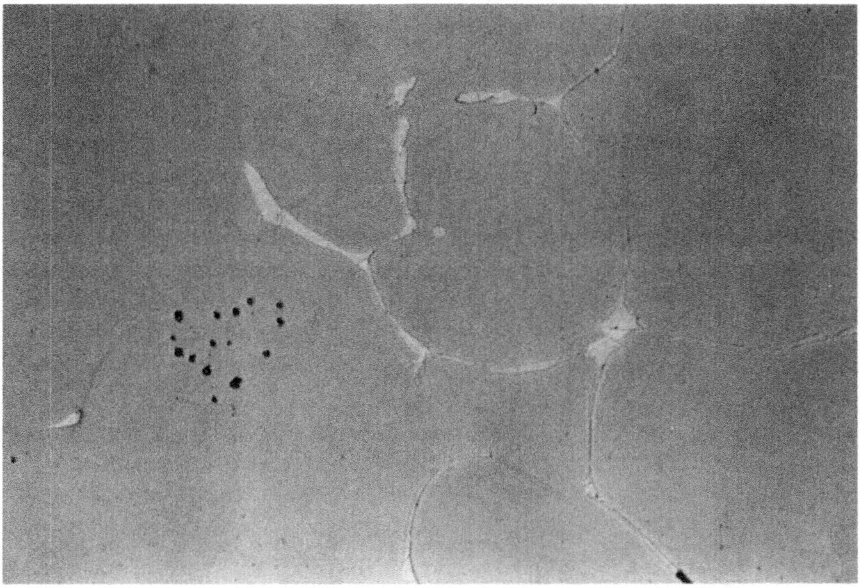

Fig. 18. Micrograph of MG-Si with crystal grains of about 100 μm and impurity precipitates mainly along the grain boundaries (taken from Dietl 1983b).

then has to be washed in a filter press to remove the acid and residual impurities dissolved therein. Chemo-mechanical grinding of MG-Si using simultaneous interaction with acid, represents an alternative hydrometallurgical procedure.

Extensive experimental work has been done in optimizing particle size of the milled MG-Si, type and concentration of acids as well as temperature and duration of the leaching treatment. Figure 21 demonstrates the influence of particle size on the purification of MG-Si powder covering a range from 10–140 μm of average particle size. Fe, Ca and Ti stand for the leaching behavior of most of the other impurities in MG-Si. The effectiveness of acid treatment is improved by grinding raw silicon to a particle size well below the average grain size of the polycrystalline structure. In case of extremely fine milling, the improvement in purification potential has, however, to be balanced, against both the difficulties in handling the material and the increasing danger of powder explosions.

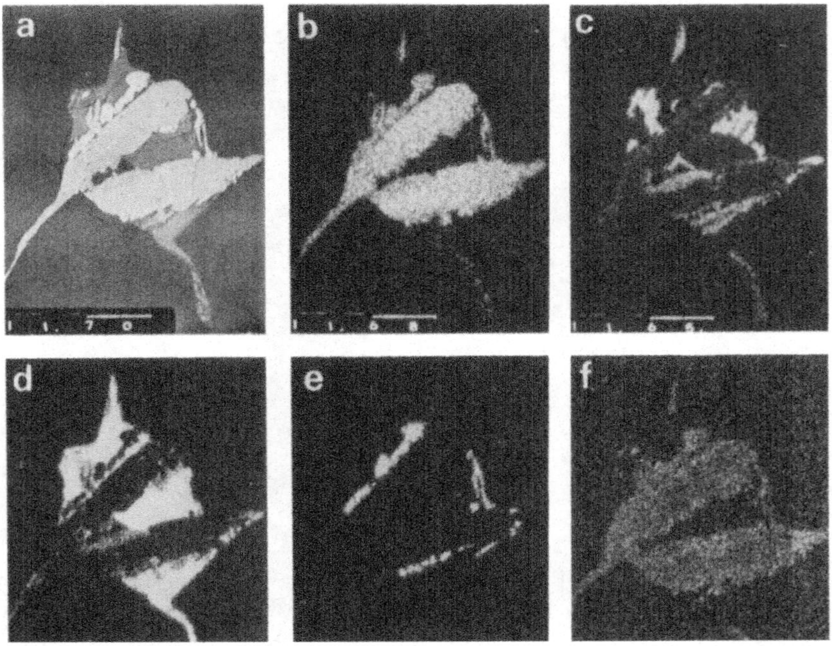

Fig. 19. Microprobe analysis of a typical impurity precipitate: (a) secondary electron image; (b) distribution of Fe; (c) distribution of Al; (d) distribution of Ca; (e) distribution of Ti; (f) distribution of Mn (after Dietl 1983b).

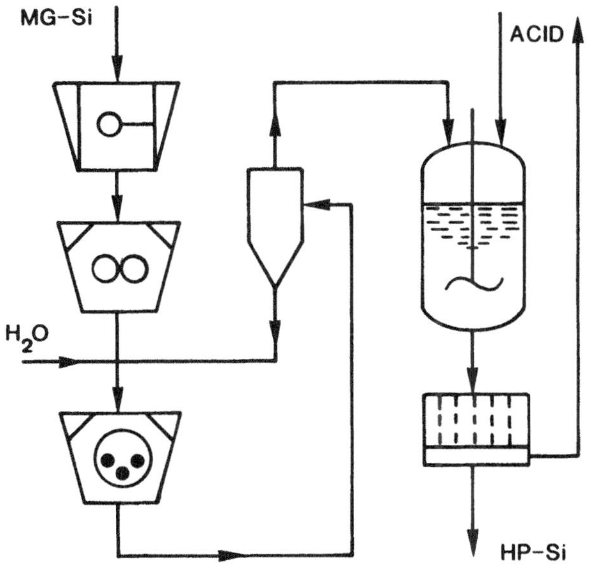

Fig. 20. Flow-chart of the hydrometallurgical refining process yielding silicon of high purity (HP-Si).

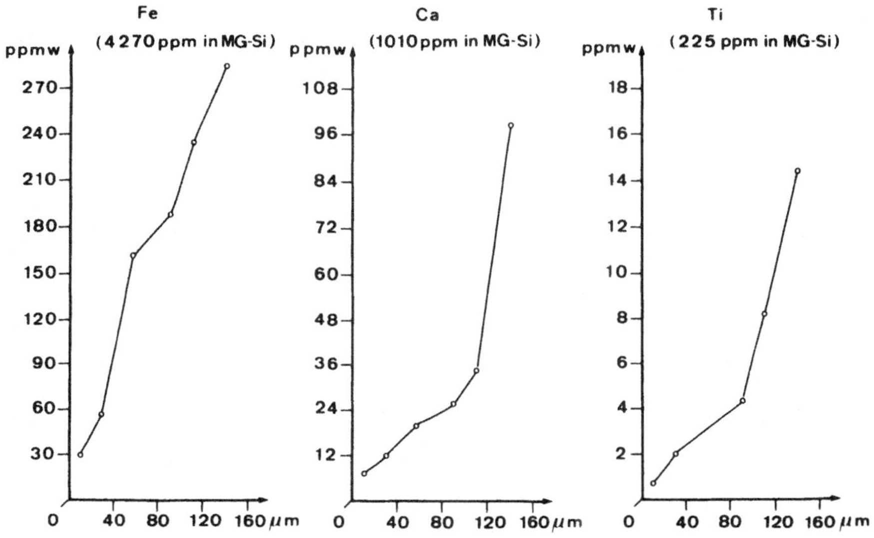

Fig. 21. Effect of particle size on the purification of MG-Si powder by acid leaching (from Dietl 1983b).

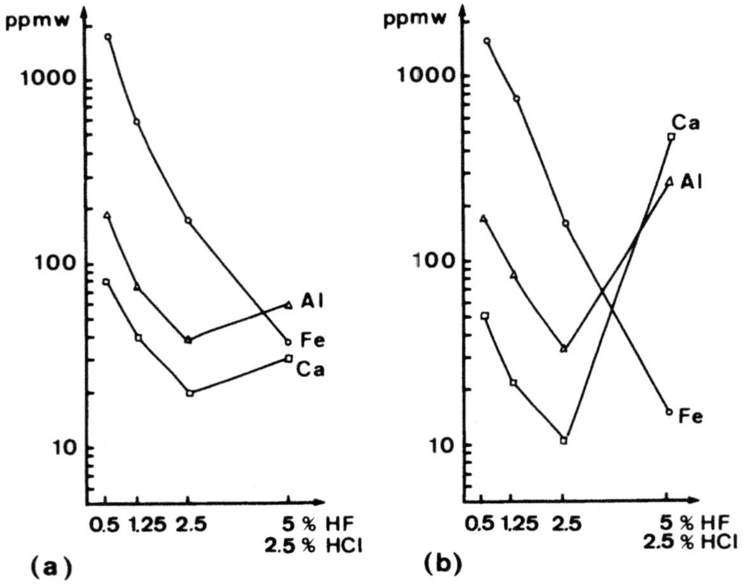

Fig. 22. Influence of the percentage of HF added to a constant 2.5% HCl on the purification of acid leaching: (a) 16 h at room temperature; (b) 2 h at 80°C (after Dietl 1983b).

Leaching with solutions of hydrochloric, hydrofluoric, sulfuric and nitric acid has been tested (see, e.g., the review by Pizzini 1982). Interesting results for most of the impurities have been reported by Dietl (1983b) when using mixtures of hydrochloric and hydrofluoric acid. In fig. 22 the influence of an increasing percentage of hydrofluoric acid added to a constant 2.5% hydrochloric acid is demonstrated for the impurities Fe, Al and Ca. The leaching parameters were 16 h at room temperature in fig. 22a and 2 h at 80°C in fig. 22b. In both cases, the Fe level is continuously reduced using increasing HF concentrations. In comparison with Fe, the removal of Al and Ca is optimum at medium HF concentrations. Whether the decrease in purification at higher HF concentrations is due to the formation of insoluble fluorides is not known yet.

In figs. 23a,b, the influence of leaching time on the purification effect is shown at 20°C and 80°C, respectively. At room temperature the removal of impurities increases with the time of acid treatment. At 80°C this applies only to Fe, whereas Ca and Al show the opposite behavior. The reason for this may be the same as the one assumed for the decrease in purification at higher concentrations of hydrofluoric acid.

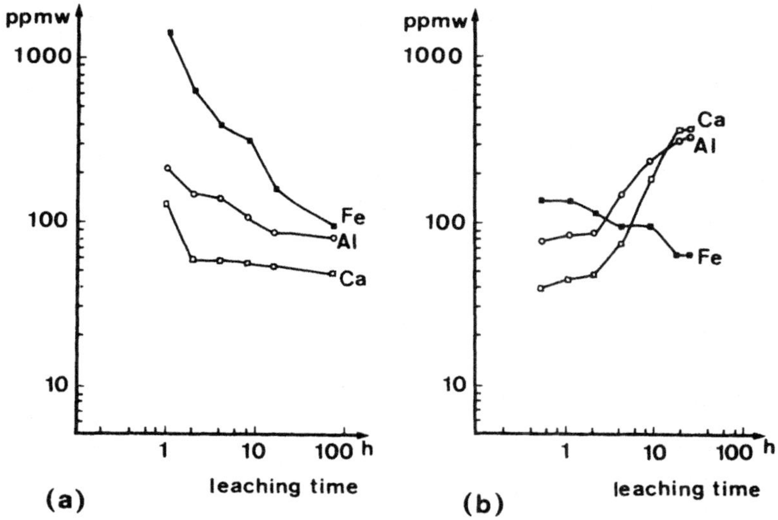

Fig. 23. Influence of leaching time on purification of a relatively coarse MG-Si powder by acid leaching at (a) 20°C and (b) 80°C (after Dietl 1983b).

Table 12
Impurity concentrations (ppma) in MG-Si and acid-leached Si (after Hurrle and Dietl 1981).

Impurity	Concentration (ppma)		Impurity	Concentration (ppma)	
	MG-Si	Leached Si		MG-Si	Leached Si
Li	12	1	Co	2.2	<0.1
Na	18	0.1	Ni	25	<0.6
Mg	45	0.5	Cu	21	1.5
Al	1500	20	Zn	14	<0.5
K	12	0.04	Zr	20	<0.2
Ca	1450	4	Mo	0.6	<0.3
Ti	250	0.2	Ag	0.5	<0.06
V	12	<0.04	Sn	2	<0.06
Cr	17	0.05	Ba	17	<0.6
Mn	350	<1	W	2.3	<0.4
Fe	1800	5	Pb	1.5	<0.04

In table 12 multi-element analysis by spark source mass spectrometry indicates that the predominant impurities in MG-Si (Fe, Al and Ca) can be eliminated down to low ppm concentrations, whereas in the case of Mn, Ti and nearly all the other metal impurities even sub-ppm levels are attainable. However, impurities with unfavourable segregation coefficients, such as the doping elements B and P, are distributed nearly homogeneously in crystalline silicon and cannot be selectively attacked by acid leaching. They have to be removed by subsequent pyrometallurgical treatment or a similar method.

5.2. *Pyrometallurgical processes*

Pyrometallurgy is used to remove impurities from metallurgical-grade silicon in the molten state. Appropriate treatments of such kind include

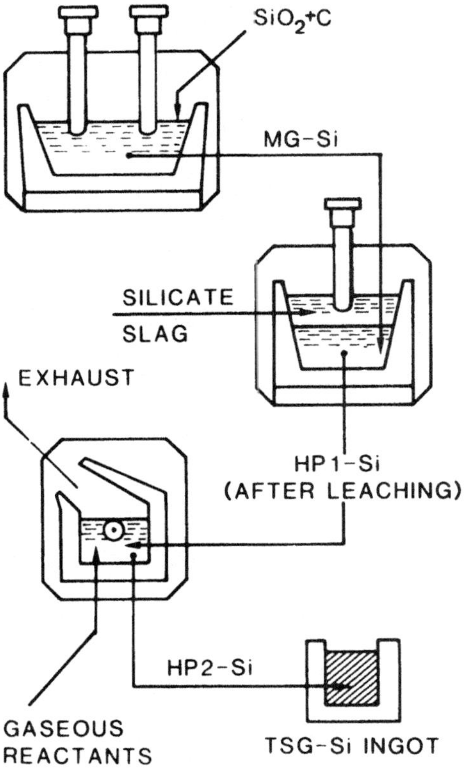

Fig. 24. Schematic sequence of pyrometallurgical refining of MG-Si (for nomenclature see fig. 4).

liquid–liquid extraction with molten slags and liquid–gas extraction using reactive gas blowing or evaporation, as schematically illustrated in fig. 24.

5.2.1. Liquid–liquid extraction

Successful utilization of liquid–liquid extraction requires a slag system which has a greater chemical affinity to impurities than silicon. The slag should be immiscible with molten silicon, which means that besides being chemically stable in the presence of molten silicon it does not dissolve major amounts of elemental silicon or, for its part, contaminate the silicon too much. Slags based on alkaline earth metal silicates have been found

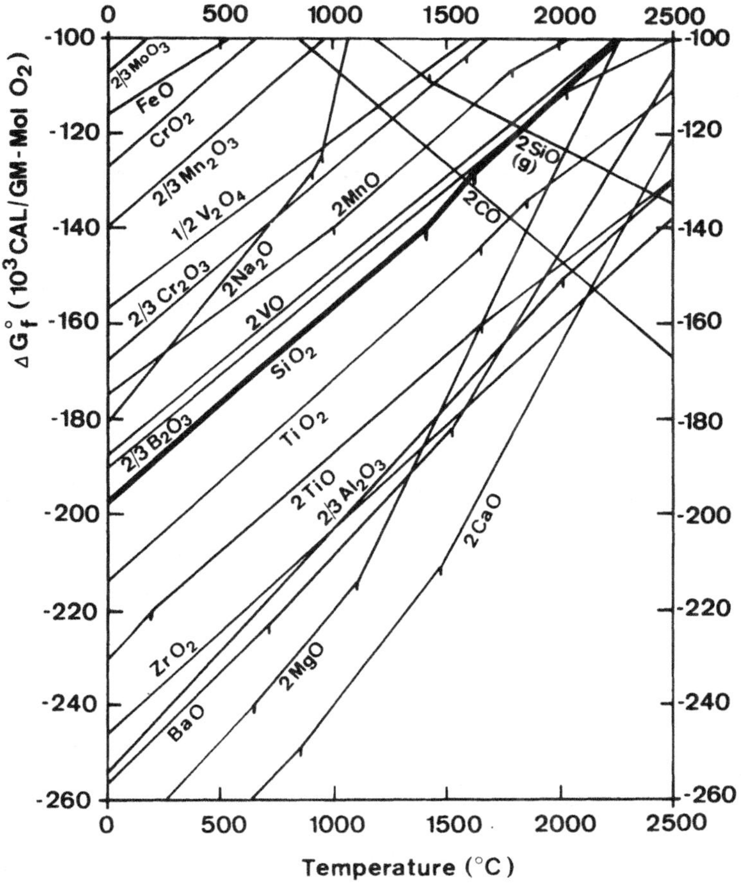

Fig. 25. Standard free-energy of oxide formation of different elements in comparison with SiO_2 (after Wilson 1973).

to fulfil these criteria (see, e.g., Secco d'Aragona 1981, Dietl and Wohlschläger 1979, 1981a, Pelosini et al. 1980).

As discussed by Pizzini (1982), optimization of slagging requires understanding of the high-temperature chemistry of liquid–liquid extraction on the basis of oxidative/reductive interactions. As the elemental impurities are reduced or oxidized at the liquid silicon–slag interface, first-order thermodynamic predictions can be made from the temperature function of standard free-energy values of oxide formation as plotted in fig. 25. Elements forming oxides of higher stability than SiO_2 are expected to get oxidized and transferred into the molten slag. This has proven to be valid for elements such as calcium, magnesium and aluminum. On the other hand, elements with lower oxidation energy such as iron, chromium or manganese tend to accumulate in the silicon melt. A more careful thermodynamic analysis has, in addition, to take into account the formation of both silicates and silicides. With regard to boron and titanium with Gibbs potentials close to that of SiO_2 (see fig. 25) deviations from the above-stated rule become understandable. Titanium as a metal forming silicides of comparatively high stability remains in the silicon melt. As reported by Dietl (1983a) boron, on the other hand, is transferred into the molten slag due to stronger bonding conditions in the silicate phase. The potentiality to remove boron makes the slagging procedure an important step of the sequence for refining metallurgical-grade silicon.

5.2.2. *Liquid–gas extraction*

Liquid–gas extraction is another pyrometallurgical treatment for upgrading MG-Si. Blowing of liquid silicon with gases such as oxygen, chlorine or carbon dioxide has been and is currently employed by several MG-Si manufacturers for achieving higher material qualities. The conceptual approach is to convert the impurities into compounds such as oxides or silicates which can be slagged or which are more volatile than silicon and can be removed by evaporation. The high reactivity of silicon itself, however, limits the effectiveness of reactive gas blowing much more than, for instance, in the case of steel refinement. As silicon is an easily oxidable element, only impurities such as aluminum, zirconium, alkaline earth metals and rare earth metals can selectively be oxidized. As shown in fig. 25, the oxidation of carbon in the presence of molten silicon can be achieved at temperatures above 1500°C only. Blowing with CO_2, for instance, is reported by Strauss (1958) to lower the carbon content in metallurgical-grade silicon.

Enk and Nickl (1961) as well as Muhamed and Fuad (1975) showed that the levels of aluminum and calcium were reduced by blowing with chlorine and oxygen, respectively. To lower the phosphorus concentration, MG-Si has been successfully treated by Ebert and Frank (1953) with oxygen, water, carbon monoxide and carbon dioxide, respectively. The gases listed above were used by Dosaj et al. (1974) in single to three component mixtures together with H_2, SiF_4 and SiF_6 in another experimental set-up. More complicated reactions, such as those with H_2/H_2O mixtures, have

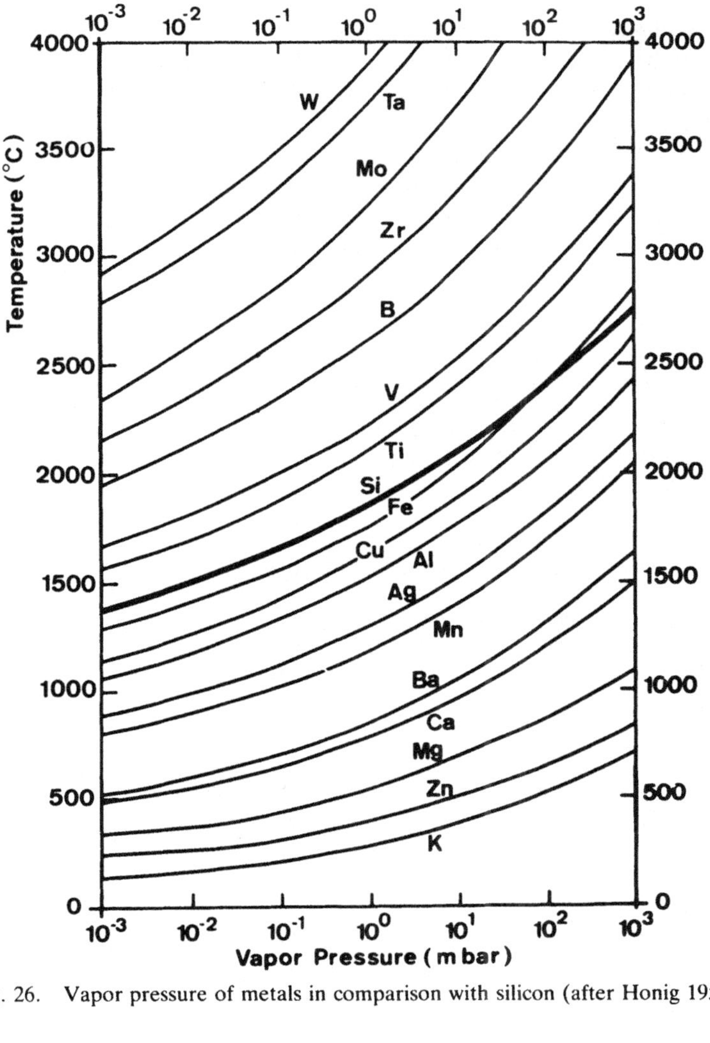

Fig. 26. Vapor pressure of metals in comparison with silicon (after Honig 1953).

been applied by Theuerer (1956) to reduce the boron content in metallurgical-grade silicon. The knowledge of both the chemistry of possible high-temperature reactions, and physical properties such as vapor pressure and solubility of the reaction products in the molten silicon phase is a prerequisite of selecting the proper experimental approach.

Direct evaporation of impurities from molten silicon represents another interesting way of liquid–gas extraction. In fig. 26 vapor pressure data of different metals are plotted versus temperature. Elements with higher vapor pressure than silicon are, in principle, expected to evaporate from the silicon melt. Impurity recovery at 1650°C in the arc furnace for impurity elements such as aluminum, calcium and especially phosphorus – as shown in fig. 10 – has in fact, been found to be inversely related to their vapor pressure. To enhance the evaporation effect it is advantageous to treat the silicon melt under vacuum conditions. Starting from metallurgical-grade levels, the doping impurity phosphorus could reproducibly be reduced by vacuum metallurgy to concentrations below 1 ppm (as claimed by Dietl and Wohlschläger 1986). This is shown in table 13, where impurity

Table 13
Impurity concentrations (ppma) in hydro- and pyrometallurgically refined MG-Si reaching HP2 quality (after Dietl 1983a).

Impurity	Concentration (ppma)		Impurity	Concentration (ppma)	
	MG-Si	Refined Si		MG-Si	Refined Si
Li	10	<0.02	Ni	25	<0.5
B	25	4	Cu	30	<0.05
C	800	<100	Zn	15	<0.5
Na	20	<0.02	Zr	20	<0.2
Mg	50	<0.2	Mo	1	<0.3
Al	1500	0.5	Ag	0.5	<0.06
P	30	0.3	Sn	2	<0.06
K	10	<0.05	Ba	20	<0.6
Ca	1500	<0.2	W	5	<0.4
Ti	250	<0.2	Pb	2	<0.04
V	10	<0.04			
Cr	20	<0.1			
Mn	350	<0.05			
Fe	2000	1.5			
Co	5	<0.1			

concentrations in hydro- and pyrometallurgically refined metallurgical-grade silicon are listed: HP2 quality – as defined in fig. 4 – is gained, with most of impurities concentrations below the detection limits of a spark source mass spectrometer.

5.2.3. Auxiliary materials

In refining silicon by treating it in the molten state, one has to consider the gradual decrease of the impurity level through extractive purification against the increase of contaminants originating from the crucible and other auxiliary materials. In fig. 27 the interaction of these two impurity situations versus time of liquid–liquid extraction and liquid–gas extraction, respectively, is schematically indicated. To limit contamination, auxiliary materials and the surface to volume ratio of the silicon melt have to be brought to optimum conditions.

A number of materials proposed to build crucibles for melting silicon were tested by Revel et al. (1983). Quartz is likely to be one of the least contaminating materials. Its utilization, however, is limited to pure silicon melts. Recrystallization of silica glass due to metal impurities and, especially, the interaction with molten slags cause rapid destruction of the crucible. Graphite with a density larger than 1.75 g/cm^3 was found to be a very attractive material because of its machinability and its multiple utilization. However, graphite reacts with silicon under formation of SiC particles. These may act as deleterious impurity inclusions and, moreover,

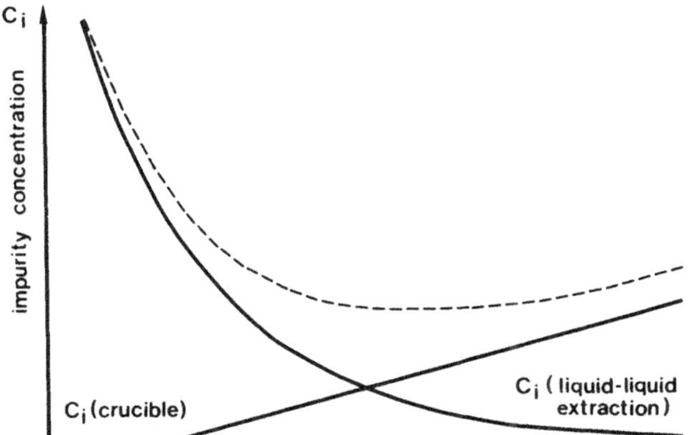

Fig. 27. Extractive purification of Si in the molten state and contamination from auxiliary materials (from Dietl et al. 1981).

disturb crystal growth by stray nucleation. As found out by Goma (1983), the formation of SiC can be largely reduced by a lamellar pyrocarbon coating without crystallographic irregularities. In the case of liquid–liquid extraction, Dietl and Kotilge (1983) claimed a layer of molten slag to protect auxiliary equipment of graphite from being attacked by the silicon melt.

Silicon nitride represents another material – proposed, e.g., by Duffy et al. (1980) – which is resistant to molten silicon. Sintered silicon nitride must be rich in β-phase and dense enough to reduce thermal-expansion mismatch and surface porosity. The reactivity of silicon towards silicon nitride of impure quality, being customary in trade, leads to sticking effects and therefore complicates the re-use of this kind of material. As shown by Sirtl and Rath (1980) and Saito et al. (1981), it is possible to rather use silicon nitride coatings fabricated by CVD or by depositing nitride powder of high purity on substrate materials such as graphite or fused silica.

Up to now all the efforts to find auxiliary materials being suitable for use in pyrometallurgical purification of silicon yielded refractory materials based exclusively on the silicon compounds silicon oxide, silicon nitride and silicon carbide. This also applies, in principle, to carbon, as this material in its Si-resistant qualities protects itself by forming a dense layer of silicon carbide.

5.3. Recrystallization processes

Purification by recrystallization represents a very effective procedure in the field of pyrometallurgical refinement of silicon. Both directional solidification of silicon from its own melt and recrystallization from metal solution make good use of the segregation behavior of most of the impurity elements.

5.3.1. Recrystallization from aluminum

While metals such as tin, zinc and silver may be employed, aluminum is the preferred material for the solvent refining technique. Recrystallization from aluminum solution is an old and well-known purification procedure (see, e.g., Vigouroux 1899) which, nevertheless, has been patented more recently by Kotval and Strock (1980) at the Union Carbide Corporation, and Dawless (1981) at the Aluminum Company of America. The complete low-cost, low-energy process for upgrading of MG-Si includes three major steps:

- Fractional crystallization of silicon-aluminum alloys containing 40–60% metallurgical-grade silicon.

- Separation of silicon crystals having the habitus of platelets from still molten aluminum–silicon eutectic.
- Removal of residual aluminum by acid leaching.

The well-known silicon–aluminum phase diagram in the right-hand figure of fig. 28 shows that essentially pure silicon can be precipitated from aluminum–silicon alloys with silicon contents greater than 12.6 wt %. Silicon platelets are formed by cooling a homogeneous silicon–aluminum alloy to a temperature well above the eutectic temperature of 577°C. Cooling rates of about 60°C/hr yield platelets of 5–10 mm. The solvent refining step is carried out in graphite crucibles, but other non-reactive materials, such as refractory oxides, can also be employed. Using a centrifugal filter is one of the different techniques described for separating the silicon platelets from the eutectic melt. The adherent material – mainly residual eutectic – is removed from the platelets by leaching with aqueous HCl solutions.

In table 14 an analysis of silicon from aluminum-based refining of MG-Si indicates transition-metal impurities of less than 5–10 ppma each, and boron and phosphorus as low as 1 ppma and 0.5 ppma, respectively. The remaining aluminum concentration of about 300 ppma corresponds to the solid solubility as depicted on the left-hand figure of fig. 28. As claimed by Kotval and Strock (1980), the leached material thereafter is melted in contact with a pure silica slag to lower the aluminum level. The slagged silicon is removed from the melting zone in a unidirectional manner, thus yielding effective further purification by directional solidification in combination with liquid–liquid extraction.

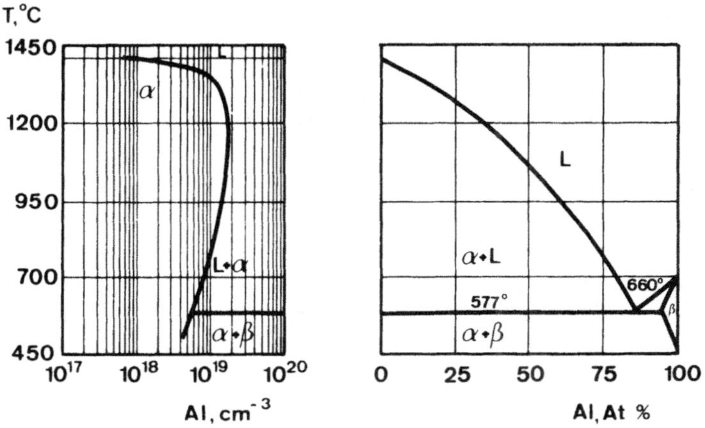

Fig. 28. Binary phase diagram of the Al–Si system (left) and solid solubility of Al in Si (right) (after Glasow and Semskow 1970).

Table 14
Impurity concentrations (ppma) in Si refined by recrystallization from Al solution (after Smith and Dawless 1981).

Impurity	Concentration (ppma)	
	MG-Si	Al-based refining
Al	1600	300
B	45	1
P	30	0.3
Cr	140	<5
Fe	2100	<10
Mn	70	<5
Ni	50	<10
Ti	165	<10
V	100	<5
Zr	–	<1
Cu	–	<5

5.3.2. Directional solidification

With regard to purification of more or less upgraded silicon by recrystallization from its own melt, two kinds of segregation should be differentiated: During solidification of polycrystalline MG-Si impurities with favourable segregation coefficients become accumulated by 'second-phase segregation' at the grain boundaries from where they can be removed by hydrometallurgical treatment. Directional solidification of partially refined silicon, on the other hand, advantageously utilizes macrosegregation.

In fig. 29 critical impurity levels in a simulated silicon meltstock are calculated from threshold values for solar cell application. In the column on the right, permissible concentrations c_s for impurity elements in p-type silicon are marked. From there the equation for impurity distribution between crystal and melt up to a solidified fraction of $g = 0.9$ leads to impurity concentrations c_L in the starting meltstock on the left. The calculations have been done with segregation coefficient values k_{eff} as obtained by Davis et al. (1980) from Czochralski crystals grown at 7 cm/hr. Bearing in mind that 5×10^{16} atoms/cm^3 equals 1 ppma, it is apparent that the calculated impurity levels in the meltstock in fig. 29 range from less than 1 ppma to somewhat above 100 ppma. The levels of nickel, chromium, copper and molybdenum even exceed those being present in typical metallurgical-grade silicon as depicted in table 2. All other impurities require at least one prior refining step to reach solar-grade levels via directional solidification when starting from metallurgical-grade quality.

Various methods have been claimed to be useful for refining silicon by directional crystallization. Techniques currently utilized such as float zoning [FZ] (as reported by Ludsteck 1981), Czochralski pulling [CZ] (as reported by McCormick 1980), heat exchanger method [HEM] (as reported by Khattak et al. 1981), Bridgman growth (as reported by Amick

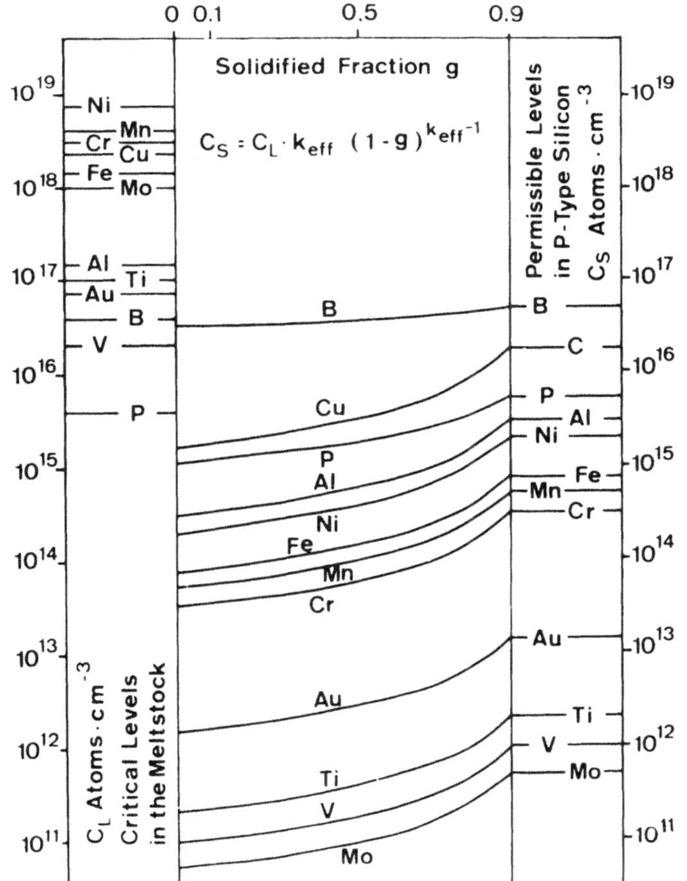

Fig. 29. Calculation of critical feedstock impurity levels from threshold values for solar cell degradation (after Dietl 1983a).

et al. 1983a), SILSO casting (as reported by Helmreich and Ast 1980) and edge-supported pulling [ESP], one of the ribbon-growth techniques (as reported by Ciszek et al. 1981), are capable of significantly purifying pre-refined silicon. The chart in fig. 30 shows the different situations possible in the field of 'solar-grade synthesis'. Starting from silicon qualities such as DAR (direct arc reacted products from high-purity arc furnace operation), HP1 and HP2 (more or less advanced stages of MG-Si refining) or from electrolytical processes, TSG-Si can be reached by different methods utilizing directional solidification. Some combinations will, however, have a limited application. By using semiconventional solidification, one has to end up with semiconventional economy. Considering the quality of partially purified starting material it remains to be seen which quality criteria may guarantee sufficiently high values of conversion efficiency. Elements with unfavourable segregation coefficients such as boron and phosphorus have to exist here at impurity levels which are acceptable to the skilled device maker.

5.4. *Refining sequences*

As pointed out a variety of potential, more or less elaborated, refining processes aimed at producing bulk-quality solar-grade silicon can be seen. The final combination of procedures to upgrade silicon will certainly not involve all the steps discussed. Differences in the sequence of refining processes are mainly caused by the different qualities of the starting material, but also by the purity requirements of both the crystallization method and cell technology finally being applied. In the following, only a small

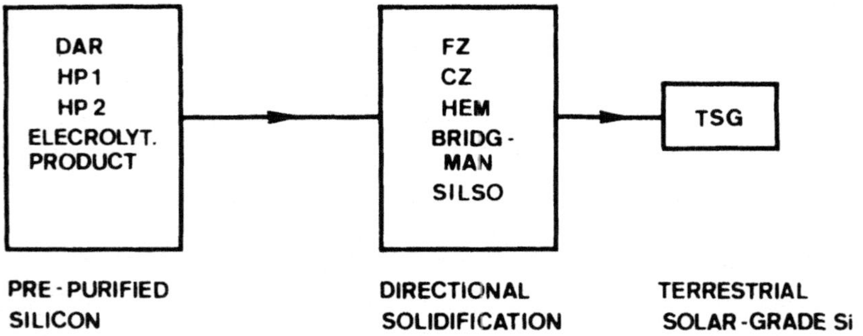

Fig. 30. 'Solar-grade synthesis' from partially refined silicon by directional solidification (after Sirtl 1982).

number of designed basic low-cost refining sequences is given as an illustration.

A process claimed by Pelosini et al. (1980) starts with liquid–liquid extraction of metallurgical-grade silicon with a barium oxide-based slag in conjunction with oxidative gas blowing and ends up with a hydrometallurgical step.

Another processing sequence, based on solvent refining, involves the following sequence (see Hanoka et al. 1981): Precipitation of silicon platelets from solution of MG-Si in aluminum, acid leaching of the silicon platelets, melting of the platelets in contact with a silica-based slag of high purity and directional solidification of the silicon–slag melt. The central part of this aluminum-based refining is also under investigation at Wacker as an integrated step in the sequence of the aluminothermic reduction of quartz (see section 4.3.1).

Moreover, differences in refining concepts may be caused by commercially available auxiliary materials or by accidently existing technical equipments. Khattak and Schmid (1983) described in situ refinement of different qualities of metallurgical-grade silicon by the heat exchanger method (HEM). Simultaneous application of vacuum treatment, slagging, moist-argon blowing through the melt and directional solidification reduced metallic impurities to less than 1 ppm levels. These procedures were also effective in reducing the boron and the phosphorus content as well as the amount of silicon carbide particles.

As a last example, a process is claimed by Dietl and Wohlschläger (1981b) for purifying metallurgical-grade silicon for the manufacture of solar cells, wherein acid treatment of silicon powder is combined with vacuum evaporation. Depending on the purity of the starting material, this may be optionally supplemented by liquid–liquid extraction and directional solidification.

6. Metallurgical routes to solar-grade silicon

To design a basic low-cost route to gain solar-grade silicon meltstock, cost, availability and purity of the materials to be used, utilization or recycling of by-products and the amount of investment as well as energy consumption have to be considered. The simpler the basic concept and the lower the number of necessary processing steps, the better, of course, are the chances to reach this goal.

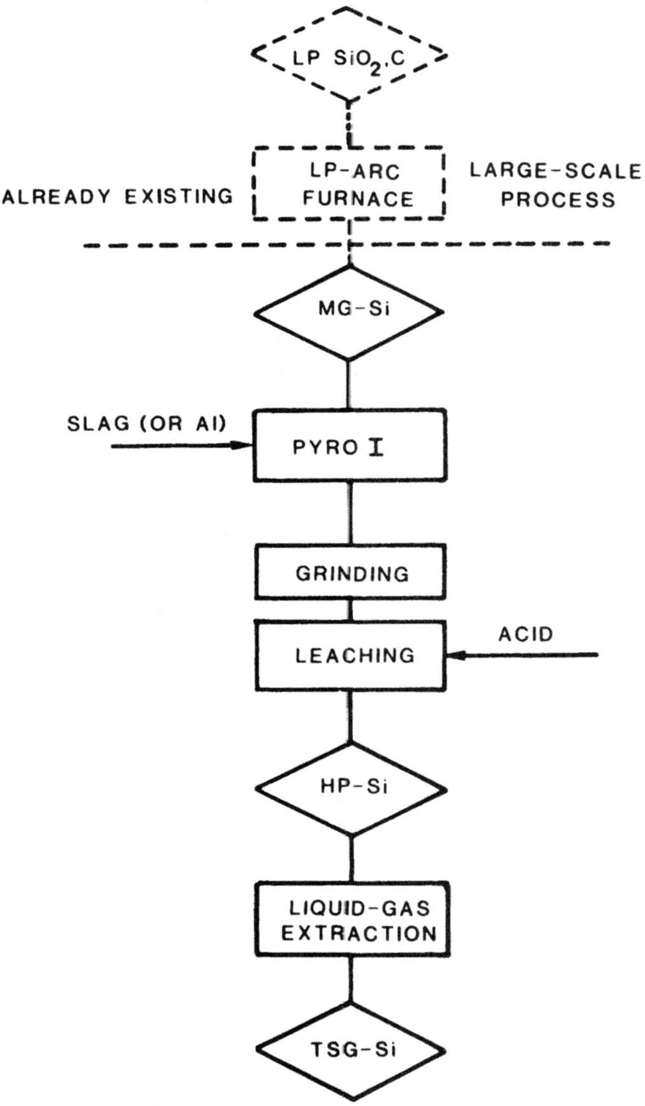

Fig. 31. Schematic flow-fiagram of the MG-Si approach (Wacker): LP, low purity; MG, metallurgical-grade; HP, high purity (advanced stage of refining); TSG, terrestrial solar-grade.

6.1. *Conceptual alternatives*

Real low-cost routes to solar-grade silicon start either from metallurgical-grade silicon or from improved qualities of quartz. This, in summary, represents the result of numerous process studies over the past decade. The most evident idea to process metallurgical-grade silicon further by metallurgical refening techniques competes with the concept to create a much higher purity situation in an improved arc furnace system.

In fig. 31 the flow-chart of an optimized refining cycle based on metallurgical-grade silicon is schematically drawn. Wacker in Germany and, somewhat modified, also Heliosil in Italy (as reported by Pizzini 1982) evaluate combinations of metallurgical refining steps. Commercially available metallurgical-grade silicon requires one first pyrometallurgical upgrading by slagging or, alternatively, by solvent extraction (Al-based refining). After grinding and acid leaching high-purity quality is gained as an advanced stage of refinement. To reach TSG quality, a second pyrometallurgical step by liquid–gas extraction is needed. Depending on the type of crystallization technique and device processing, final purification by in situ or additional directional solidification may be applied.

In figs. 32 and 33 flow-charts of sophisticated modifications of the standard arc furnace process are shown. Whereas Dow Corning/Elkem and later Elkem/Exxon (see section 4.2.2) concentrated on the use of selected high-purity quartz and carbon black, Siemens more recently developed a process based on effectively purified quartz sand and carbon (see section 4.3.2). Since in both cases finely powdered starting materials are applied, the specific reactions in the arc furnace require pelletizing or pressing into briquettes of the powders with the aid of high-purity binders.

A comparison of the three different flow-charts in figs. 31–33 reveals a shift of purification expenditure from the reduction product towards the starting materials. Whereas in the conception of MG-Si refinement, the first pyrometallurgical step is discussed to be at most accomplished in connection with the reduction process, in the Siemens approach upgrading of the materials predominantly takes place before being utilized in the arc furnace. The Elkem/Exxon approach represents an intermediate situation in terms of preceding and subsequent refinement. As a consequence, purity requirements concerning auxiliary materials such as pellet binder, refractory materials of construction and electrodes for the arc process are the more stringent the less the actual conception provides purification in the subsequent processing steps.

As far as information is available, the approaches of strongly improved

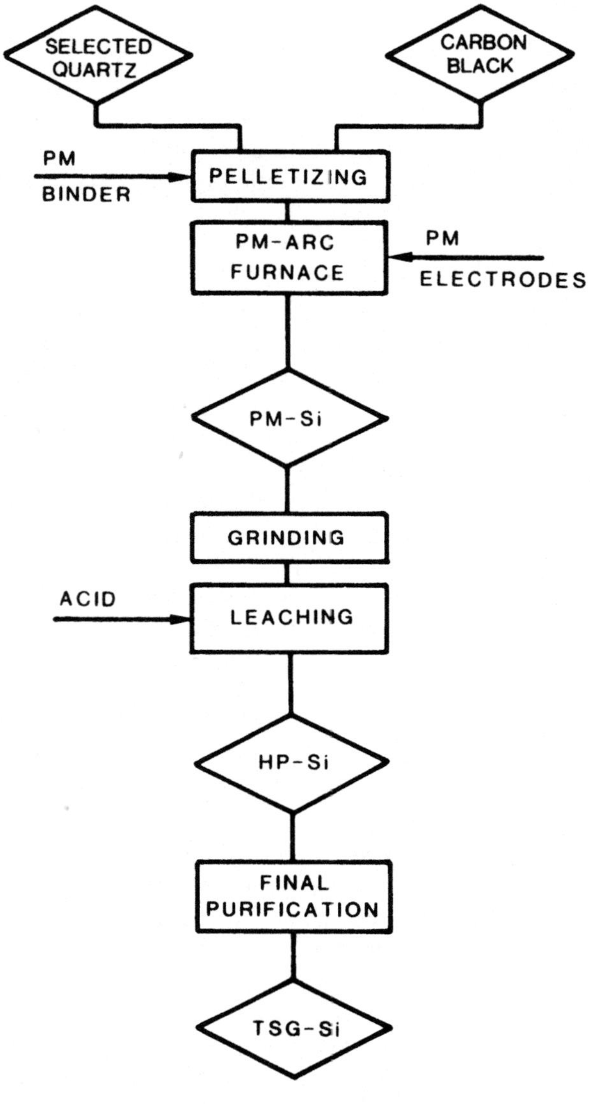

Fig. 32. Schematic flow-diagram of the Elkem/Exxon approach: PM, pure metallurgical; HP, high purity; TSG, terrestrial solar-grade.

versions of the arc furnace process seem to offer an attractive alternative to at least the major part of the refining sequences of standard metallurgical-grade silicon as described in section 5.4. It ought to be mentioned, however, that the carbon content in silicon is above saturation level, irrespective of the way the arc furnace process is modified. The answer to the

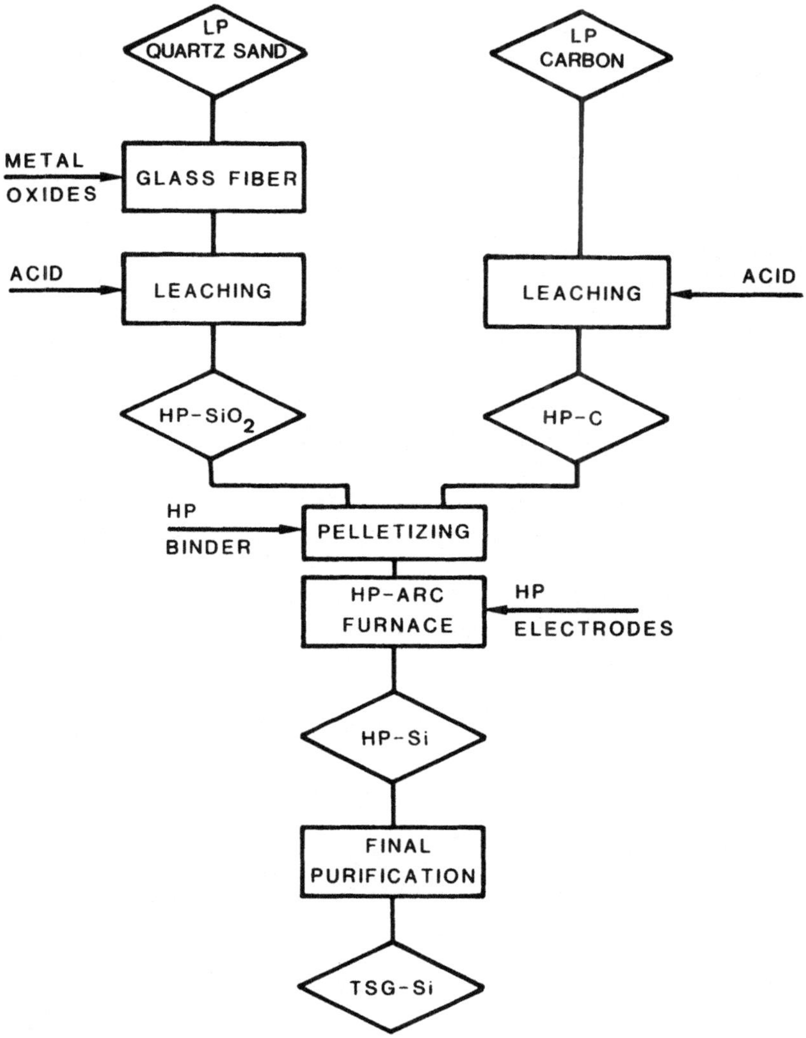

Fig. 33. Schematic flow-diagram of the Siemens approach: LP, low purity; HP, high purity; TSG, terrestrial solar-grade.

question, whether preceding purification of the starting materials quartz and carbon or extensive refining of commercially available metallurgical-grade silicon will be the better alternative, in terms of material quality and production costs, remains open. That is, in particular, caused by the fact that silicon products obtained from both improved arc furnace processes still need additional upgrading, at least by means of directional solidification, to achieve solar-grade quality.

6.2. *Economical aspects and future trends*

The meltstock material used and being used up to around 1985 exclusively has its origin in the conventional chemical vapor deposition (CVD) process. As predicted by Sirtl (1983) semiconventional feedstock material in the 20–30 $/kg price range could be available in the future in medium quantities through a, more or less, modified CVD technology. Whereas Wacker plans to obtain this semiconventional material by scaling-up deposition rates, Dow Corning and Union Carbide try to utilize higher conversion rates of catalytically synthesized silanes. To have such a starting material available will be of great importance for the developing photovoltaic industry. Silicon from an optimized refining cycle based on metallurgical-grade silicon or a sophisticated high-purity arc furnace operation should be seen in the price range of 5–10 $/kg. No process of this type, however, has a chance to reach industrial maturity prior to the end of the decade.

Silicon at costs of 10 $/kg and below will only be realized by applying unconventional methods in every particular subtechnology. To design a basic low-cost material technology the following guidelines have to be considered:

- Costs and availability of both starting materials and auxiliary materials;
- Minimization of the number of refining steps at lowest possible energy consumption;
- Maximization of throughput in low-cost technology equipments;
- Utilization of by-products and environmental compatibility.

With regard to by-products, at least three different situations have to be assessed: The first one is to obtain valuable by-products of great demand. The second situation is one of increasing importance, namely complete recycling. The third one is silicon production generating worthless by-products which may be discarded without major environmental restrictions. Many a process has looked quite different in terms of cost balance when treated in such a manner.

The minimum amount of investment per annual unit output is another factor of major impact on total costs. The simpler the design and the lower the number of necessary processing steps, the better the chance for producing real low-cost solar-grade silicon. With certainty, proper choice of both heating systems and cycles as well as time of residence during high-temperature reactions will be of considerable influence. The various forms of the silicon product to be transferred to the crystallization area is of additional importance. From merely economical considerations, the simple transport of liquid silicon to the crystallization equipment appears to be the optimum case unless the process of directional solidification is required for the final purity level. Chunks from there or powder from different refining steps may be two more acceptable product forms.

The time necessary to operate the solar cell so as to pay back the energy required to its manufacture is roughly 10 years for cells based on 'conventional terrestrial' technologies. Improvements in crystal growth technology and ingot sawing, and, in particular, replacement of the Siemens process by a basic low-cost process were predicted by Hunt (1976) to reduce energy pay-back time to at most twelve months including a material's share of about six months. This prognosis can be within reach over different routes if cost-consciousness is observed in the design of every particular processing step, since energy consumption and economics have to be widely considered of the same value.

7. Conclusion

Providing a low-cost base material is a fundamental requirement of large-scale solar cell fabrication as an option for future energy generation. Regarding favorable physical properties and practically unlimited availability, silicon has unambiguously shown its superiority for photovoltaic application. As pointed out in this survey, a limited number of high-potential processes aimed at producing terrestrial solar-grade silicon can be seen.

Upgrading of metallurgical-grade silicon represents one of the most promising ways of silicon meltstock processing. Silicon from metallothermic processes or from improved arc furnace processes utilizing raw materials of high purity turned out to be interesting alternatives. All products of different processes for quartz reduction require more or less severe further purification to reach solar-grade quality. To design a basic low-cost refining concept, cost and availability of starting and auxiliary materials to be used as well as amount of investment, throughput and energy consumption have to be considered.

The main advantage of the metallurgical ways of silicon meltstock processing is to produce impurity optimized silicon at lowest possible costs. To transfer this initial advantage to the final goal of large-scale terrestrial solar cell production, crystallization processes and cell technologies of adequate low-cost character are demanded. Certainly, the photovoltaic industry will develop multi-process/multi-purpose systems and, thus create a large spectrum of material specifications. This, in turn, will lead to competition, a sound basis for any kind of growing business. At the time being, technical feasibility as well as economy of the different metallurgical routes have to be investigated by further research and development work via gradually enlarged pilot stages.

8. Summary

In this chapter a survey is given on the present status of silicon meltstock processing for terrestrial solar cell application. After dealing with detrimental effects of typical impurities, specification of terrestrial solar-grade silicon and its dependence on subsequent processes are discussed. Material characterization is considered in the light of diagnostic needs, leading to methods of skilled impurity analysis. Various technical conceptions based on metallurgical processes are examined, considering their long-range assessment of fundamental low-cost potentiality. Approaches in chemical preparation of solar silicon by means of metallothermic processes are sketched as well as electrowinning and electrorefining. Major emphasis is put on currently investigated methods in both the production of silicon of higher purity from improved arc furnace operation and the refining of commercially available metallurgical-grade silicon. In this framework, the family of real low-cost refining processes consisting of hydrometallurgical treatment and different means of pyrometallurgy is accentuated.

Acknowledgement

The author is strongly indebted to Prof. Dr. E. Sirtl for many valuable suggestions and stimulating discussions. Furthermore, acknowledgement is directed to my colleagues Dr. D. Helmreich, Dr. C. Holm, Dr. D. Seifert and Dr. H. Seiter for helpful contributions to this survey article.

References

Aas, H.H., and J.A. Kolflaath, 1974, US Patent 3,809,548 (May 7, 1974) assigned to Christiana Spiegerverk (Norway).

Abe, T., S. Kikuchi, S. Shirai and S. Muroaka, 1981, Semiconductor Silicon (Electrochemical

Society, Pennington, NJ) p. 54.
Amick, J.A., J.P. Dismukes, R.W. Francis, L.P. Hunt, P.S. Ravishankar, M. Schneider, K. Mathei, R. Sylvain, K. Larsen and A. Schei, 1983a, Proc. Symp. Materials and New Processing Technologies for Photovoltaics, eds J.A. Amick, V.P. Kapur and J. Dietl (Electrochemical Society, Pennington, NJ) p. 67.
Amick, J.A., J.P. Dismukes, P.S. Ravishankar, M. Schneider, J. Blake, K. Mathei, R. Sylvain, G. Halvorsen, K. Larsen and A. Schei, 1983b, Proc. 5th EC Photovoltaic Solar Energy Conf., Kovouri, Athens (Reidel, Dordrecht) p. 1068.
Aulich, H.A., 1983, Proc. 5th EC Photovoltaic Solar Energy Conf., Kovouri, Athens (Reidel, Dordrecht) p. 936.
Aulich, H.A., W. Dietze, K.H. Eisenrith, J. Schäfer, F.-W. Schulze and H.-P. Urbach, 1982a, 4th EC Photovoltaic Solar Energy Conf., Stresa (Reidel, Dordrecht) p. 868.
Aulich, H.A., K.H. Eisenrith, H.-P. Urbach and J.G. Grabmaier, 1982b, Proc. 3rd Symp. Materials and New Processing Technologies for Photovoltaics, eds J.P. Dismukes, E. Sirtl, P. Ray-Choudhury and L.P. Hunt (Electrochemical Society, Pennington, NJ) p. 177.
Breneman, W.C., E.G. Farrier and H. Morihara, 1978, Proc. 13th IEEE Photovoltaic Spec. Conf., Washington DC (IEEE, New York) p. 339.
Camanzi, A., G. Capannesi, G. Grillo, M. Vittoris and F. Alessandrini, 1983, Proc. 5th EC Photovoltaic Solar Energy Conf., Kavouri, Athens (Reidel, Dordrecht) p. 60.
Ciszek, T.F., 1985, Crystal Growth of Electronic Materials, ed. E. Kaldis (North-Holland, Amsterdam) p. 185.
Ciszek, T.F., M. Schietzelt, L.L. Kazmerski, J.L. Hurd and B. Fernelius, 1981, Proc. 15th IEEE Photovoltaic Spec. Conf., Orlando (IEEE, New York) p. 581.
Craven, R.A., 1981, Semiconductor Silicon (Electrochemical Society, Pennington, NJ) p. 254.
Davis, J.R., A. Rohatgi, R.H. Hopkins, P.D. Blais, P. Rai-Choudhury, J.R. McCormick and H.C. Mollenkopf, 1980, IEEE Trans. Electron Devices **ED-24** (4), 677.
Davis, J.R., R.H. Hopkins and A. Rohatgi, 1982, Proc. Symp. Materials and New Processing Technologies for Photovoltaics, eds J.P. Dismukes, E. Sirtl, P. Ray-Choudhury and L.P. Hunt (Electrochemical Society, Pennington, NJ) p. 177.
Dawless, R.K., 1981, US Patent 4,246,249 (Jan. 20) assigned to Aluminium Company of America.
De Mattei, R.C., D. Elwell and R.S. Feigelson, 1981, Proc. Symp. Electrocrystallization, eds R. Weil and R.G. Barradas (Electrochemical Society, Pennington, NJ) p. 323.
Dietl, J., 1981a, Proc. Symp. Materials and New Processing Technologies for Photovoltaics, eds J.A. Amick, E. Sirtl, P. Ray-Choudhury and J.P. Dismukes (Electrochemical Society, Pennington, NJ) p. 48.
Dietl, J., 1981b, Extend. Abstr. Electrochemical Society **81-2**, Abstr. No. 475, 1143.
Dietl, J., 1983a, Proc. Symp. Materials and New Processing Technologies for Photovoltaics, eds J.A. Amick, V.K. Kapur and J. Dietl (Electrochemical Society, Pennington, NJ) p. 52.
Dietl, J., 1983b, Sol. Cells **10**, 145.
Dietl, J., and J. Kotilge, 1983, DE (Germ. Pat. Appl.) 32 08 877 (Sept. 22) assigned to Wacker.
Dietl, J., and M. Wohlschläger, 1979, DOS (Germ. Pat. Appl.) 27 29 464 (Nov. 1) assigned to Wacker.
Dietl, J., and M. Wohlschläger, 1981a, DOS (Germ. Pat. Appl.) 29 33 164 (Feb. 26) assigned to Wacker.
Dietl, J., and M. Wohlschläger, 1981b, US Patent 4,304,763 (Dec. 8) assigned to Wacker.

Dietl, J., D. Helmreich and E. Sirtl, 1981, Solar silicon, in: Crystals: Growth, Properties and Applications, Vol. 5 (Springer, Berlin) p. 43.
Dietl, J., C. Holm and E. Sirtl, 1982a, US Patent 4,312,850 (Jan. 26) assigned to Wacker.
Dietl, J., C. Holm and E. Sirtl, 1982b, Proc. 4th EC Photovoltaic Solar Energy Conf., Stresa (Reidel, Dordrecht) p. 941.
Dietze, W., H.J. Frenzl and J. Schäfer, 1982, Extend. Abstr. Electrochemical Society **82-2**, Abstr. No. 245, 394.
Dosaj, V.D., L.P. Hunt and L.D. Crossman, 1974, Proc. 11th IEEE Photovoltaic Spec. Conf., Scottsdale (IEEE, New York) p. 275.
Duffy, M.T., S. Berkman, G.W. Gullen, R.V D'Aiello and H.J. Moss, 1980, J. Cryst. Growth **50**, 347.
Ebert, H., and K. Frank, 1953, DBP (Germ. Pat.) 10 29 163 (May 8) assigned to Knappsack AG.
Elwell, D., and R.S. Feigelson, 1982, Sol. Energy Mater. **6**, 123.
Enk, E., and J. Nickl, 1961, DBP (Germ. Pat.) 10 98 931 (Oct. 19) assigned to Wacker.
ERDA Report, 1976, Photovoltaic Conversion Program, Summary Report, Energy Research & Development Administration, Washington, DC (ERDA – 76 – 161).
Fischler, S., 1962, J. Appl. Phys. **33**, 1615.
Freiesleben, W., 1980, Proc. 3rd EC Photovoltaic Solar Energy Conf., Cannes (Reidel, Dordrecht) p. 166.
Glasow, W.M., and W.S. Semskow, 1970, Die Germanium- und Silicium-Zweistofflegierungen (VEB Deutscher Verlag der Wissenschaften, Berlin) p. 84.
Goma, J., 1983, Doctorat d'Etat Orléans (Thesis) May 5.
Gutsche, W.H., and D.E. Hill, 1976, Final Report DOE/JPL - 954 338.
Hamakawa, Y., 1981, Proc. 3rd EC Photovoltaic Solar Energy Conf., Cannes (Reidel, Dordrecht) p. 22.
Hanoka, J.I., H.B. Strock and P.S. Kotval, 1981, J. Appl. Phys. **52**, 5829.
Helmreich, D., and G. Ast, 1980, Proc. Symp. Electronic and Optical Properties of Polycrystalline or Impure Semiconductors and Novel Silicon Growth Methods, eds K.V. Ravi and B. O'Mara (Electrochemical Society, Pennington, NJ) p. 233.
Helmreich, D., and H. Seiter, 1979, Proc. 2nd EC Photovoltaic Solar Energy Conf., Berlin (Reidel, Dordrecht) p. 742.
Holm, C., 1982, unpublished results.
Holm, C., and E. Sirtl, 1981, J. Cryst. Growth **54**, 253.
Holm, C., and E. Sirtl, 1986, to be published.
Honig, R.E., 1953, RCA-Rev. 1953, p. 195.
Hopkins, R.H., R.G. Seidensticker, J.R. Davis, P. Rai-Choudhury, P.D. Blais and J.R. McCormick, 1977, J. Cryst. Growth **42**, 493.
Hopkins, R.H., J.R. Davis, A. Rohatgi, R.B. Campbell, P.D. Blais, P. Rai-Choudhury, R.E. Stapleton, H.C. Mollenkopf and J.R. McCormick, 1980, Progr. Rep. DOE/JPL – 954 331 – 80/9.
Hopkins, R.H., J.R. Davis, A. Rohatgi and J.R. McCormick, 1982, Final Report, Silicon materials Task, DOE/JPL – 954331 – 78/3.
Hunt, L.P., 1974, Proc. 11th IEEE Photovoltaic Spec. Conf., Scottsdale (IEEE, New York) p. 259.
Hunt, L.P., 1976, Proc. 12th IEEE Photovoltaic Spec. Conf., Baton Rouge (IEEE, New York) p. 347.
Hunt, L.P., 1983, Proc. Symp. Materials and New Processing Technologies for Photovoltaics, eds J.A. Amick, V.K. Kapur and J. Dietl (Electrochemical Society, Pennington, NJ) p. 154.

Hunt, L.P., and V.D. Dosaj, 1979, Final Report DOE/JPL – 954 559–78/7 (Oct. 1979).
Hunt, L.P., V.D. Dosaj, J.R. McCormick and A.W. Rauchholz, 1978, Proc. 13th IEEE Photovoltaic Spec. Conf., Washington DC (IEEE, New York) p. 333.
Hunt, L.P., J.P. Dismukes and J.A. Amick, 1983, Proc. Symp. Materials and New Processing Technologies for Photovoltaics, eds J.A. Amick, V.K. Kapur and J. Dietl (Electrochemical Society, Pennington, NJ) p. 106.
Hurrle, A., and J. Dietl, 1981, Fresenius Z. Anal. Chem. **309**, 277.
Jasper, H.-P., 1983, Int. Ferro Alloys Conf. (INFACON), Techn. Session II, Tokyo.
Kapur, V.K., L. Nanis and A. Sanjurjo, 1979, Extend. Abstr. Electrochem. Soc. **79-1**, Abstr. No. 60, 164.
Kazmerski, L.L., 1983, Proc. 5th EC Photovoltaic Solar Energy Conf., Kavouri, Athens (Reidel, Dordecht) p. 40.
Khattak, C.P., and F. Schmid, 1983, Proc. Symp. Materials and New Processing Technologies for Photovoltaics, eds J.A. Amick, V.K. Kapur and J. Dietl (Electrochemical Society, Pennington, NJ) p. 478.
Khattak, C.P., M. Basaran, F. Schmid, R.V. D'Aiello, P.H. Robinson and A.H. Firester, 1981, Proc. 15th IEEE Photovoltaic Spec. Conf., Orlando (IEEE, New York) p. 1432.
Kotval, P.S., and H.B. Strock, 1980, US Patent 4,195,067 (March 25) assigned to Union Carbide Corporation.
Ludsteck, A., 1981, Extend. Abstr. Electrochem. Soc. **81-2**, Abstr. No. 473, 1139.
Lutwack, R., 1981, Proc. 3rd EC Photovoltaic Solar Energy Conf., Cannes (Reidel, Dordrecht) p. 718.
McCormick, J.R., 1980, Proc. 14th IEEE Photovoltaic Spec. Conf., San Diego (IEEE, New York) p. 298.
McCormick, J.R., 1982, Proc. 16th IEEE Photovoltaic Spec. Conf., San Diego (IEEE, New York) p. 57.
Miller, G.L., D.V. Lang and L.C. Kimerling, 1977, Ann. Rev. Mater. Sci. **7**, 377.
Millet, E.J., 1980, J. Cryst. Growth **48**, 666.
Motzfeld, K., 1982, Extend. Abstr. Electrochem. Soc. **82-2** Abstr. No. 240, 385.
Muhamed, F., and M. Fuad, 1975, Proc. Int. Symp. Ferro Alloys and Silicon Metal, Sibenik (Yu) p. 124.
Ohl, R.S., 1946, US Patent 2,402,839 (June 25) assigned to Bell Telephone.
Olsen, J.M., 1983, Proc. Symp. Materials and New Processing Technologies for Photovoltaics, eds J.A. Amick, V.K. Kapur and J. Dietl (Electrochemical Society, Pennington, NJ) p. 119.
Olsen, J.M., and K.L. Carleton, 1981, J. Electrochem. Soc. **128**, 2698.
Olsen, J.M., and A. Kibbler, 1981, Abstracts 5th Am. Conf. Cryst. Growth, San Diego, p. 248.
Pelosini, L., A. Parisi and S. Pizzini, 1980, US Patent 4,241,037 (Dec. 23) assigned to Montedison (Italy).
Pizzini, S., 1982, Sol. Energy Mater. **6**, 253.
Pizzini, S., L. Giarda, A. Parisi, A. Solmi and G. Soncini, 1980, Proc. 14th IEEE Photovoltaic Spec. Conf., San Diego (IEEE, New York) p. 902.
Pizzini, S., C. Calligarich, C. Chemelli, M. Gasparini, P. Rava and L. Sardix, 1983, Proc. Symp. Materials and New Processing Technologies for Photovoltaics, eds J.A. Amick, V.K. Kapur and J. Dietl (Electrochemical Society, Pennington, NJ) p. 200.
Rao, G.M., D.E. Elwell and R.S. Feigelson, 1980, J. Electrochem. Soc. **127**, 1940.
Rao, G.M., D.E. Elwell and R.S. Feigelson, 1981, J. Electrochem. Soc. **128**, 1708.
Redfield, D., 1981, 15th IEEE Photovoltaic Spec. Conf., Orlando (IEEE, New York) p. 1179.

Revel, G., J.L. Pastol, D. Hania and H. Nguyen Dinh, 1983, Proc. 5th EC Photovoltaic Solar Energy Conf., Kavouri, Athens (Reidel, Dordrecht) p.1037.
Roskill Information Services Ltd., 1976, Silicon and Ferrosilicon, 2nd Ed (London).
Rozgony, G.A., and C.W. Pearce, 1978, Appl. Phys. Lett. **37**, 747.
Rustioni, M., R. Garavaglia and S. Pizzini, 1982, Extend. Abstr. Electrochem. Soc. **82-2**, Abstr. No. 244, 393.
Saito, T., A. Shimura and S. Ichikawa, 1981, Proc. 15th IEEE Photovoltaic Spec. Conf., Orlando (IEEE, New York) p. 576.
Salama, A.M., 1978, Proc. 13th IEEE Photovoltaic Spec. Conf., Washington DC (IEEE, New York) p. 496.
Sanjurjo, A., 1982, Proc. Symp. Materials and New Processing Technologies for Photovoltaics, eds J.P. Dismukus et al. (Electrochemical Society, Pennington, NJ) p. 165.
Sanjurjo, A., L. Nanis, K. Sancier, R. Bartlett and V.K. Kapur, 1981, J. Electrochem. Soc. **128**, 179.
Secco d'Aragona, F., 1981, Proc. Symp. Materials and New Processing Technologies for Photovoltaics, eds J.A. Amick, E. Sirtl, P. Ray-Choudhury and J.P. Dismukes (Electrochemical Society, Pennington, NJ) p. 119.
Sigmund, H., 1982, J. Electrochem. Soc. **129**, 2809.
Sigmund, H., R. Braungart and Ch. Höpfl, 1980, Verhandlg. DPG (IV) **15**, 176.
Sirtl, E., 1979, Proc. 2nd EC Photovoltaic Solar Energy Conf., Berlin (Reidel, Dordrecht) p. 84.
Sirtl, E., 1981, Proc. 3rd EC Photovoltaic Solar Energy Conf., Cannes (Reidel, Dordrecht) p. 236.
Sirtl, E., 1982, Proc. 4th EC Photovoltaic Solar Energy Conf., Stresa (Reidel, Dordrecht) p. 858.
Sirtl, E., 1983, Sol. Cells **10**, 101.
Sirtl, E., and H.-J. Rath, 1980, US Patent 4,194,028 (March 19) assigned to Wacker.
Smith, N.F., and R.K. Dawless, 1981, Extend. Abstr. Electrochem. Soc. **81-2**, Abstr. No. 477, 1147.
Storti, G.M., 1981, Proc. 15th IEEE Photovoltaic Spec. Conf., Orlando (IEEE, New York) p. 442.
Strauss, J., 1958, US Patent 2,866,701 (Dec. 30) assigned to Vanadium Corporation of America.
Theuerer, H.C., 1956, J. Met. **8**, 1316.
Trumbore, F.A., 1960, Bell Syst. Tech. J. **39**, 205.
Tschöpel, P., 1980, Fresenius Z. Anal. Chem. **302**, 1.
Tucker, N.P., 1927, J. Iron and Steel Inst. **15**, 412.
US DOE Program, 1978, Program Summary, January 1978 (DOE/ET – 0019/1), Division of Solar Technology (US Department of Energy, Washington, DC).
Van Overstraeten, R.J., 1981, Proc. Symp. Materials and New Processing Technologies for Photovoltaics, eds J.A. Amick, E. Sirtl, P. Ray-Choudury and J.P. Dismukes (Electrochemical Society, Pennington, NJ) p. 37.
Vigouroux, E., 1899, Compt. Rend. **129**, 334.
Weber, J., and P. Wagner, 1980, J. Phys. Soc. Jpn. Suppl. A 263, **49**, 263.
Wilson, W.G., 1973, 31st Electric Furnace Conf., Cincinnati, Dec. 5–7, 1973.
Wolf, M., and H. Rauschenbach, 1976, Solar Cells (IEEE, New York) p. 146–170.
Wünstel, K., and P. Wagner, 1982, J. Appl. Phys. **A27**, 207.
Zehaf, M., H. Amzil, J.P. Crest, G. Mathian, S. Martinuzzi and J. Oualid, 1982, Proc. Symp. Materials and New Processing Technologies for Photovoltaics, eds J.P. Dismukes et al. (Electrochemical Society, Pennington, NJ) p. 273.

CHAPTER 7

BIFACIAL SOLAR CELLS

ANTONIO LUQUE

Instituto de Energia Solar
Universidad Politécnica de Madrid
Madrid, Spain

Silicon Processing for Photovoltaics II, edited by C.P. Khattak and K.V. Ravi

Contents

List of symbols

a_i	Extinction coefficient of monochromatic light of wavelength λ_i
A	Cell area
b	μ_e/μ_h ratio
d	Half of the metal grid finger width (see fig. 5)
D	Grid busbar width
D_e	Diffusion coefficient for electrons
D_{eff}	High-injection diffusion coefficient [see eq. (22)]
D_h	Diffusion coefficient for holes
E	Electric field in base
g_L	Rate of light-induced generation per unit of volume of electron–hole pairs
h	Injection parameter at the high–low junction (see section 4)
H	Abbreviation for $\exp(q\phi_P/2kT)$
H_0	Value of H for $\phi_P = V_{HI}$
J	Current density
J_e	Electron current density
J_h	Hole current density
J_{qe}	Electron quasi-current density [see eq. (39]
J_{qh}	Hole quasi-current density [see eq. (31)]
J_{B0}	Base recombination current (see fig. 1)
J_{BNL}	Base illumination current into the p–n junction (see fig. 1)
J_{BPL}	Base illumination current into the high–low junction (see fig. 1)
J_{C0}	Base transferred current at the p–n junction (see fig. 1)
J_{NL}	Total illumination current at the p–n junction (see fig. 1)
J_{NNL}	Emitter illumination current at the p–n junction (see fig. 1)
J_{N0}	Front emitter recombination current (see fig. 1)
J_{PL}	Total illumination current at the high–low junction (see fig. 1)
J_{P0}	Back-emitter recombination current (see fig. 1)
J_{PPL}	Emitter illumination current of the high–low junction (see fig. 1)
J_T	Dipole layer recombination current [see eq. (12)]
k	Boltzmann constant, 1.38×10^{-23} J/K
l	Half of the metal grid fingers separation (see fig. 5)

L	Metal grid finger length
L_c	Diffusion length of electrons in a p-type base
L_{eff}	Effective diffusion length in high injection, $L_{eff} = (D_{eff}\tau_{eff})^{1/2}$
n	Electron density
n_i	Intrinsic carrier density in silicon (see table 3)
n_{B0N}	Dark electron profile in base if $n(W) = 0$
n_{B0P}	Dark electron profile in base if $n(0) = 0$
n_{PL}	Illuminated electron profile in the p^+-emitter for zero density at its boundary
n_{P0}	Dark electron profile in the p^+-emitter for zero density at its boundary
N_c	Equivalent density of states in the conduction band
N_B	Base doping
N_p	p-emitter doping
$N_{ph\,i}$	Photon flux
p	Hole density
p_{NL}	Illuminated hole profile in the n^+-emitter for zero density at its boundary
p_{N0}	Dark hole profile in the n^+-emitter for zero density at its boundary
q	Electronic charge, 1.60×10^{-19}C
r	Net thermal recombination rate (recombination – generation)
R_{eb}	Sheet resistance due to electrons in base under high injection [see eq. (74)]
R_{hb}	Sheet resistance due to holes in base under high injection [see eq. (74)]
R_B	Vertical series resistance [see eq. (67)]
R_G	Lateral series resistance due to the grid [see eqs. (69) or (73)]
$R_{\square m}$	Sheet resistance at the metal fingers
$R_{\square n}$	Sheet resistance due to the n-emitter
$R_{\square p}$	Sheet resistance due to the p-emitter
$R_{\square B}$	Sheet resistance due to the base in low injection
S_{eff}	Effective recombination velocity at the p–p^+ interface [see eq. (16)]
t_L^n	J_{PL}^n / J_{NL}^n ratio for illumination on the n-side
t_L^p	J_{NL}^p / J_{PL}^p ratio for illumination on the p-side
T	Absolute temperature
u	Injection parameter at the p–n junction (see section 4)
U	Abbreviation for $\exp(q\phi_n/2kT)$
U_0	Value of U for $\phi_N = V_{HI}$ $(U_0 = H_0)$
V_i	Voltage of the cell without series resistance effects
V_n	Separation of the conduction band and Fermi level in the n^+-emitter in eV (see fig. 2)

V_D Dember voltage in the base [see eq. (39)]
V_H Voltage at the high–low junction [see eq. (37)]
V_{HI} High-injection voltage [see eq. (4)]
V_U Voltage at the p–n junction [see eq. (34)]
W Base thickness
α_F Direct current gain [see eq. (48)]
α_R Reverse current gain [see eq. (49)]
θ *U*/*H* ratio
μ_e Electron mobility in the base
μ_h Hole mobility in the base
τ_e Electron lifetime in the base
τ_h Hole lifetime in the base
τ_{eff} High-injection lifetime in the base, $\tau_{eff} = \tau_e + \tau_h$
ϕ_N Separation of quasi-Fermi levels at the p–n junction
ϕ_P Separation of quasi-Fermi levels at the high–low junction
ϕ_{THH} Thermodynamic barrier of potential at the high–low junction [see eq. (35)]
ϕ'_{THH} Barrier of potential at the high–low junction in high injection [see eq. (36)]
ϕ_{THU} Thermodynamic barrier of potential at the p–n junction (see fig. 2)

1. Introduction

The present interest of bifacial solar cells (i.e., cells active on both faces) for terrestrial applications lies in the collection of the earth's albedo. Flat panels made with bifacial cells encapsulated between two transparent covers can increase the power output by over 50% when placed in very simple adequate surroundings, such as a set of two white-painted wooden panels (Cuevas et al. 1982) or, simply, a large area of bifacial panels placed on a white-painted ground. Such an albedo-collecting panel has recently been developed by our team, and is today available commercially.

There is an additional and fundamental reason to develop bifacial cells; this is linked to the fabrication of static concentrators. These concentrators must be designed so that they cast onto the cells the light coming from a large region of the sky, in particular from the region where the sun can be found at some moment during the year. In fact, static concentrators should be designed to collect the highest possible energy from an extended solar source which may be the yearly averaged radiance of the sky.

Regardless of the specific nature of a given concentrator there is an upper limit of the effective concentration that can be achieved from extended sources. This upper limit is based on the Fermat principle (Miñano and Luque 1983) for optical concentrators (i.e., concentrators based on reflection and refraction) or on physical optics considerations for holographic concentrators (Winston 1982).

A result of these works is that the limit of the effective concentration is reached when:

(a) the concentrator is able to cast isotropic illumination on the collector if the concentrator itself is illuminated isotropically (this concentrator is said to be optimal), and

(b) all the rays in the region of the sky bound by a line of iso-radiance are collected and the remaining ones are rejected (this concentrator is said to be adapted to the source).

The upper limit of concentration is proportional to the second power of the refractive index of the medium surrounding the cell.

Condition (a) clearly implies that the collector has to be a bifacial cell. A monofacial cell leads to concentrations which are, at most, one half of its upper limit.

The upper limit of concentration is a decreasing function of the fraction of the total colleted energy available. If 80% of the available energy is collected, the effective concentration upper limit is 8.0 for a sunny place, such as the outskirts of Madrid (Spain) and 9.2 if no diffuse radiation exists (as in the outer atmosphere). In a cloudy climate with an isotropic sky this upper limit is 4.5. The preceding values correspond to a bifacial cell immersed in a medium with a refractive index $n = 1.5$. If a monofacial cell is used, the preceding values have to be divided by two. If the cell is surrounded by air, they have to be divided by $n^2 = 2.25$. On the other hand, if two, yearly, positions are accepted, the preceding values are, roughly, doubled.

The moderate values of concentration achievable by a static concentrator seem to indicate that bifacial cells are a key aspect of their economic viability.

A third application of bifacial cells can be related to increasing the efficiency of luminiscent concentrators. Such devices, due to their frequency changing mechanisms do not present an absolute upper bound of concentration from extended sources (the optical pumping of lasers is a practical proof of this). However, their efficiency is low enough as to benefit from the use of bifacial cells in the way indicated elsewhere (Cuevas et al. 1979a).

Today, practical bifacial cells are back surface field (BSF) cells presenting metallization grids on both faces. As far as we know this bifacial BSF structure has been independently proposed by Soviet (Borodine et al. 1976) and Spanish (Luque et al. 1978a) scientists. Other bifacial structures like n^+–p–n^+ (Luque et al. 1978b, Chambouleyron and Chevalier 1978) or the vertical multi-junction (VMJ) cells, intrinsically bifacial, present poorer performances than the present BSF structure (Cuevas et al. 1979a,b).

The principle involved in manufacturing bifacial cells are very similar to those used for cells wich have all the contact fingers at the back side in order to avoid front-side shadowing. Among these structures we should mention the interdigitated back contact cell (Lammert and Swartz 1977) the tandem cell (Chiang et al. 1978) of n^+–p–n^+ structure, with a floating p–n junction, or the front surface field cell (Von Ross and Anspaugh 1978) with a p^+–p–n^+ structure.

Bifacial cells are either in a low- or in a high-injection regime. Indeed, very good results are obtained within a wide range of base resistivities of both types (n and p), therefore working in a variable regime of injection. In this chapter we present the low-injection theory first, and afterwards a model of cells working under p–i–n (i for intrinsic base) representing the limit of very high injection. Both cases allow an analytical treatment. An unitary treatment of both cases that can qualitatively account for intermediate levels of injection is presented thereafter. This qualitative model is based on the Ebers–Moll equations, and is, therefore, familiar to device scientists. This model is very useful to provide physical insight into the behaviour of a bifacial cell. We discuss the bifacial cell properties based on this model in short-circuit as well as open-circuit conditions, and we give a general description of the curve shape excluding series-resistance effects. The latter are analyzed in section 5, including the specific conditions set by the double grid used in bifacial cells. Finally, we devote a section (section 8) to discuss applications of the bifacial cell.

The belief that high base resistivity cells give poor open-circuit voltage is very extended (e.g., Iles and Soclof 1975). This is true for p–n cells but not for BSF ones, and the reason is that these last cells, at open-circuit, are usually in high injection, so that a substantial voltage is developed at the high–low junction. In fact, the open-circuit voltage of a cell depends very little on the base resistivity (Ruiz et al. 1983) (it is somewhat higher for bases of 0.5 Ω cm) and even in the case of p–i–n cells it is close to 600 mV (at AM1 and 25°C).

This active role of the high–low junction, which in the extreme case of a p–i–n cell becomes a half-junction, justifies the nomenclature used here, calling emitters both highly doped layers (p^+ or n^+). Therefore, the classical concept of effective surface recombination velocity usually used for the study of BSF cells is abandoned here (some references have been made for the connection on to classical works). Actually, we consider that the reduction of effective surface recombination is not the main role of the high–low junction, because this effect can also be achieved by a proper passivation of the surface, as is the case in the different structures of interdigitated back contact cells. Of those cells the best is probably the one with a carefully passivated surface. The role that we consider as the main feature of the high–low junction in bifacial cells is the build-up of voltages.

2. Low-injection theory of bifacial cells

The unidimensional low-injection minority profile equation for the base of a n^+–p–p^+ back surface field cell is given by

$$D_e \frac{d^2 n}{dx^2} = \frac{n}{\tau_e} - g_L \tag{1}$$

where D_e and τ_e are the diffusion constant and lifetime of the electrons, and g_L is the term of light generation. In this equation we have neglected the equilibrium concentration in the base with respect to the actual concentration of electrons n. Since this is a linear differential equation, its solution can be considered as the sum of a profile $n_{BL}(x)$ with zero boundary conditions at both junctions (n^+–p and p–p^+) and two dark profiles (with $g_L = 0$) $n_{BON}(x)$ and $n_{BOP}(x)$ with zero concentration at the opposite junction. Those arbitrary concentration profiles $n_{BON}(0)$ and $n_{BOP}(W)$ are linked to the quasi-Fermi level splitting through

$$n_{BON}(0) = \frac{n_i^2}{N_B}\exp\left(\frac{q\phi_N}{kT}\right) = n_i\frac{\exp(q\phi_N/kT)}{N_B/n_i} \equiv n_i\frac{U^2}{U_0}, \tag{2}$$

and

$$n_{BOP}(W) = \frac{n_i^2}{N_B}\exp\left(\frac{q\phi_P}{kT}\right) = n_i\frac{\exp(q\phi_P/kT)}{N_B/n_i} \equiv n_i\frac{H^2}{H_0}, \tag{3}$$

where U, H and $U_0 = H_0$ are defined by eqs. (2) and (3). Besides

$$U_0 = H_0 = \frac{N_B}{n_i} = \exp\left(\frac{qV_{HI}}{2kT}\right), \tag{4}$$

where V_{HI} is the separation of quasi-Fermi levels causing the onset of high-injection conditions ($n = N_B$). The use of U_0 and H_0 is intended to for achieving an unitary nomenclature for low- and high-injection conditions, and its usefulness will be clearer later on. Under low injection ϕ_N, $\phi_P < V_{HI}$ and $U, H < U_0, H_0$.

In our calculations, ϕ_P and ϕ_N are supposed to assume known values, providing a set of boundary conditions for the calculation of the minority-carrier profile. We will analyze later-on the way to calculate ϕ_P and ϕ_N as functions of the current density in the cell.

The calculation of $n_{BON}(x)$ and $n_{BOP}(x)$ is an elementary one for which we refer to the classical work by Wolf (1960). These profiles are shown in fig. 1 (see this figure for nomenclature).

For the calculation of n_{BL} we have to define the generation function g_L. In monocrystalline silicon this is simply the absorption of light per unit of length

$$g_L = \sum_i N_{phi} a_i \exp(-a_i x), \tag{5}$$

where N_{phi} is the flux of photons of a given wavelength i entering the base, and a_i is the absorption constant for each wavelength. The solution of $n_{BL}(x)$ for each wavelength is analytical, and for a polychromatic radiation the solution is the superposition of the solutions corresponding to each monochromatic component.

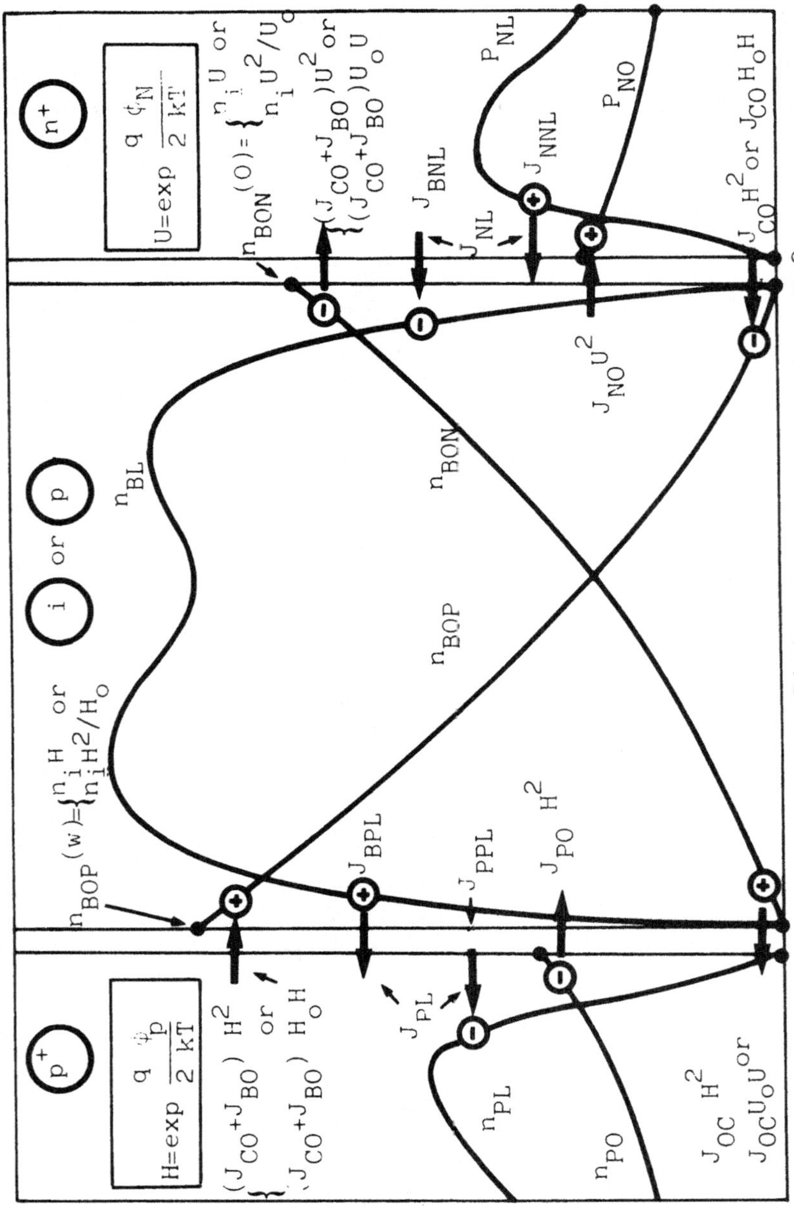

Fig. 1. Qualitative photogenerated and dark carrier profiles, and current and quasi-current components.

A widely accepted AM1 spectrum is the one measured by Thomas et al. (1976). In table 1 we present the number of photons corresponding to eight wavelengths that we use at our Institute to implement eq. (5). The details of the elaboration of this table can be found elsewhere (Cuevas 1980). The reflective coefficient of bare silicon, taken from several sources (Gray et al. 1972, Michel 1978, Hovel 1975), has been introduced there, as well as those with a 600 Å thick antireflecting coating of titanium oxide (Michel 1978). The absorption coefficient of silicon has also been taken from the literature (Volfson and Subashiev 1967, Dash and Newman 1955).

We should remember that, after traversing the emitter, the flux of photons, reaching the base has been reduced by a factor $\exp(-ax_j)$. The value of a should be multiplied by the inverse of the cosine of the incidence angle if the light is not normal to the cell. The absolute value of the flux of photons is seldom important in our calculations.

The electron current densities in the base are mainly electron currents. The diffusion currents associated with each profile at the ends of the base are obtained with the well-known formula

$$J_e = qD_e \,\mathrm{d}n/\mathrm{d}x. \tag{6}$$

According to the nomenclature used in fig. 1 we obtain

$$J_{C0} = q\,\frac{n_i^2}{N_B}\,\frac{D_e}{L_e}\,\frac{1}{\sinh(W/L_e)}, \tag{7}$$

Table 1
AM1 spectrum, silicon absorption coefficient, and reflection coefficient with and without TiO_x coating, together with the number of photons entering the silicon in both reflectance cases.

λ(μm)	$N(\lambda)/10^{14}$ photons $(\mathrm{cm}^2\,\mathrm{s}\,0.1\,\mu\mathrm{m})^{-1}$	$\alpha(\lambda)$ (cm^{-1})	$R(\lambda)$	$R_{TiO_x}(\lambda)$	$N(1-R)$	$N(1-R_{TiO_x})$
0.4	209	7×10^4	0.45	0.25	115	157
0.5	407	1.35×10^4	0.38	0.09	252	370
0.6	450	4.7×10^3	0.33	0.022	301	440
0.7	451	2.1×10^3	0.31	0.039	311	433
0.8	400	950	0.32	0.0775	272	369
0.9	200	375	0.32	0.105	136	179
1.0	314	95	0.32	0.125	213	275
1.1	241	9	0.335	0.15	160	205

and

$$J_{B0} = q \frac{n_i^2}{N_B} \frac{D_e}{L_e} \frac{\cosh(W/L_e)}{\sinh(W/L_e)}, \tag{8}$$

where $L_e = (D_e \tau_e)^{1/2}$ is the diffusion length of electrons in the base.

For the current associated to illumination profiles we obtain, based on Wolf's work and further developed by us (Luque et al. 1978),

$$J_{BNL}^{n} = \sum_i q N_{phi}^{n} \frac{a_i L_e}{a_i^2 L_e^2 - 1} \left[\frac{\exp(a_i W) - \cosh(W/L)}{\sinh(W/L)} + a_i L_e \right], \tag{9}$$

and

$$J_{BPL}^{n} = \sum_i q N_{phi}^{n} \frac{a_i L_e}{a_i^2 L_e^2 - 1} \times \left[\frac{1 - \exp(-a_i W)\cosh(W/L)}{\sinh(W/L)} - a_i L_e \exp(-a_i W) \right]. \tag{10}$$

The superscript n means that the illumination is on the n-side. If an additional illumination is produced on the p-side, two sets of terms similar to those of eqs. (10) and (9) are added, respectively, to eqs. (9) and (10). No superscripts means an arbitrary illumination without specification about the side.

Neither the profiles in fig. 1, nor the current densities, whose expressions we have written, correspond to any physical situation. For instance, eq. (9) does not represent the short-circuit cell current. Only a proper superposition of currents, corresponding to a superposition of profiles, will lead to physically meaningful results.

Let us analyze now the profiles at the heavily doped (n^+ or p^+) zones corresponding to the quasi-Fermi levels splittings ϕ_N and ϕ_P, which are considered to be constant throughout the space-charge region at the p–n or high–low junctions, respectively. In these regions a surface recombination velocity (SRV) is assumed at each surface. At high doping levels this SRV is constant. However, the differential equation of the minority-carrier profiles is a complex one. This complexity arises from several facts. First, the variable doping produces an electric field (Tanenbaum and Thomas 1956) that opposes the flux by diffusion of majority carriers. Yet, the bandgap shrinkage produced at the heavily doped zones, which is due to a variety of causes and was analyzed in detail by a number of authors (Volfson and Subashiev 1967, Dumbar and Hauser 1973, Slotboom and de Graaf 1976, Lanyon and Tuft 1979), generates an additional driving force (Van Overstraeten et al. 1973) – a quasi-field depending solely on the local doping density – that must be taken into account in the current balance mentioned above.

To calculate the space charge giving rise to an electric field, we have to analyze the fraction of ionized doping impurities, and here again, the ionization energy also depends on the doping concentration (Li and Thurber 1977, Li 1978) (it is reduced to zero for concentrations above $5 \times 10^{18} cm^{-3}$).

Finally, we have to consider the variation of mobility with the doping level, as well as the minority-carrier lifetime, sharply affected by Auger mechanisms (Sheng 1975).

The preceding current balance has to produce a horizontal majority-carrier quasi-Fermi level, and induces a specific bending of the band edge of the minority carriers that affects the transport of those cariers through a position-dependent electric field plus quasi-field. As a result, the minority-carrier profile differential equation is linear, with position-dependent coefficients whose values are not readily known. Nevertheless, it is possible to consider the solution as a superposition of a dark profile and an illumination one. The dark profiles p_{N0} or n_{P0} correspond to a surface recombination velocity at the surface's end, and to a boundary condition connected to the quasi-Fermi splitting ϕ_N or ϕ_P at the space-charge zone, respectively.

Indeed, the relationship of the carrier concentration and the quasi-Fermi level is not simple, since the bandgap shrinkage and the degeneration (Lindholm and Fossum 1981, Fossum and Shibib 1981) of the majority-carrier band lead to a little known value of an activity factor f^+ to multiply the doping concentration N^+ at the edge of the space-charge zone of the heavily doped zone either at the p–n or at the high–low junctions in the boundary conditions equations, similar to eqs. (2) or (3). Yet, it can be assured that there is a proportionality between carrier concentrations and $\exp(q\phi / kT)$. This implies that the currents corresponding to those profiles at the edge of the space-charge layers are proportional to $\exp(q\phi / kT)$, and their constants of saturation, J_{N0} and J_{p0}, include all the high-doping effects mentioned here. To further analyze these constants is out of the scope of the present work.

Including illumination in the diffusion equation introduces additional factors in the generation term due to the variation of the bandgap (Volfson and Subashiev 1967), and, therefore, in the variation of the absorption coefficients a_i with distance. The solutions p_{NL} and n_{PL} correspond to surface recombination velocity conditions at the surface end of the heavily doped zone and to zero boundary conditions at the junction (p–n or high–low).

The minority-carrier currents associated to each profile at the p–n or high–low junction edges are shown in fig. 1. The actual profiles of minority carriers in the base for a given set of quasi-Fermi level splittings, ϕ_N and ϕ_P, are $n_{BL}(x) + n_{BON}(x) + n_{BOP}(x)$ in the base, $p_{NL}(x) + p_{N0}(x)$ in the n$^+$-emitter and $n_{PL}(x) + n_{P0}(x)$ in the p$^+$-emitter. The currents at each point are the sum of the currents produced at that point by each individual profile.

By neglecting the generation and recombination produced at the space-charge zone in the p–n junction, the total current of electrons plus holes traversing this junction is given by (see fig. 1)

$$J = J_{NL} - [(J_{N0} + J_{B0} + J_{C0})U^2 - J_{C0}\, H^2], \tag{11}$$

where $J_{NL} = J_{BNL} + J_{NNL}$.

We can consider, in eq. (11), the current photogenerated at the space-charge zone as included in J_{NL}. We can also add a term due to recombination in the space-charge zone that is approximately proportional to $\exp(q\phi/2kT)$ (Sah et al. 1957). Actually, this term appears because of the high value of the recombination at a point where the electron and hole concentrations are such that $n/\tau_e = p/\tau_h$. In the space-charge zone this recombination is important in the space region where the voltage varies as kT/q. If the electric field, in which this relationship among electrons and holes holds, is E we can write the total recombination as

$$J_T U_0 U = n_i \frac{kT}{qE} \frac{1}{2(\tau_e \tau_h)^{1/2}} U, \tag{12}$$

where the equivalent lifetime in this region is twice the geometrical mean of those of electrons and holes. As a result we can write eq. (11) as

$$J = J_{NL} - [(J_{N0} + J_{B0} + J_{C0})U^2 + J_T U_0 U - J_{C0} H^2], \tag{13}$$

where the separation of quasi-Fermi levels of the high–low junction appears through the term H^2.

An additional relation between U^2 and H^2 can be written, taking into account the continuity of current at the high–low junction. Since the condition $n/\tau_e = p/\tau_h$ is never fulfilled, because $p \gg n$, the space-charge zone recombination is negligible. As a consequence, we can state that the electron current is the same at both ends of the space high–low junction. Therefore,

$$0 = J_{PL} - [(J_{P0} + J_{B0} + J_{C0})H^2 - J_{C0}\, U^2], \tag{14}$$

where, to be strict, any photogeneration contribution of the high–low transition region, if it exists, has been added to J_{PL}, which otherwise is $J_{PPL} + J_{BPL}$. We have to remember the close resemblance of eq. (14) to an open-circuited p–n junction [see eq. (11) with $J = 0$].

The voltage at the cell in the absence of series resistance is

$$V_i = \frac{kT}{q} \ln U^2$$

To obtain the J–V curve we eliminate U^2 and H^2 from eqs. (13), (14) and (15) which is a very simple task. Actually, we have to solve in H^2 eq. (14) and substitute its value as a function of U^2 in eq. (13). In this way we obtain a function $J(U)$ that is easily converted into $J(V_i)$ using eq. (15), leading to a two-exponential I–V_i relationship. The exponential in $\exp(qV_i/2kT)$ is caused by the term in J_T.

Some authors (e.g., Madelkorn and Lammeck 1972) prefer to refer to the high–low junction as a surface of effective recombination velocity S_{eff}. In our formalism

$$S_{eff} = J_{P0} N_B / q n_i^2 \tag{16}$$

where all the heavy doping effects affecting this parameter are included in J_{P0}.

The values of J_{P0} and J_{N0} are calculated according to the analytical model of Fossum and Shibib (1981), modified by our own analysis of the n^+- and p^+-emitters (Eguren et al. 1981). These calculated values have been compared to the measurements we have done on our best cells (Araújo et al. 1982, 1984). We conclude that a good lower limit for J_{P0} and J_{N0} in diffused junctions is, respectively, 2×10^{-13} and 10^{-12} A/cm^2, and we use these values in our calculations here.

With respect to the base lifetime we use the values obtained by Fossum and Lee (1982) for silicon, which are variable but for pure silicon with doping below 10^{15}cm^{-3} τ_e is 0.7×10^{-3}s and $\tau_h = 0.5 \times 10^{-3}$s. The diffusion constant and the mobility of carriers are related through the Einstein relationship $\mu = qD/kT$, valid at low doping concentration. The mobility is obtained from Thurber et al. (1980a,b) and, for low-doping concentrations, gives values of $D_e = 36.6$ for electrons and $D_h = 12.2$ for holes. We present, in table 3, a set of values for several base resistivities.

Finally, acording to measurements by G.L. Araújo (Araújo et al. 1984) we assume that $J_T U_0$ is of about 2×10^{-8}A/cm^2 for good cells.

3. High injection p–i–n model of bifacial cells

In this section we present a summary of our model presented elsewhere (Luque and Eguren, 1982).

3.1. Profile differential equation

In an intrinsic base in thermal equilibrium placed between two heavily doped n^+ and p^+ regions, a quasi-neutral region appears at its center, bound by two space-charge zones which extend about one intrinsic Debye length, beside the metallurgical junctions. In silicon the intrinsic Debye length is about 20 μm. If the base is not intrinsic there also appear two space-charge layers, whose lengths decrease when the base doping increases. In this case the p–n junction side has a space-charge layer much longer than that of the high–low junction. In some cases no space-charge layer appears in the last junction (Dharival and Kulshreshtha 1981).

In the quasi-neutral region the one-dimensional transport equations are

$$J_e = q\mu_e nE + qD_e \frac{dn}{dx}, \tag{17}$$

$$J_h = q\mu_h nE - qD_h \frac{dp}{dx}. \tag{18}$$

Under very high injection conditions, or if the substrate is intrinsic. The condition

$$n = p \tag{19}$$

is accurately fulfilled, confirming our assumption of quasi-neutrality through the base. Adding eqs. (17) and (18), and using eq. (19) and the Einstein relationship we obtain

$$E = \frac{J/q}{\mu_e + \mu_h}\frac{1}{n} - \frac{kT}{q}\left(\frac{\mu_e - \mu_h}{\mu_e + \mu_h}\right)\frac{1}{n}\frac{dn}{dx}, \tag{20}$$

where J is the total current density $J_e + J_h$. In the same way, by subtracting eq. (18) from eq. (17) and using eq. (20)

$$\frac{J_e - J_h}{q} = \left(\frac{\mu_e - \mu_h}{\mu_e + \mu_h}\right)\frac{J}{q} + 2D_{eff}\frac{dn}{dx}, \tag{21}$$

where

$$D_{\text{eff}} = \frac{2D_e D_h}{D_e + D_h}. \tag{22}$$

On the other hand, the continuity equations are

$$\frac{1}{q}\frac{dJ_e}{dx} = (r - g_L), \tag{23}$$

$$\frac{1}{q}\frac{dJ_h}{dx} = -(r - g_L). \tag{24}$$

By subtracting eq. (24) from eq. (23) and using eq. (21), we obtain

$$D_{\text{eff}}\frac{d^2 p}{dx^2} = r - g_L. \tag{25}$$

Under very high injection conditions, with $n = p$ the Read–Hall–Shockley equations lead to (Van Vliet 1982)

$$r = \frac{n}{\tau_e + \tau_h} = \frac{n}{\tau_{\text{eff}}}, \tag{26}$$

where there is a small difference with the conditions at the space-charge region because there the lifetime is twice the geometrical mean of τ_e and τ_h, which here is twice the arithmetical mean. This is due to the condition that the highest recombination achieved in the junction occurs for $n/\tau_e = p/\tau_h$, but this condition cannot be sustained in the whole base where the neutrality of charge imposes $n = p$. As a consequence τ_{eff} is somewhat higher than the value in the preceding section, even though the physical phenomenon is very similar.

The effective diffusion length implied by the lifetime in eq. (26) and the diffusion constant in eq. (25), is $L_{\text{eff}} = (D_{\text{eff}}\tau_{\text{eff}})^{1/2}$.

The term of optical generation in eq. (25) is the same as discussed in the preceding section for the low-injection case. The solutions for this equation are the same as indicated there, but the relationship among profiles and quasi-Fermi levels at the boundaries is in this case ($n = p$)

$$n_{\text{B0N}}(0) = n_i \exp(q\phi_N/2kT) = n_i U, \tag{27}$$

$$n_{\text{B0P}}(0) = n_i \exp(q\phi_P/2kT) = n_i U, \tag{28}$$

Any physically meaningful profile is a superposition of profiles similar to that of the low-injection case, as indicated in fig. 1. In the emitters the profiles are exactly the same as those analyzed in the preceding section.

3.2. Current components

If the electron current is known at a point x, the electric field can be obtained from eq. (18). That field can now be introduced in the expression of the total current obtained from the addition of eqs. (17) and (18). If condition (19) is taken into account we finally obtain

$$J = \left(-qD_{\text{eff}}\frac{dp}{dx} + J_e\right)\frac{b+1}{b} = (J_{qh} + J_e)\frac{b+1}{b}, \tag{29}$$

where b is the ratio μ_e/μ_h, and we call J_{qh} hole quasi-current. Equation (29) is quite mnemonic since it states that the current at any point is the sum of the current of a carrier plus the quasi-current of the other, multiplied by a factor that takes into account high-injection phenomena. The quasi-current is obtained from the known profiles as if it were a diffusion current with diffusion constant D_{eff}. The corresponding formula, if the hole current is known, is

$$J = \left(qD_{\text{eff}}\frac{dn}{dx} + J_h\right)(b+1) = (J_{qe} + J_h)(b+1). \tag{30}$$

These equations are of particular interest in the base boundaries where we know the minority-carrier currents of the heavily doped regions.

3.3. Voltages on the cell

The total voltage on the cell is the sum of the voltages across the base. Let us consider now a n^+–p–p^+ structure of which the n^+–i–p^+ one can be considered as a particular case. If the n^+–p junction has a direct bias V_U, the thermal barrier ϕ_{THU} is reduced by that amount, as represented in fig. 2. The relationship of this voltage to the electron concentration in the base edge n_B (if the quasi-Fermi levels are constant through the dipole layer) is

$$n_B(0) = N_c \exp\left[-\frac{q(\phi_{THU} + V_n - V_U)}{kT}\right], \tag{31}$$

where V_n is defined in fig. 2. The term $\phi_{THU} + V_n$ can be easily calculated as a function of the doping concentration in the base, and is independent from the emitter and, therefore, from heavy doping effects, even though ϕ_{THU} or V_n alone do depend on those effects. If its value is introduced in eq. (31) we obtain

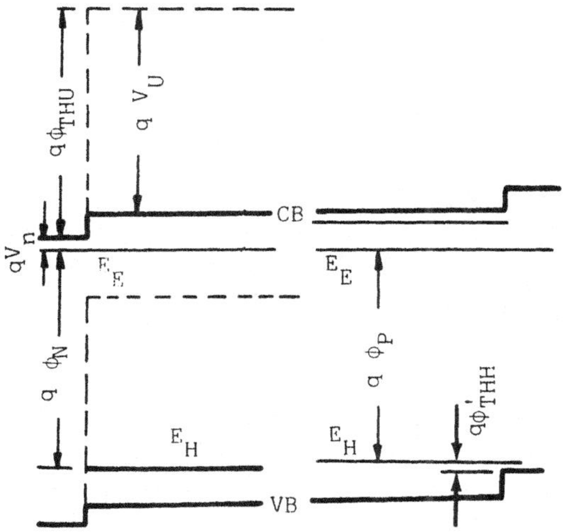

Fig. 2. Electron-energy-band diagram for a p–n junction and a high–low junction with the base under high injection.

$$n_B(0) = \frac{n_i^2}{N_B} \exp\left(\frac{qV_U}{kT}\right). \tag{32}$$

This equation is the same as that obtained for the low-injection case. Taking into account eqs. (19) and (32),

$$[n_B(0)]^2 = n_i^2 \exp\left(\frac{q\phi_N}{kT}\right) = \frac{n_i^4}{N_B^2} \exp\left(\frac{2qV_U}{kT}\right), \tag{33}$$

so that the relationship between ϕ_N and V_U is

$$V_U = \frac{\phi_N + V_{HI}}{2}, \tag{34}$$

V_{HI} is defined in eq. (4).

Under low-injection conditions a potential barrier ϕ_{THH} is built at the high–low junction to balance the flow of majority carriers through it. For holes the quasi-Fermi level is constant, and, therefore, this barrier is given by

$$\phi_{THH} = \frac{kT}{q} \ln\left(\frac{N_P f_P^+}{N_B}\right), \tag{35}$$

where f_P^+ is an activity factor taking into account, if applicable, the partial ionization, degeneracy and bandgap shrinkage in the heavily doped zone. Under high-injection conditions $p = n$, so that

$$\phi'_{THH} = \frac{kT}{q} \ln\left(\frac{N_P f_P^+}{n}\right). \tag{36}$$

The voltage V_H across the high–low junction is the subtraction of ϕ_{THH} and ϕ'_{THH}. Its relationship to ϕ_P is given, according to eqs. (28) and (4), by

$$V_H = \frac{\phi_P - V_{HI}}{2}. \tag{37}$$

If the base is intrinsic, eqs. (34) and (37) are still valid by considering $N_A = n_i$, so that $V_{HI} = 0$.

The voltage drop in the base is obtained by integration of eq. (20). This voltage is formed by two terms: the ohmic drop

$$V_R = \int_0^w \frac{J/q}{\mu_e + \mu_h} \frac{1}{n} \, dx, \tag{38}$$

and the Dember potential V_D caused by the different mobility of electrons and holes. This term, once integrated, is

$$V_D = \frac{b-1}{b+1} \frac{\phi_P - \phi_N}{2}. \tag{39}$$

Therefore, the total voltage is

$$V = V_U + V_H + V_D - V_R = V_i - V_R, \tag{40}$$

where V_i is the cell voltage if the ohmic drop V_R is not considered. V_i is related to ϕ_P and ϕ_N by

$$V_i = \frac{b\phi_P + \phi_N}{b+1} = \frac{kT}{q} \ln[(U^2)^{1/(b+1)}(H^2)^{b/(b+1)}]. \tag{41}$$

3.4. J–V_i characteristics

The p–i–n current can be obtained for given values of the quasi-Fermi level splittings at the end of the base with the help of eqs. (29) and (30). We know the profiles at the base and, therefore, we can calculate the quasi-currents of electrons and holes at the end of the base, respectively adjacent to the n^+- or p^+-emitters. These quasi-currents are represented in fig. 1, where we use the same symbol used for the minority currents in the low-injection case.

Doubtlessly, the differential equation in the base is linear also in p–i–n conditions and allows the superposition in an identical way as under low injection. The only difference here lies in the fact that the values of the concentration at the boundaries are given by eqs. (27) and (28) instead of eqs. (2) and (3). Also the lifetime, diffusion constant and diffusion length must be changed from τ_e, D_e and L_e into τ_{eff}, D_{eff} and L_{eff} in the constants J_{BPL}, J_{BNL}, J_{C0} and J_{B0}. Apart from this, the current at the n–i and p–i junctions, which in both cases is the total current J, can be written, according to eqs. (30) and (29), as, respectively

$$\frac{J}{b+1} = J_{NL} - [J_{N0}U^2 + (J_{B0} + J_{C0})U_0 U - J_{C0}H_0 H] \tag{42}$$

$$\frac{Jb}{b+1} = J_{PL} - [J_{P0}H^2 + (J_{B0} + J_{C0})H_0 H - J_{C0}U_0 U] \tag{43}$$

which play a similar role as eqs. (11) and (14) in the low-injection case. The terms in U_0U and H_0H are quasi-currents and not real currents, and, therefore, they are proportional to U and H instead of to U^2 and H^2, as can be deduced by comparing eqs. (2) and (3) to eqs. (27) and (28).

These equations do not take into account the generation of the space-charge layers. However, they can be added to the constants J_{NL} and J_{PL} which, otherwise, are simply $J_{NNL} + J_{BNL}$ or $J_{PPL} + J_{BPL}$, including a sum of a current and a quasi-current. Also the recombination term in the space-charge zone can be added to these equations. In this case, the point where $n/\tau_n = p/\tau_p$ can appear either on the p–n junction or on the high–low one, depending on the type of cell (n^+–p–p^+ or p^+–n–n^+) and on the relative value of τ_n and τ_p. In most CZ-grown silicon samples $\tau_n > \tau_p$, so that the highest recombination appears where $n > p$, i.e. in the n–i junction, therefore we want to note that in a p^+–n–n^+ cell the space-charge recombination term linked to the p–n junction in low injection will appear linked to the n–n^+ junction (and to the separation of quasi-Fermi levels there) in high injection, while in n^+–p–p^+ cells this term is always placed at n^+–p junction. Accordingly, eq. (42) becomes

$$\frac{J}{b+1} = J_{NL} - [J_{N0}U^2 + (J_{B0} + J_T + J_{C0})U_0 U - J_{C0}H_0 H] \tag{44}$$

where the term $J_T U_0 U$ is not a quasi-current but a real current.

The J–V_i curve is obtained by eliminating V and H using eqs. (41), (43) and (44). To do this eq. (44) is multiplied by b and substracted from eq. (43). After rearranging terms, we obtain

$$J_{N0}bU^2 + [J_{B0}b + J_Tb + J_{C0}(b+1)]U_0U$$
$$= J_{P0}H^2 + + [J_{B0} + J_{C0}(b+1)]H_0H + (J_{NL}b - J_{PL}) = 0 \tag{45}$$

and giving an arbitrary value to H we obtain the value for U after solving second order equation (45). With these values of U and H, we obtain J from eqs. (44) or (43), and V_i from eq. (41). We repeat this operation for a different value of H, and in this way we obtain the J–V_i curve. We note that while the procedure based on setting H is appropriate for a n-illuminated cell, we advise to set U and solve in H for a p-illuminated one. This is due to the fact that in the latter case it is impossible to find a real solution for eq. (45) for H close to zero.

The calculation of the vertical ohmic drop requires the knowledge of the profile, which can be obtained once the quasi-Fermi level splitting is known at both base ends. The procedure indicated above for calculating the quasi-Fermi levels ϕ_N and ϕ_P allows calculation of the profile, and therefore V_R, but in this case we need a numerical integration procedure.

4. The unitary Ebers–Moll model of bifacial cells

In this section we present a formalism presented in more detail elsewhere (Luque and Araújo 1984).

The low-injection equations, eqs. (13) and (14), and the high-injection ones, eqs. (44) and (43), can be written in an unitary form

$$J_{NL} - \frac{J}{b+1} = \frac{(J_{C0}/u)}{\alpha_F} U^2 - (J_{C0}/h)H^2, \tag{46}$$

$$J_{PL} - \frac{Jb}{b+1} = -(J_{C0}/u)U^2 + \frac{(J_{C0}/h)}{\alpha_R} H^2, \tag{47}$$

where

$$\alpha_F = \frac{J_{C0}/u}{(J_{B0} + J_{C0})/u + J_TU_0/U + J_{N0}} = \frac{J_{C0}}{J_{B0} + J_{C0} + J_TuU_0/U + J_{N0}u}, \tag{48}$$

$$\alpha_R = \frac{J_{C0}/h}{(J_{B0} + J_{C0})/h + J_{P0}} = \frac{J_{C0}}{J_{B0} + J_{C0} + J_{P0}h}, \tag{49}$$

where u and h are specially defined parameters to present the level of injection. These equations reduce to the low-injection case for $u = h = 1$ and $b = 0$, and reduce to the high-injection case for $u = U/U_0$ and $h = H/H_0$. Setting $b = 0$ has an intuitive interpretation. In the high-injection case the electron and hole concentrations are similar, but in the low-injection case (for the n^+–p–p^+ structure) the hole concentration is higher. If we admit that only a fraction of this, equal to the electron concentration, participates in the conduction process, we have to assign a much higher mobility to these carriers, leading to small values of b. In a p^+–n–n^+ structure the proper setting will be $b = \infty$.

The use of u, h, and the proper setting of b ($b = 0$ or $b \simeq 3$) allows to analyze the cases of low and very high (p–i–n) levels of injection. Cases of intermediate injection could also be interpreted by using these parameters. We are further studying this possibility in an attempt to give an intuitive interpretation of the intermediate injection phenomena with more insight than the one obtained from pure calculations.

Going back to the p–i–n case, the α's in eqs. (48) and (49) decrease with the level of injection. This behaviour is due to the odd dependence of the collector current, and is the opposite to the one expected in the low-injection case, caused by the space-charge recombination terms there. In the model presented here α_R is constant, and α_F increases with the level of injection for the low-injection case.

The formulation of eqs. (46) and (47) is similar to that of the Ebers–Moll model for bipolar transistors if the quasi-Fermi level splittings are considered as the emitter and collector voltages. The terms on the left-hand side are the emitter and collector currents. Their sum (with reversed sign) is the base current, and represents the net recombination at the cell due to dark profiles.

It is obvious that eqs. (46) and (47) cannot be solved for U^2 and H^2 in the high-injection case, since U and H appear in the coefficients. The actual procedure for obtaining U and H is explained in detail in section 2 for the low-injection case, and in section 3 for the high-injection one. Nevertheless the equations presented in this section allow a complete physical insight of the bifacial cell behaviour. In particular in short-circuit and open-circuit conditions the behaviour of bifacial cells is easily interpreted when based on these equations.

4.1. Short-circuit conditions

An important aspect of bifacial cells is that they present a good photo-response for illumination impinging on either of their two sides. Let J^{n}_{NL} and $J^{n}_{PL} = t^{n}_{L} J^{n}_{NL}$ be the values of the photogenerated currents in eqs. (17) and (18) under front illumination only (the superscript n corresponds to illumination on the n-side). For back illumination we use J^{p}_{PL} and $J^{p}_{NL} = t^{p}_{L} J^{p}_{PL}$. The transference parameters t are, in general, rather small.

The short-circuit current is obtained by setting U or $H = 0$. Under such condition one of the junctions represents a value of $\phi = -\infty$, and the same occurs for V_i. As a result, the short-circuit currents that we are calculating here are actually the uncompensated photogenerated currents. In practice, there is no difference for good cells.

Under low injection, the p–n junction is the one withstanding the reverse bias, and therefore $U = 0$. We can easily eliminate H, and we obtain

$$J^{n}_{sc} = J^{n}_{NL}(1 + \alpha_R t^{n}_{L}), \tag{50}$$

$$J^{p}_{sc} = J^{p}_{PL}(\alpha_R + t^{p}_{L}). \tag{51}$$

These equations have an intuitive interpretation. Under front illumination the short-circuit current consists of two terms: J^{n}_{NL}, collected directly at the front junction, and $J_{NL} t^{n}_{L} \alpha_R$, which is the fraction of photogenerated current transmitted to the high–low junction (factor t^{n}_{L}) and reflected there towards the front junction (fractor α_R). Under back illumination the current collected at the high–low junction J^{p}_{PL} is reflected towards the p–n junction (factor α_R) or transmitted there directly (factor t^{p}_{L}).

Under p–i–n conditions the half-junction (p–i or n–i) that becomes reversed-biased is, usually, the one opposite to the illuminated face. The corresponding equations are

$$J^{n}_{sc} = J^{n}_{NL} \frac{1 + b}{\alpha^{n}_{Fsc} + b} (\alpha^{n}_{Fsc} + t^{n}_{L}), \tag{52}$$

$$J^{p}_{sc} = J^{p}_{PL} \frac{1 + b}{1 + b\alpha^{p}_{Rsc}} (\alpha^{p}_{Rsc} + t^{p}_{L}). \tag{53}$$

The use of indices in eqs. (52) and (53) is intended to remind us that the α's are variable. n or p refers to which side is illuminated. sc refers to short-circuit conditions.

The factors $(\alpha + t)$ in these equations are characteristic of the low-injection case for back illumination. The high-injection factor containing b is slightly higher than one. In consequence, for n^+–p–p^+ where the α's are similar for low and high injection, the low-injection case gives a slightly higher front-illuminated response but a lower back-illuminated one.

The fact that the dark junction becomes reverse-biased requires some explanation. From a formal point of view this is explained by analysis of eq. (45). As we mentioned before, if the condition $J_{NL}b > J_{PL}$ is approximately fulfilled, setting $U = 0$ leads to imaginary values of H. This means that in fact we have to set $H = 0$, in agreement with our previous statement. The opposite is true if the condition mentioned before is reversed.

Let us consider a front-illuminated n^+–p–p^+ cell with a high base resistivity so that a p–i–n behaviour can be ascribed to this cell. When we approach short-circuit conditions the concentration at the p–p^+ junction decreases until it reaches the value N_B. At this moment the cell enters a transition regime; concentration at its front side begins to be reduced and a drastic process of redistribution of carriers in the base results. This behaviour has been shown through numerical calculations (Weaver and Nasby 1981). However, above a certain base resistivity (1000 Ω cm in p-type silicon, or 500 Ω cm in n-type) the current is saturated before the base concentration has been reached. In these cases the short-circuit current obtained from the p–i–n model coincides with the actual one.

In back-illuminated cells this rearrangement of carriers does not occur. The high-injection predicts a decrease of the carriers near the p–n junction, and this junction can actually allow this decrease by taking a negative value of polarization. Of course, the high-injection conditions cease to prevail near the junction, but this results in negligible changes in the recombination, and therefore in the I–V curve.

In general, a n^+–p–p^+ cell presents its highest electron concentration at the p–p^+ junction (remember that it acts as an open-circuited p–n junction). If the illumination is on the n^+-side this value is lower than if the illumination is on the p^+-side. Depending on the base resistivity the cell can be in low injection for both cases or can enter into a transition regime towards high injection in the back-illuminated case. For base resistivities below 50 Ω cm, the low-injection regime applies at AM1 conditions, and the symmetry factor defined as J^p_{sc}/J^n_{sc} is given by the low-injection formulas

$$\frac{J^p_{sc}}{J^n_{sc}} = \frac{J^p_{PL}}{J^n_{NL}} \frac{\alpha_R + t^p_L}{1 + \alpha_R t^n_L}. \tag{54}$$

The second factor on the right-hand side is always smaller than one, accounting for the transport of a large number of carriers through the base in the back-illuminated case.

For higher values of base resistivity, it is more appropriate to use the formula obtained by using the high-injection model for back illumination and the low-injection one for front illumination

$$\frac{J^{\mathrm{p}}_{\mathrm{sc}}}{J^{\mathrm{n}}_{\mathrm{sc}}} = \frac{J^{\mathrm{p}}_{\mathrm{PL}}}{J^{\mathrm{n}}_{\mathrm{NL}}} \frac{1+b}{1+b\alpha^{\mathrm{p}}_{\mathrm{Rsc}}} \frac{\alpha^{\mathrm{p}}_{\mathrm{Rsc}} + t^{\mathrm{p}}_{\mathrm{L}}}{1+\alpha_{\mathrm{R}} t^{\mathrm{n}}_{\mathrm{L}}}. \tag{55}$$

The transition regime can be represented by setting values of b between zero and the real one (≈ 3). We observe that the high-injection factor is greater than one and therefore improves the cell symmetry.

Finally, for cells of very high base resistivity (above $1000\,\Omega\,\mathrm{cm}$) we can consider, for the reasons given above, that the short-circuit currents are the one given by the high-injection formulas. The symmetry factor is then

$$\frac{J^{\mathrm{p}}_{\mathrm{sc}}}{J^{\mathrm{n}}_{\mathrm{sc}}} = \frac{J^{\mathrm{p}}_{\mathrm{PL}}}{J^{\mathrm{n}}_{\mathrm{NL}}} \frac{\alpha^{\mathrm{n}}_{\mathrm{Fsc}}+b}{1+b\alpha^{\mathrm{n}}_{\mathrm{Rsc}}} \frac{\alpha^{\mathrm{p}}_{\mathrm{Rsc}} + t^{\mathrm{p}}_{\mathrm{L}}}{\alpha^{\mathrm{n}}_{\mathrm{Fsc}} + t^{\mathrm{n}}_{\mathrm{L}}}. \tag{56}$$

The high-injection factor, i.e. the second term on the right-hand side, is always slightly higher than one. This leads to a high symmetry factor that, in some cases, must be higher than one. We should remember that a cell working under p–i–n conditions has lost any memory of the type of conductivity at the base, and therefore does not know where the p–n junction is located.

In tables 2, 3 and 4, we present some calculations corresponding to good cells of several base resistivities to support the range of validity of each approach. The recombination parameters J_{N0}, J_{P0}, τ_{e} and τ_{n} have been selected according to the best achievable values as discussed in section 2. In this case we have not included the term $J_{\mathrm{T}}U_0$, because in low injection it almost does not affect V_{oc} or J_{sc}. In high-injection it is hidden by the terms of recombination in the base that have the same dependence on the voltage. For the photogenerated currents we have been rather moderate and have used values more in accordance with those commonly obtained in our industrial process with a certain grid opacity and a fast-deposited AR coating. We observed very good values of the symmetry factors for all situations, but the tendency is always the one indicated in the preceding theoretical discussion. In table 2 we present a verification of the validity of the high- or low-injection models.

From an experimental point of view we have to mention that the lack of homogeneity of the AR coatings affects the symmetry factor (affecting $J^{\mathrm{p}}_{\mathrm{PL}}/J^{\mathrm{n}}_{\mathrm{NL}}$) so that values above one are observed sometimes, even in low injection. This has been clearly shown by diffusing a p–n junction through a bright-etched surface and a p–p$^+$ junction through a polished surface. A

Table 2a
Comparison between low-injection and high-injection calculations for different base resistivities for n^+–p–p^+ structures under short-circuit and open-circuit operation.

Low-injection model calculations						Low or high injection	High-injection model calculations							
Illuminated side	N_B (cm^{-3})	J (mA/cm^2)	V (mV)	$n_B(0)$ (cm^{-3})	$n_B(W_B)$ (cm^{-3})		h	u	$n_B(W_B)$ (cm^{-3})	$n_B(0)$ (cm^{-3})	V (mV)	J (mA/cm^2)	ϱ_B (Ωcm)	Illuminated side
n	4.0×10^{16}	31.4	SC	–	1.0×10^{13}	L							0.3	n
p	4.0×10^{16}	24.8	SC	–	1.67×10^{14}	L							0.3	p
n	1.3×10^{16}	31.6	SC	–	9.68×10^{12}	L							1.0	n
p	1.3×10^{16}	29.3	SC	–	1.61×10^{14}	L							1.0	p
n	2.3×10^{15}	31.7	SC	–	8.85×10^{12}	L							5.0	n
p	2.3×10^{15}	30.9	SC	–	1.48×10^{14}	L							5.0	p
n	8.0×10^{14}	31.8	SC	–	8.60×10^{12}	L							15.0	n
p	8.0×10^{14}	31.1	SC	–	1.43×10^{14}	L							15.0	p
n	2.3×10^{14}	31.8	SC	–	8.46×10^{12}	L							50.0	n
p	2.3×10^{14}	31.2	SC	–	1.41×10^{14}	L↑, H↓							50.0	p
n	2.0×10^{13}	31.8	SC	–	8.37×10^{12}	L↑, H↓		9.3		1.86×10^{14}	315	31.2	600.0	n
p	2.0×10^{13}	31.2	SC	–	1.40×10^{14}	H	2.96		5.91×10^{13}		304	31.6	600.0	p
n	1.3×10^{13}	31.8	SC	–	8.35×10^{12}	L↑, H↓		14.3		1.86×10^{14}	315	31.2	1000.0	n
p	1.3×10^{13}	31.2	SC	–	1.39×10^{14}	H	4.55		5.91×10^{13}		304	31.6	1000.0	p
n	4.0×10^{16}	OC	600	1.34×10^{14}	1.13×10^{14}	L							0.3	n
p	4.0×10^{16}	OC	594	1.06×10^{14}	2.48×10^{14}	L							0.3	p
n	1.3×10^{16}	OC	599	3.88×10^{14}	3.65×10^{14}	L							1.0	n
p	1.3×10^{16}	OC	597	3.59×10^{14}	4.90×10^{14}	L							1.0	p
n	2.3×10^{15}	OC	590	1.54×10^{15}	1.50×10^{15}	L↑, H↓	0.86	0.83	1.92×10^{15}	1.98×10^{15}	591	OC	5.0	n
p	2.3×10^{15}	OC	589	1.50×10^{15}	1.60×10^{15}	L↑, H↓	0.92	0.86	2.12×10^{15}	1.92×10^{15}	594	OC	5.0	p
n	8.0×10^{14}					H	2.38	2.44	1.95×10^{15}	1.90×10^{15}	591	OC	15.0	n
p	8.0×10^{14}					H	2.60	2.38	1.90×10^{15}	2.08×10^{15}	594	OC	15.0	p
n	2.3×10^{14}					H							50.0	n
p	2.3×10^{14}					H							50.0	p
n	1.3×10^{13}					H	144.6	149.2	1.94×10^{15}	1.88×10^{15}	590	OC	1000.0	n
p	1.3×10^{13}					H	140.8	154.3	1.89×10^{15}	2.07×10^{15}	593	OC	1000.0	p

Other parameters used: those of tables 2 and 3, where necessary; J_L (illuminated side) = 30 mA/cm^2; J_L (non-illuminated side) = 1.8 mA/cm^2.

Table 2b
Comparison between low-injection and high-injection calculations for different base resistivities for p^+–n–n^+ structures under short-circuit and open-circuit operation.

Low-injection model calculations						Low or high injection	High-injection model calculations							
Illuminated side	N_B (cm^{-3})	J (mA/cm^2)	V (mV)	$p_B(0)$ (cm^{-3})	$p_B(W_B)$ (cm^{-3})		h	u	$p_B(W_B)$ (cm^{-3})	$p_B(0)$ (cm^{-3})	V (mV)	J (mA/cm^2)	ϱ_B (Ωcm)	Illuminated side
p	1.3×10^{16}	30.8	SC	–	1.33×10^{13}	L							0.3	p
n	1.3×10^{16}	14.7	SC	–	2.21×10^{14}	L							0.3	n
p	4.5×10^{15}	31.2	SC	–	1.82×10^{13}	L							1.0	p
n	4.5×10^{15}	22.5	SC	–	3.04×10^{14}	L							1.0	n
p	9.0×10^{14}	31.6	SC	–	2.10×10^{13}	L							5.0	p
n	9.0×10^{14}	27.8	SC	–	3.51×10^{14}	L							5.0	n
p	3.0×10^{14}	31.6	SC	–	2.15×10^{13}	L				6.02×10^{13}	305	31.7	15.0	p
n	3.0×10^{14}	28.9	SC	–	3.58×10^{14}	L↑, H↓			1.89×10^{14}	–	319	31.2	15.0	n
p	8.5×10^{13}	31.6	SC	–	2.15×10^{13}	L↑, H↓			–	5.87×10^{13}	329	31.2	50.0	p
n	8.5×10^{13}	29.3	SC	–	3.59×10^{14}	H	2.18		1.86×10^{14}	–	329	31.2	50.0	n
p	4.0×10^{12}					H		14.79		5.91×10^{13}	304	31.6	1000.0	p
n	4.0×10^{12}					H	46.44		1.86×10^{14}		315	31.2	1000.0	n
p	1.3×10^{16}	OC	606	5.22×10^{14}	2.38×10^{14}	L							0.3	p
n	1.3×10^{16}	OC	588	1.50×10^{14}	3.28×10^{14}	L							0.3	n
p	4.5×10^{15}	OC	594	9.18×10^{14}	6.52×10^{14}	L							1.0	p
n	4.5×10^{15}	OC	586	6.61×10^{14}	7.59×10^{14}	L↑							1.0	n
p	9.0×10^{14}					H	2.08	2.28	1.88×10^{15}	2.05×10^{15}	594	OC	5.0	p
n	9.0×10^{14}					H	2.14	2.07	1.93×10^{15}	1.87×10^{15}	589	OC	5.0	n
p	3.0×10^{14}					H	6.33	6.95	1.90×10^{15}	2.08×10^{15}	594	OC	15.0	p
n	3.0×10^{14}					H	6.51	6.33	1.95×10^{15}	1.90×10^{15}	590	OC	15.0	n
p	8.5×10^{13}					H	22.12	24.23	1.88×10^{15}	2.06×10^{15}	593	OC	50.0	p
n	8.5×10^{13}					H	22.82	22.12	1.94×10^{15}	1.88×10^{15}	590	OC	50.0	n
p	4.0×10^{12}					H	472.5	517.5	1.89×10^{15}	2.07×10^{15}	593	OC	1000.0	p
n	4.0×10^{12}					H	485.0	470.0	1.94×10^{15}	1.88×10^{15}	590	OC	1000.0	n

Other parameters used: those of tables 2 and 3, where necessary; J_L (illuminated side) = 30 mA/cm^2; J_L (non-illuminated side) = 1.8 mA/cm^2.

Table 3
Low-injection parameters of different solar cell structures.

Solar cell structure	ϱ_B (Ω cm)	N_B (cm^{-3})	τ^* (μs)	L^* (μm)	J_{B0} (A/cm^2)	J_{C0}	α_F	α_R	$U_0 = H_0$
p^+–n–n^+	0.3	1.3×10^{16}	152	386	1.96×10^{-13}	9.03×10^{-13}	0.6951	0.4302	9.169×10^5
	1.0	4.5×10^{15}	236	510	3.72×10^{-13}	3.04×10^{-13}	0.8414	0.6887	3.174×10^5
	5.0	9.0×10^{14}	287	585	1.54×10^{-12}	1.66×10^{-11}	0.9052	0.8673	6.348×10^4
	15.0	3.0×10^{14}	296	600	4.47×10^{-12}	5.07×10^{-11}	0.9157	0.9027	2.116×10^4
	50.0	8.5×10^{13}	299	606	1.56×10^{-11}	1.81×10^{-10}	0.9196	0.9159	5.995×10^3
	1000.0	4.0×10^{12}	300	609	3.30×10^{-10}	3.73×10^{-9}	0.9188	0.9186	2.820×10^2
p^+–p–n^+	0.3	4.0×10^{16}	890	1390	1.12×10^{-14}	6.92×10^{-13}	0.4062	0.7661	2.821×10^6
	1.0	1.3×10^{16}	746	1410	4.15×10^{-14}	2.63×10^{-12}	0.7166	0.9160	9.169×10^5
	5.0	2.3×10^{15}	557	1320	3.11×10^{-13}	1.73×10^{-11}	0.9294	0.9713	1.622×10^5
	15.0	8.0×10^{14}	520	1290	9.69×10^{-13}	5.14×10^{-11}	0.9631	0.9778	5.643×10^4
	50.0	2.3×10^{14}	506	1290	3.43×10^{-12}	1.82×10^{-10}	0.9763	0.9805	1.622×10^4
	1000.0	1.3×10^{13}	500	1289	6.17×10^{-11}	3.26×10^{-9}	0.9812	0.9814	7.053×10^2

Other parameters used: $J_{P0} = 2.0 \times 10^{-13}$ A/cm^2; $J_{N0} = 1.0 \times 10^{-12}$ A/cm^2; $n_i^2 = 2.01 \times 10^{20}$ cm^{-3}; $W_B = 250$ μm.

* τ, the minority-carrier lifetime is taken from Fossum and Lee (1982) and L, the diffusion length is calculated using the μ's (mobilities) found by Thurber et al. (1980a,b).

Table 4
High-injection model calculations for a p^+–i–n^+ structure.

Illuminated side	J (mA/cm^2)	V (mV)	α_F	α_R	$p_{BP} = n_{BP}$ (cm^{-3})	$p_{BN} = n_{BN}$ (cm^{-3})	θ
p	OC	593	0.9059	0.9617	2.07×10^{15}	1.89×10^{15}	1.0956
	9.0	583	0.9182	0.9648	1.69×10^{15}	1.55×10^{15}	1.0927
	15.3	573	0.9286	0.9673	1.38×10^{15}	1.27×10^{15}	1.0931
	22.9	554	0.9445	0.9709	9.37×10^{15}	8.49×10^{14}	1.1040
	27.4	530	0.9577	0.9738	5.87×10^{15}	5.15×10^{14}	1.1395
	29.5	507	0.9658	0.9655	3.77×10^{14}	3.12×10^{14}	1.2081
	31.6	304	0.9787	0.9782	5.91×10^{13}	–	–
n	OC	590	0.9039	0.9032	1.88×10^{15}	1.94×10^{15}	0.9710
	10.4	578	0.9182	0.9668	1.45×10^{15}	1.55×10^{15}	0.9352
	16.5	566	0.9286	0.9693	1.14×10^{15}	1.27×10^{15}	0.9005
	23.9	543	0.9445	0.9729	6.93×10^{14}	8.49×10^{14}	0.8161
	28.2	510	0.9577	0.9758	3.42×10^{14}	5.15×10^{14}	0.6640
	30.2	469	0.9658	0.9776	1.32×10^{14}	3.12×10^{14}	0.4233
	31.2	198	0.9711	0.9787	9.76×10^{10}	1.83×10^{14}	0.0005
$1p + \frac{1}{2}n$	OC	604	0.8877	0.9578	1.56×10^{15}	2.41×10^{15}	1.0654
	16.4	592	0.9058	0.9625	1.98×10^{15}	1.89×10^{15}	1.0468
	31.7	571	0.9286	0.9680	1.29×10^{15}	1.27×10^{15}	1.0200
	39.2	550	0.9445	0.9717	8.44×10^{14}	8.49×10^{14}	0.9947
	45.6	497	0.9658	0.9763	2.84×10^{14}	3.12×10^{14}	0.9103
	47.6	107	0.9773	0.9787	1.51×10^{10}	3.36×10^{13}	0.0004
$1n + \frac{1}{2}p$	OC	603	0.8868	0.9585	2.47×10^{15}	2.43×10^{15}	1.0164
	17.2	589	0.9058	0.9634	1.86×10^{15}	1.89×10^{15}	0.9824
	29.5	573	0.9236	0.9678	1.31×10^{15}	1.40×10^{15}	0.9393
	39.7	544	0.9445	0.9727	7.22×10^{14}	8.49×10^{14}	0.8508
	46.0	476	0.9658	0.9773	1.62×10^{14}	3.12×10^{14}	0.5179
	47.2	97	0.9723	0.9787	–	1.54×10^{14}	0.0000

Other parameters used: $J_{P0} = 2 \times 10^{-13}$ A/cm^2; $J_{N0} = 1.0 \times 10^{-12}$ A/cm^2; $\tau_B = 800$ μs; $L_B = 1200$ μm; $w_B = 250$ μm; $n_i^2 = 2.01 \times 10^{20}$ cm^{-3}; $J_{B0}U_0 = 3.53 \times 10^{-8}$ A/cm^2; $J_{C0}U_0 = 1.62 \times 10^{-6}$ A/cm^2; $b = 2.7$; $kT/q = 0.025$ V.

symmetry factor of 1.05 has been obtained (Eguren et al. 1981). As a result, it is more relevant to look at the statistical behaviour when both faces have similar polishing (bright-etched). With this objective in mind we have analyzed a set of laboratory made cells (Calleja et al. 1984a) with base p-resistivity of 5 and 37 Ω cm and n-resistivity of 1000 Ω cm. Also, we have compared those results with a group of industrial bifacial n^+–p–p^+ cells of 16 and 48 Ω cm (Calleja et al. 1984b). In table 5 we present the symmetry factor (J_{sc} back-illuminated/J_{sc} front-illuminated) for the averaged values of the short-circuit current densities. We observe that the symmetry factor increases with the base resistivity, ranging from values of 0.84 up to nearly 1.

Even though the 1000 Ω cm cells are p^+–n–n^+, we have not interpreted (Calleja et al. 1984a) the slightly higher current under p^+-illumination as due to the presence of the p–n junction at the p^+-side. Yet, we consider that this is due to the fact that the B-diffused p^+-layer collects more current than the P-diffusion n^+ one. This has been unequivocally shown by measuring the spectral response (Eguren et al. 1981) at $\lambda = 0.40\,\mu m$, which is higher under back illumination in a n^+–p–p^+ cell than under front illumination. This is coherent with the fact that p^+-emitters have been shown to be transparent to electrons, while n^+-emitters are not transparent to holes. This has been shown by measuring V_{oc} in partially metallized bifacial solar cells (Eguren et al. 1981). The better behaviour of the p^+-emitters adds to the high-injection behaviour, shown in eq. (56), to produce higher short-circuit currents in p^+-illuminated p–i–n cells.

Therefore we have to conclude that very high resistivity n^+–p–p^+ cells should give values of the symmetry factor, not affected by the surface

Table 5
Symmetry factor (J_{sc} front-illuminated/J_{sc} back-illuminated) for several base resistivities for the averaged values of the short-circuit currents in groups of cells. Data from Calleja et al. (1984a) and Eguren et al. (1981).

Number of cells	Cell type	Base ϱ (Ωcm)	$\langle J_{sc}\rangle_{back}/\langle J_{sc}\rangle_{front}$
45	n^+–p–p^+	5	0.839
34	n^+–p–p^+	15	0.843
32	n^+–p–p^+	35	0.878
25	n^+–p–p^+	48	0.915
23	p^+–n–n^+	1000	0.962

roughness, higher than one. However, this effect has not yet been experimentally shown.

Still we have to consider the case of bifacial illumination. Under low injection strict superposition occurs. Under p–i–n behaviour the half-junction sustaining negative voltage can change with respect to that under monofacial illumination. If $J_{NL}b > J_{PL}$, which in the bifacial case becomes $(J^{n}_{NL} + t^{p}_{L}J^{p}_{PL})b > J^{p}_{PL} + t^{n}_{L}J^{n}_{NL}$, this results in $H = 0$ i.e., the p–p^{+} junction is the one withstanding the negative bias. In this case

$$J^{np}_{sc} = J^{n}_{NL}\frac{1+b}{\alpha^{np}_{Fsc}+b}(\alpha^{np}_{Fsc}+t^{n}_{L}) + J^{n}_{PL}\frac{1+b}{\alpha^{np}_{Fsc}+b}(1+t^{p}_{L}\alpha^{np}_{Fsc}), \tag{57}$$

which shows that in this case there is a back collecting junction (factor $\alpha + t$) and a front collecting one (factor $1 + \alpha t$), both affected by the high-injection factor.

In the p–i–n case, the superposition is not fully respected. The presence of the p^{+} front collecting junction increases the current and the decrease of the α's under high injection levels causes opposite effects on both terms of eq. (57). The overall effect is usually an overlinearity in the superposition. However, this non-linear effect is small, due to the fact that the total current is near the total photogenerated one, which follows a superposition rule. We have not yet carried out careful measurements to analyse this effect.

A final remark is pertinent here. According to the calculations in table 1 and 2 the term J_{B0} is 22 × higher than J_{P0} for the n^{+}–p–p^{+} cells of 15 Ω cm and 78 × higher for the cells of 50 Ω cm. This means that α_{R} is not affected by the emitter recombination in good n^{+}–p–p^{+} cells and the same happens with the symmetry factor [see eqs. (54) and (55)]. Besides the J^{n}_{NL}/J^{p}_{PL} uncertainties, this factor is mainly affected by the base material properties. In p^{+}–n–n^{+} cells or p–i–n cells this is also true, but to a smaller extent (the base influence is about 5 times that of the n^{+}-emitter).

4.2. Open-circuit voltage

The open-circuit voltage is obtained by setting $J = 0$ in eqs. (43) and (44). In this way we obtain U^{2} and H^{2}, and we substitute these in eq. (41). For the case of front illumination we obtain

$$V^{n}_{oc} = \frac{kT}{q}\ln\left\{\left[\frac{J^{n}_{NL}}{J_{C0}/u^{n}_{oc}}\frac{\alpha_{Foc}(1+t^{n}_{L}\alpha^{n}_{Rcc})}{1-\alpha^{n}_{Foc}\alpha^{n}_{Roc}}\right]^{1/(b+1)} \times\left[\frac{J^{n}_{NL}}{J_{C0}/h^{n}_{oc}}\frac{\alpha_{Roc}(\alpha^{n}_{Foc}+t^{n}_{L})}{1-\alpha^{n}_{Foc}\alpha^{n}_{Roc}}\right]^{b/(b+1)}\right\}. \tag{58}$$

For back illumination

$$V_{\rm oc}^{\rm p} = \frac{kT}{q} \ln\left\{\left[\frac{J_{\rm PL}^{\rm p}}{J_{\rm C0}/u_{\rm oc}^{\rm p}} \frac{\alpha_{\rm Foc}^{\rm p}(\alpha_{\rm Foc}^{\rm p} + t_{\rm L}^{\rm p})}{1 - \alpha_{\rm Foc}^{\rm p}\alpha_{\rm Roc}^{\rm p}}\right]^{1/(b+1)} \times \left[\frac{J_{\rm PL}^{\rm p}}{J_{\rm C0}/h_{\rm oc}^{\rm p}} \frac{\alpha_{\rm Roc}^{\rm p}(1 + t_{\rm L}^{\rm p}\alpha_{\rm Foc}^{\rm p})}{1 - \alpha_{\rm Foc}^{\rm p}\alpha_{\rm Roc}^{\rm p}}\right]^{b/(b+1)}\right\}. \tag{59}$$

In these equations n or p refers to the illuminated face, and oc refers to open-circuit conditions. For low injection the same equations are valid if $b=0$ and $u_{\rm oc} = h_{\rm oc} = 1$ and in this case the indices can be dropped from the α's, since they are constant.

The difference $\Delta V_{\rm oc}^{\rm n-p} = V_{\rm oc}^{\rm n} - V_{\rm oc}^{\rm p}$ takes a complex expression due to the small differences of the α's, u and h for front and back illumination. If these differences are disregarded we obtain the approximate formula (exact for low injection if $b=0$)

$$\Delta V_{\rm oc}^{\rm n-p} = \frac{kT}{q} \ln\left[\frac{J_{\rm NL}^{\rm n}}{J_{\rm PL}^{\rm p}} \left(\frac{1 + t_{\rm L}^{\rm n}\alpha_{\rm R}}{\alpha_{\rm R} + t_{\rm L}}\right)^{1/(b+1)} \times \left(\frac{\alpha_{\rm F} + t_{\rm L}^{\rm n}}{1 + t_{\rm L}^{\rm p}\alpha_{\rm F}}\right)^{b/(b+1)}\right]. \tag{60}$$

If $J_{\rm NL}^{\rm n} = J_{\rm PL}^{\rm p}$, and $\alpha = 1$, this difference is zero. If $\alpha_{\rm F}$, $\alpha_{\rm R} < 1$ the first term in parentheses is greater and the second one smaller than one. As a consequence, if $b=0$ (the low-injection case) we obtain a positive difference as it is to be expected by the superposition principle. Even more, if we consider the ratio $J_{\rm sc}^{\rm p}/J_{\rm sc}^{\rm n}$ from eq. (54) for the case $b=0$, we obtain

$$\Delta V_{\rm oc}^{\rm n-p} = \frac{kT}{q} \ln\left(\frac{J_{\rm sc}^{\rm n}}{J_{\rm sc}^{\rm p}}\right), \tag{61}$$

showing the exact verification of the superposition principle.

Under high injection, with $b=3$, the second factor in parentheses in eq. (60), which is smaller than one, is preponderant. As a result, $\Delta V_{\rm oc}^{\rm n-p}$ tends to be negative. In intermediate cases $(0<b<3)$ an intermediate situation is produced. In conclusion, $\Delta V_{\rm oc}^{\rm n-p}$ is expected to take positive or negative values in a continuous form, the more positive being the ones corresponding to a clear low-injection behaviour (implying the verification of the superposition principle).

If we consider eqs. (43) and (44) for the open-circuit case $(J=0)$, we note that they are fairly symmetrical and they do not depend on b (i.e.,

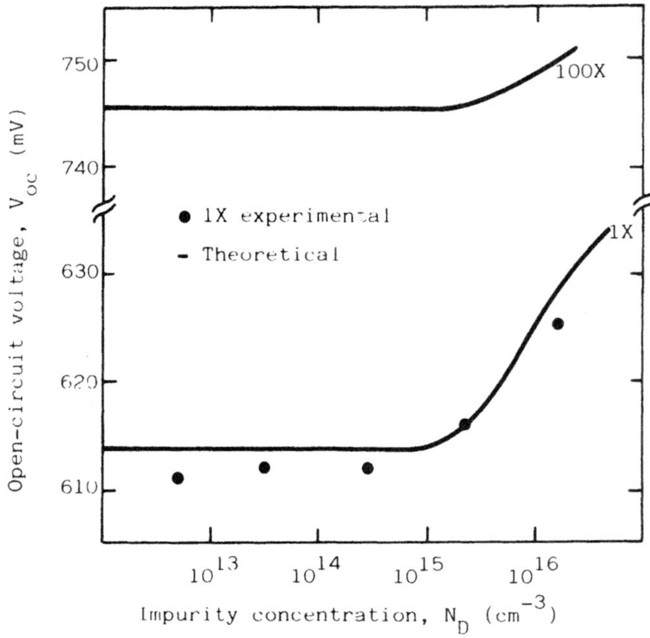

Fig. 3. Open-circuit voltage versus base doping in p–n–p$^+$ cells for two concentration levels. The data are experimental points. (After Ruiz et al. 1983)

on the high-injection parameter). So, the main reason for the lack of symmetry lies in eq. (41), showing that for low injection ($b = 0$) the voltage is caused by ϕ_N at the junction and therefore only U^2 is important, while under high injection the voltage is created mainly by ϕ_P, so that the value of H^2 at the high–low junction is very important.

An exact calculation of V_{oc} can be done using the procedure indicated in section 3. We present in tables 2, 3 and 4 the values of α, u, h and V_{oc} for good cells and various values of base resistivity. To check the validity of the high- or low-injection model the highest or the lowest values of carriers concentration in the base are to be compared to the base doping. We conclude that above 15 Ω cm all the cells are in high injection for open-circuit conditions. The high values of V_{oc} in high injection are caused, as observed in eqs. (58) and (59), by the high-injection parameters u and h, which are greater than one. These approximate calculations can be compared to the exact model by Ruiz, Cuevas and Cid (Ruiz et al. 1983) showing the dependence of the open-circuit voltage of a p$^+$–n–n$^+$ cell versus the base resistivity, as shown in fig. 3.

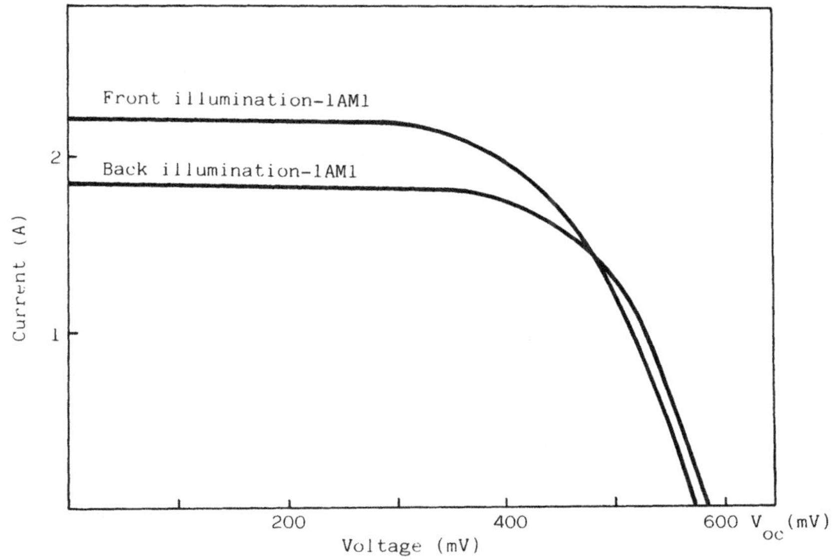

Fig. 4. Front and back illuminated *I–V* curves of the same n^+–p–p^+ industrial cell showing a cross-over that implies a violation of the superposition principle.

As we mentioned above, most industrial bifacial cells have base resistivities between 15 and 50 Ω cm. They are in low injection in short-circuit conditions and in high injection in open-circuit ones. In general, higher short-circuit current for n-illumination and higher open-circuit voltage for p-illumination are to be expected. This means that the front- and back illuminated *I–V* curves of a given cell cross one another and we present this behaviour in fig. 4. This is a clear violation of the superposition principle so commonly invoked when dealing with silicon cells (see for an analysis Lindholm et al. 1979). This crossing effect is not expected to occur in the case of p^+–n–n^+ cells, where either in high or low injection the voltage is higher when the cell is p-illuminated.

From an experimental point of view the open-circuit voltage of bifacial cells (industrially manufactured) is close to the theoretical values calculated in tables 1, 2 and 3 for the best achievable cells. This can be seen by looking at the figures presented in section 6. Indeed, for the best industrial cells, the values of V_{oc}^{n} and V_{oc}^{p} are 590 and 587 mV, and 594 mV under bifacial illumination (AM1 on the n-side and 0.5 AM1 on the p-side) to be compared with values of 590, 593 and 603 mV calculated, respectively, for n, p and bifacial illumination. All these data correspond to measurements or calculations at 25°C.

Finally, we have to mention that under open-circuit conditions, in high injection, $J_{N0}u_{oc}^{n}$, $J_{N0}u_{oc}^{p}$, $J_{P0}h_{oc}^{n}$ and $J_{P0}h_{oc}^{p}$ are of the same order of magnitude as J_{B0}, so that they seriously affect the value of the α's, and therefore the front to back illuminated open-circuit voltage. In other words: the fabrication of the emitter is an important step in obtaining high V_{oc} voltages under either n- or p-illumination.

4.3. *Curve shape*

The shape of the curve of a bifacial solar cell is affected by three types of mechanisms: the two-exponential curve shape, the series resistance effects and the parallel resistance effects.

The third effect is quite unknown, and we do not believe that it is very specific of bifacial cells. We do not treat this aspect in this work. The series resistance is considered in next section.

For the low-injection case the ideal cell curve shape is very easily obtained if we get H^2 versus U from eq. (14), and substitute this into eq. (13), and using eq. (15) and the definition of α_R,

$$J = (J_{NL} + J_{PL}\alpha_R) - [J_{N0} + \alpha_R J_{P0} + (1 + \alpha_R)J_{B0}] \exp\left(\frac{qV_i}{kT}\right) - J_T U_0 \exp\left(\frac{qV_i}{2kT}\right) \tag{62}$$

which, for good cells ($\alpha_R \rightarrow 1$), is equal to

$$J = (J_{NL} + J_{PL}) - (J_{N0} + J_{P0} + 2J_{B0}) \exp\left(\frac{qV_i}{kT}\right) - J_T U_0 \exp\left(\frac{qV_i}{2kT}\right), \tag{63}$$

showing a two-exponential shape that verifies the principle of superposition (except, possibly, for the constant J_T which might be somewhat different in dark, as compared with under illumination).

For the p–i–n case, we add eqs. (46) and (47) after substituting the α's as given by eqs. (48) and (49). We obtain

$$J = (J_{NL} + J_{PL}) - [J_{N0}U^2 + J_{P0}H^2 + J_{B0}(H_0U + U_0H)] - J_T U_0 U. \tag{64}$$

If we consider the parameter θ defined by

$$\theta = U/H, \tag{65}$$

we find that this parameter is a slowly variating function of the ideal cell voltage, as shown in the example presented in table 4. We can write eq. (64) in the form

$$J = (J_{NL} + J_{PL}) - [J_{N0}\theta^{2b/(b+1)} + J_{P0}\theta^{-2/(b+1)}] \exp\left(\frac{qV_i}{kT}\right)$$

$$- [(J_{B0} + J_T)\theta^{b/(b+1)} + J_{B0}\theta^{-1/(b+1)}] U_0 \exp\left(\frac{qV_i}{2kT}\right), \quad (66)$$

which is a two-exponential equation slightly modified by the dependence of θ on V_i.

When considering eq. (66) we are tempted to set $J_{sc} = J_{NL} + J_{PL}$. As discussed in detail in section 4.1 the short-circuit current is given by different expressions. This is a consequence of the variation of θ with V, which is particularly strong in short-circuit conditions. However, under maximum power point conditions taking θ constant is a good approximation.

In low injection, due to the superposition principle, the coefficients of the two exponentials are the same for dark-front or back-illuminated operations or in the J_{sc} (V_{oc}) curve. In the p–i–n case, θ and therefore these coefficients are not exactly the same. These points have been studied in some detail in previous papers (Araújo et al. 1982a, 1983) and we plan to analyze them further. However, the differences among the various ways of determining these constants are small. In all cases a fitting procedure has been used (Araújo et al. 1982b). In good cells the value of the first exponential, related to the emitters recombination and to the base in low injection, is of about 1×10^{-12}A/cm^2, while the second exponential in low-injection cells is 5×10^{-8}A/cm^2, and in high-injection cells is 3×10^{-7}A/cm^2. The difference is related to the fact that the whole base recombines near the maximum of the R–H–S law. Still, the difference is smaller than what could be expected from the large size of the base region, and merits to be examined in more detail.

5. Series resistance

The series resistance comes from two different mechanisms. The first one is related to the unidimensional flow of carriers through the wafer from the front to the back face. In the p–i–n case this resistance is given by

$$R_B = \int_0^W \frac{1}{q(\mu_e + \mu_n)A} \frac{1}{p} \, dx, \quad (67)$$

where A is the cell area.

According to this term, the shape of the curve in eq. (66) is now obtained by changing V into $(V - R_B J)$. The procedure to calculate the profiles mentioned in section 3 is used to calculate R_B, as discussed in detail elsewhere (Luque and Eguren 1982). We observe that both carriers take part in the conduction mechanism. This series resistance is characteristic of p–i–n cells regardless whether they are monofacial or bifacial. It decreases (Araújo et al. 1983) with the level of current, according to $(J_L + J)^{-1/2}$ for cells dominated by the $\exp(qV/2kT)$ term and to $(J_L + J)^{-1}$ for cells dominated by the $\exp(qV/kT)$ term. In both cases $J_L = J_{NL} + J_{PL}$. This feature is important, since it permits the operation of p–i–n cells with a maximum fill factor of 0.81 at 10 suns and a maximum efficiency of 21% at 400 AM1.5 suns. Efficiencies of 19.5% with 600 Ω cm base p–n–n$^+$ cells (calculated with respect to the uncovered area) have been experimentally obtained at 120 AM1.5 suns (Ruiz et al. 1982).

For calculations in cells for flat panels (Luque and Eguren 1982) the reduction of the maximum power point voltage due to this resistance is 3.6% in a p-illuminated p–i–n at AM1 and 2.3% in a n-illuminated cell. For a cell under bifacial illumination (AM1 on the n-side) this reduction is 1.9%.

In the low-injection case this vertical resistance is approximately

$$R_B = \int_0^W \frac{1}{q\mu_h A}\frac{1}{N_B}\,dx = \varrho_B \frac{W}{A}, \tag{68}$$

leading to a maximum power point reduction of 1.6% for 10 Ω cm cells and of 0.16% for 1 Ω cm cells. We conclude that this effect of series resistance is of a minor concern for the efficiency of cells working under p–i–n conditions.

A second term related to series resistance is the one based on the lateral flow of carriers across the cell in order to reach the metal fingers of the contact grid, and then along the fingers to reach the cell busbar. In low-injection cells this effect has been carefully analyzed (Luque et al. 1980) and for properly designed grids the series resistance corresponding to an elementary cell (the small cell strip adjacent to a metal finger, according to fig. 5)

$$R_G = \frac{1}{3}\left(R_{\square n} + \frac{R_{\square p}R_{\square B}}{R_{\square p} + R_{\square B}}\right)\frac{l}{L} + \frac{2}{3}R_{\square m}\frac{L}{d}. \tag{69}$$

We refer to fig. 5 for nomenclature. The $R_\square$ correspond to the sheet resistance of the n$^+$- and p$^+$-layers. To compare them with monofacial

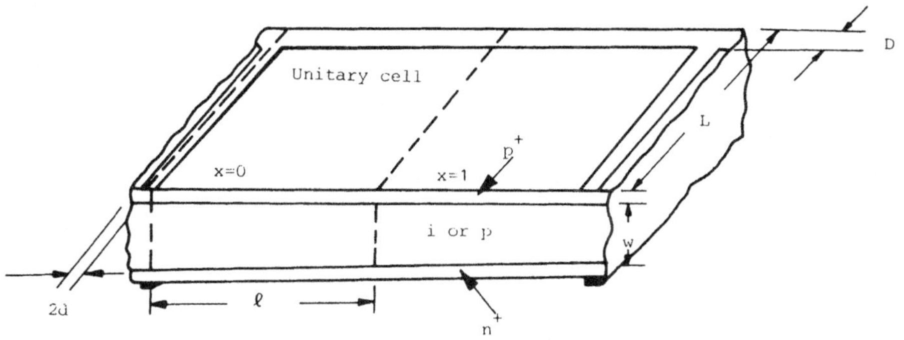

Fig. 5. Cell's element for bidimensional series resistance studies.

cells we must set $R_{\square p} = 0$ (because this layer is shorted by the back metal layer) and we must divide the last term by two, due to the presence of only one metal grid. The total resistance is obtained by dividing R_G by the number of elementary cells that constitute the actual cell.

With the typical values of sheet resistance the series resistance of a bifacial cell doubles that of a monofacial one. The effect of the series resistance is to reduce the fill factor according to a law that has been the object of a number of approximations (Luque et al. 1978c, Green 1984, Araújo and Sánchez 1982). Using a very dense metallization grid increases the fill factor but introduces a certain opacity, thus reducing the photogenerated current. As a result, there is a trade-off leading to a maximum efficiency for a given value of the irradiance on the cell. This optimum is achieved when the separation between fingers (see fig. 5) takes a specific value. However, the variation in the efficiency when l is somewhat different from the best value is small, allowing to use an approximate calculation based on optimizing the equation

$$\eta = \eta_i\left(1 - \frac{l}{L}\right)\left(1 - \left(\frac{I_{sc}R}{V_{oc}}\right)\right), \tag{70}$$

where η_i is the resistance-free efficiency.

This is done when the fraction $x = l/d$ is a solution of eq. (54)

$$x^3 - \left(\frac{2}{3}\frac{J_{sc}L^2R_{\square m}}{V_{oc}}\right)x - \frac{2}{3}\frac{J_{sc}d^2}{V_{oc}}\left(R_{\square n} + \frac{R_{\square p}R_{\square B}}{R_{\square p} + R_{\square B}}\right) = 0, \tag{71}$$

which can be solved either by analytical standard methods for solving algebraic third-order equations or by using numerical methods.

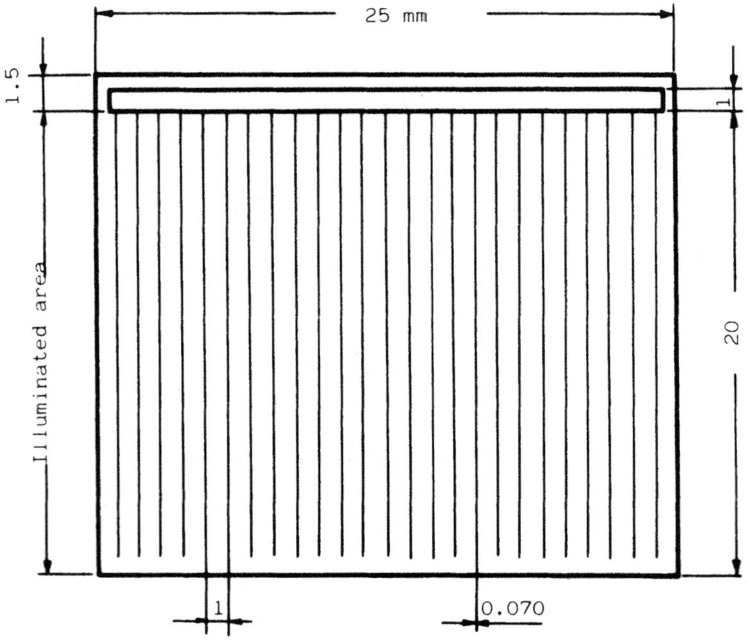

Fig. 6. Lay-out of the metal grids of a bifacial cell optimized for operation at 7 AM1 suns.

As observed in this equation the thinner the metal lines the closer they have to be spaced, but thinner lines lead to higher efficiency values and at the limit of $d \to 0$ no efficiency reduction is produced by lateral series resistance.

In fig. 6 we present an optimally designed grid for operation at 7 AM1 suns (Cuevas et al. 1983). In this case the lateral series resistance is 0.3 $\Omega\,cm^2$ (6 mΩ for the whole cell), and the resulting measurements for a p^+–n–n^+ cell with 10 Ω cm base resistivity are given in fig. 7. We can observe a maximum efficiency of 16.5% at 7 AM1 suns under front illumination. In this case the efficiency has been calculated with respect to the cell area not covered by the busbar (which is hidden by the concentrator for which this cell was designed) and therefore with an opacity factor d/l of 7%. The fill factor at 7 AM1 suns is 0.75. This shows that there is no major problem caused by series resistance to obtain good bifacial cells for the moderate value of concentration compatible with static concentration.

If the busbar is to be exposed to illumination, as happens with cells in

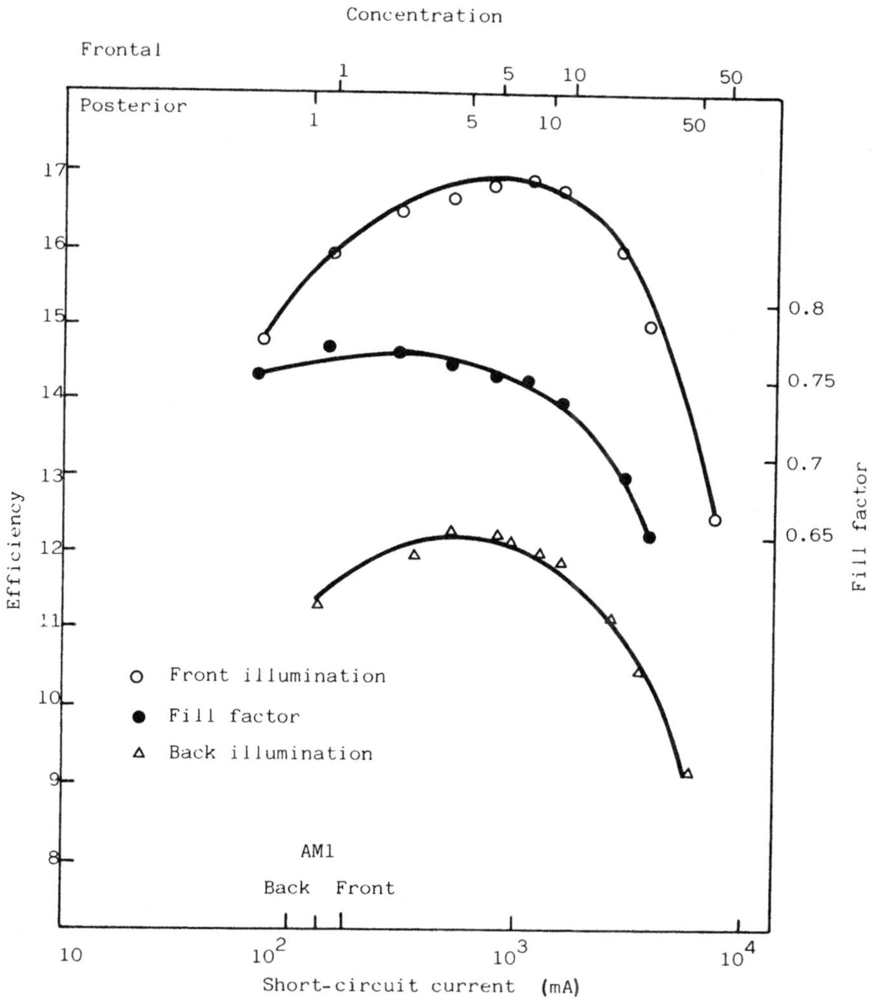

Fig. 7. Efficiency and fill factor of the cell in this figure versus concentration.

flat panels, its optimum width is given by (Luque et al. 1978c)

$$D = A\left(\frac{2}{3}\,\frac{J_{sc}R_{\square m}}{V_{oc}}\right)^{1/2}, \tag{72}$$

valid for bifacial cells (two similar busbars) and where it has to be considered that $R_{\Box m}$ here is the sheet resistance of the metal used in the busbar which sometimes is greatly reduced by soldering a copper wire over it.

For cells with a conventional non-optimized grid for flat panels the measured (Calieja et al. 1984a) series resistance is $1.36\,\Omega\,\mathrm{cm}^2$. This value includes all vertical and lateral effects. The resistance represents a maximum power point voltage reduction of 8.6%. In a monofacial cell this amount is reduced by a factor of two.

The lateral resistance in p–i–n cells is somewhat reduced by conductivity modulation (Luque et al. 1984b). Equation (69) can be rewritten as

$$R_G = \frac{1}{3}\left(\frac{R_{\Box n}R_{eb}}{R_{\Box n}+R_{eb}} + \frac{R_{\Box p}R_{hb}}{R_{\Box p}+R_{hb}}\right)\frac{l}{L} + \frac{2}{3}R_{\Box m}\frac{L}{d}\,, \tag{73}$$

where the lateral transport of electrons and holes in the intrinsic base is represented by R_{eb} and R_{hb}

$$R_{eb}\mu_e = R_{hb}\mu_n \simeq \left[q\int_0^W p\,\mathrm{d}x\right]^{-1} \tag{74}$$

In practice, for low concentration levels, these terms are much higher than $R_{\Box n}$ or $R_{\Box p}$ and they do not affect the value of the grid series resistance.

6. Fabrication of bifacial cells

A bifacial cell is, in essence, a BSF cell, and as such, it can be manufactured in many ways. However, there are some principles that should be observed.

First of all we should use a base with high diffusion length, and the subsequent steps must be careful in order to preserve it. Actually, normal float-zone or Czochralski material grown from electronic grade silicon fulfills this requirement, provided that it has a base resistivity above 5 Ω cm, for either conduction type, p or n. This statement is based on the well-known dependence of the base lifetime from the doping as measured by Iles and Soclof (1975), or in the model proposed by Fossum and Lee (1982) to explain these facts. We have obtained poor symmetry, J_{sc}-back/J_{sc}-front is 73%, by using 3 Ω cm n-type bases (Eguren et al. 1981). We have obtained also poor values of symmetry (30%) when using lower resistivity bases. This result is also found theoretically in table 2.

From a theoretical point of view, higher symmetry (J_{sc}-back/J_{sc}-front) is to be expected when using p-type bases, due to the higher value of the diffusion length of the minority carriers there, based on the higher values and higher diffusion constants of the electrons as compared to those of the holes in the dual (p^+–n–n^+) structure. Indeed, the electron diffusion length of low-injection bases in n^+–p–p^+ structure is similar to the effective one in p–i–n structures (or even higher in some ideal cases) if the Fossum–Lee model is to be used. However, we have not yet confirmed experimentally this behaviour through a large number of p^+–n–n^+ cells. Some experiments show typical symmetry values of 75% for 10 Ω cm p^+–n–n^+ cells, while for n^+–p–p^+ cells the symmetry is in the range of 85–90% for base resistivities from 5–40 Ω cm.

For very high base resistivity of about 1000 Ω cm, we were able to confirm experimentally (Calleja et al. 1984a) (with p^+–n–n^+ cells) that we can obtain a fully symmetrical behaviour (J_{sc}-back/J_{sc}-front≈1) in agreement with our theoretical discussion in the preceding sections.

The second condition is to form a high–low junction of high quality, able to:

(a) permit the light to pass through,
(b) to sustain a voltage with low leakage, and
(c) to reduce the surface recombination.

In fact, these conditions are exactly the same as those needed for p–n junctions. Boron or phosphorus layers, diffused in the conventional way to produce junctions, are good choices. Aluminum-alloyed high–low junctions are not likely to work well if some voltage is to be sustained in high injection, due to the formation of alloyed spikes (Roberts and Wilkinson 1971) which cause leakage. This does not contradict the well-known ability of thick aluminum-alloyed layers to reduce the surface recombination at their back side in low-injection BSF cells (del Alamo et al. 1981). The problem with Al-alloyed bifacial cells is more the voltage than the symmetry of the cell itself, as it has been shown experimentally (Luque et al. 1978b, del Alamo et al. 1981).

In the deposition of metal grids on both faces we again have to take the same care with both faces, to avoid perforating the shallow p–n or high–low junctions.

Standard industrial cells are done on p-type CZ substrates of 10 cm diameter and variable resistivity (above 8 Ω cm). Laboratory cells are done on a variety of substrates of 5 cm diameter including CZ and FZ monocrystals and HEM, or cast multicrystals. A standard double diffusion technique is then used to obtain either n^+–p–p^+ or p^+–n–n^+ structures. A nickel solder metallization procedure of simultaneous self delineating grid deposition on both faces is used in the industrial process. A classical Ti–Pd–Ag,

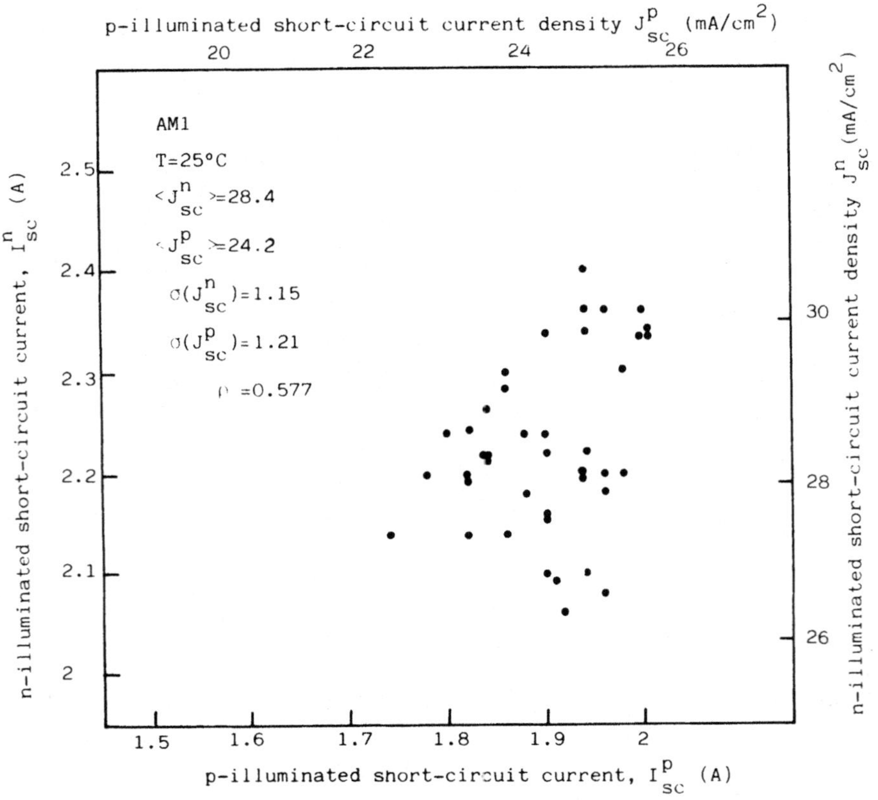

Fig. 8. n-illuminated versus p-illuminated short-circuit currents for 10 cm diameter industrial cells.

either by evaporation or sputtering with an additional electrolytic Ag regrowth, is used in most of our laboratory cells. In both cases a TiOx–SiOx antireflecting layer is spun-on on both sides of the cell.

We present now the results of a typical group of industrially produced n^+–p–p^+ bifacial solar cells with arbitrary base resistivities, usually between 15–50 Ω cm. The plot of J^n_{sc} versus J^p_{sc} is presented in fig. 8, and that of V^n_{oc} versus V^p_{oc} is shown in fig. 9. That of η^n versus η^p in fig. 10. Figure 11 shows the plot of η^{np} versus J^{np}_{sc}, and fig. 12 that of η^{np} versus V^{np}_{oc}. In the last two cases the cells are measured under bifacial illumination equivalent to 1 AM1 sun (100 mW/cm²) on the n-side and 0.5 AM1 sun (50 mW/cm²) on the p-side. The temperature is 25°C. For the bifacial case, together

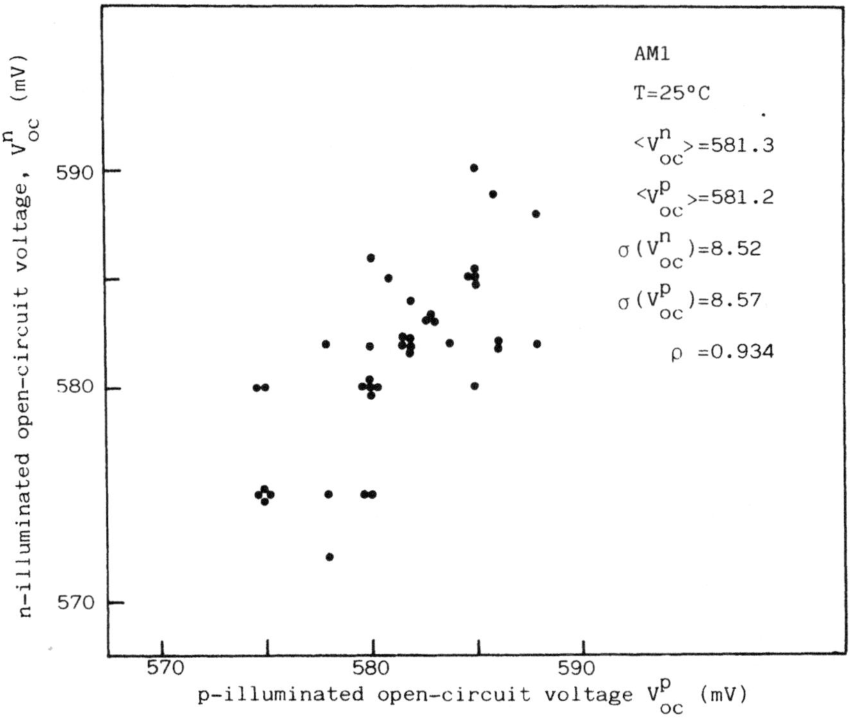

Fig. 9. n-illuminated versus p-illuminated open-circuit voltages for the same cells as in fig. 8.

with the conventional efficiency obtained by, dividing by a $150\,mW/cm^2$ irradiance, we also give the efficiency for engineering purposes, based on dividing by a $100\,mW/cm^2$ irradiance. In all cases the values are referred to the total area of a 10 cm diameter cell.

We see that industrial bifacial cells can be fabricated with performances essentially similar to those of conventional monofacial ones.

7. Multicrystalline bifacial cells

The possibility of manufacturing bifacial cells with multicrystalline silicon (Calleja et al. 1984a) is of great interest because it allows to use a lower cost substrate. Some preliminary experiments have been carried out to show the ability of heat-exchange method (HEM) grown or cast silicon to

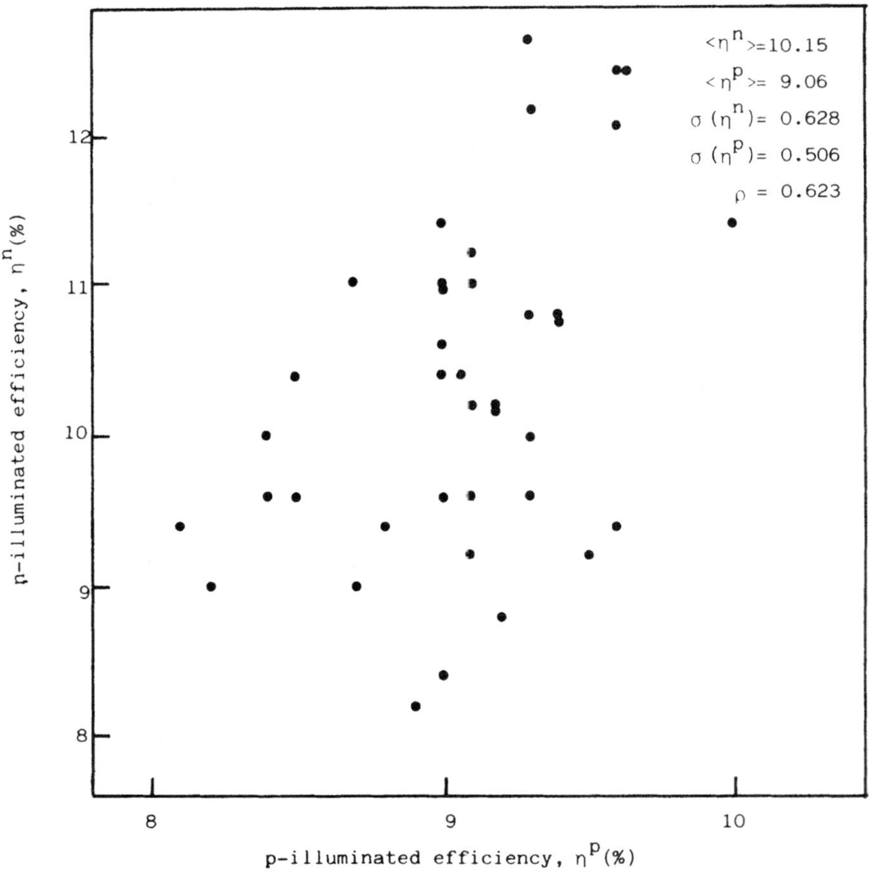

Fig. 10. n-illuminated versus p-illuminated efficiency for the same cells as in fig. 8. The efficiency is calculated for the total cell area.

produce good bifacial cells. For these experiments specially grown wafers with high base resistivity of about 13 Ω cm (n-material) for HEM and 35 Ω cm (p-material) for cast were used. In most of our experimental cells we have not used any AR coating to remove the indetermination that this coating introduces in the cell symmetry. Good cells have been obtained with HEM, with an efficiency of 9.5% for front illumination and 8.11% for back illumination, to be compared with 11.01% and 10.04% efficiencies obtained with a float-zone single crystal with the same technological pro-

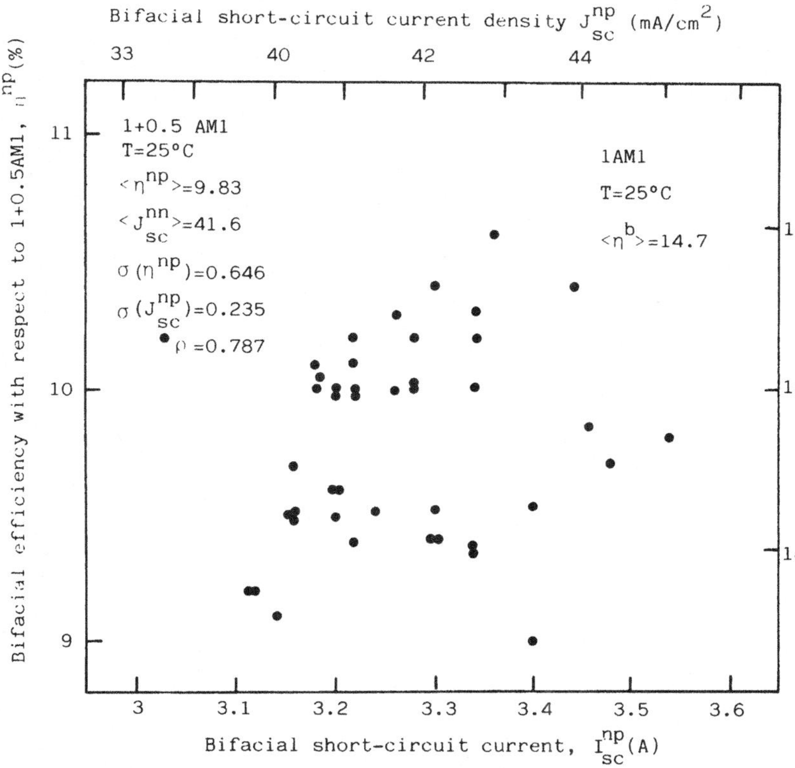

Fig. 11. Efficiency versus short-circuit current for cells illuminated simultaneously on both sides with AM1 conditions on the n-side and 0.5 AM1 conditions on the p-side. The latter irradiance can be easily collected from the albedo. The efficiency on the left-hand side of the figure is calculated for (1 + 0.5) AM1 irradiance and the one on the right-hand side is calculated with respect to the front irradiane only, i.e., the energy coming directly from the sun.

cess. With AR coating this will lead to cells with efficiencies over 13% in the best cases.

These cells follow the two-exponential law, as can be seen in fig. 13 where a dark *I–V* curve is presented, as well as its fitting. The cells present a rather high value of the second exponential, typical of a high-injection process, or more likely to an increased area of space-charge recombination (possibly the grain boundaries), but no odd current appears in the cell. The first exponential [$\exp(qV/kT)$] has a high correlation with the cell symmetry, as can be seen in fig. 14, where this cell symmetry is presented

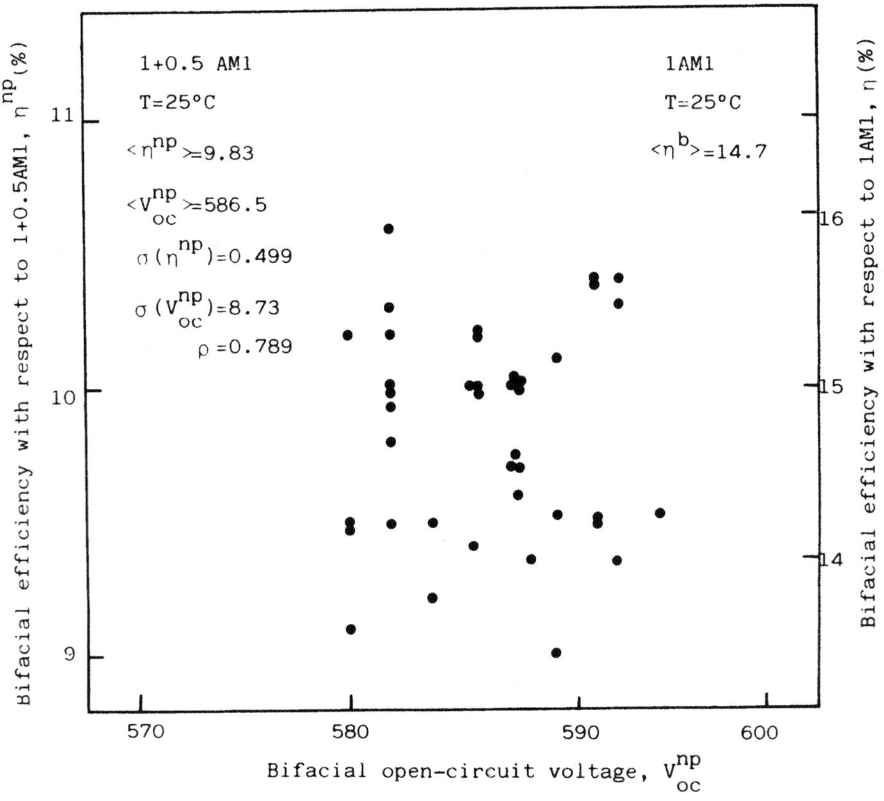

Fig. 12. Efficiency versus open-circuit voltage for the same cells as in fig. 8 and with the same illumination as that of fig. 11.

versus the first exponential coefficient for float-zone monocrystals, HEM and cast multicrystals. We see that the HEM material gives a first exponential component that, in the best cases, is similar to that of the FZ material, but in general is one order of magnitude higher. Accordingly, the symmetry is very high for cells with small first exponential coefficient and becomes poorer for higher values of this parameter, showing that in some cells the diffusion length of the HEM material is, in order of magnitude, about 1000 μm and that, in most cases, is reduced to lower values. The diffusion length of cast silicon is much smaller and does not produce good bifacial cells. We can conclude that HEM material is potentially promising to make good bifacial cells.

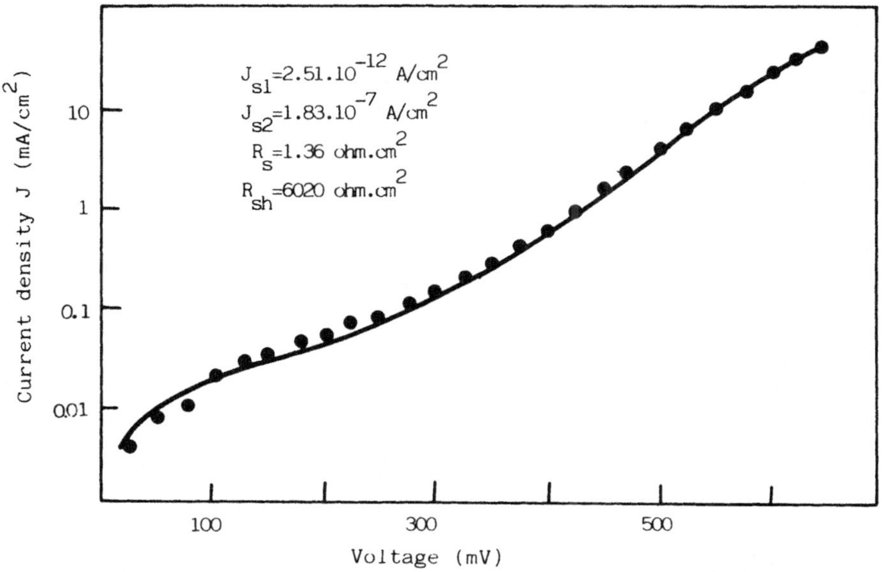

Fig. 13. Dark current versus voltage for a multicrystalline bifacial cell. The solid line is obtained from a best fitting with a two exponential model, series and parallels resistances. The dots are experimental data. A good fitting can be observed.

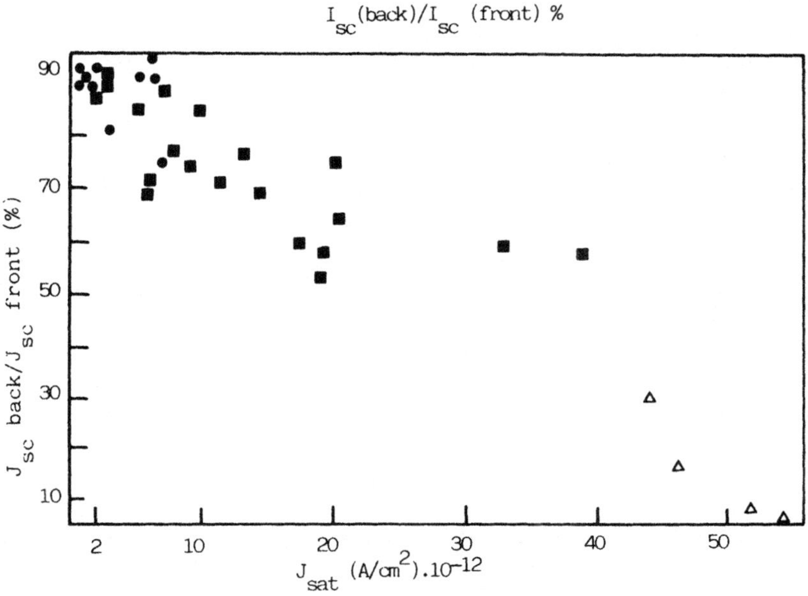

Fig. 14. Symmetry (J_{sc}-back/J_{sc}-front) versus the first exponential [term preceding $\exp(qV/kT)$] saturation current for three types of silicon. ● single crystal (FZ); ■ HEM; △ cast multicrystal.

8. Applications

Two basic practical applications exist today for terrestrial bifacial cells: albedo-collecting panels and static concentrators, either optical or, possibly, holographic. A third application, mentioned in the introduction, is the luminiscent concentrator. Of all those, only the albedo-collecting panels are already produced industrially.

Albedo-collecting bifacial panels are flat panels using bifacial cells mounted as in normal flat panels but covered with a transparent material on both sides. They are installed in the conventional way, and collect not only the energy from the sun reaching the front side, but also the energy from the surroundings, impinging mainly on the back side. If the area surrounding the panels is painted white, the amount of energy reaching the back side of the panel can be considerable.

A model has been developed (Luque et al. 1984b) based on the assumption that the white-painted surroundings are Lambertian reflectors. This assumption, as well as the results obtained with this model have been verified experimentally. We present in fig. 15 the results of comparing the model to the experimental measurements on a given day. In this case, only the ground was painted white in an area of 10 m^2 and a group of six panels were placed together (to include the shadows cast on the ground by an array of panels), and front and back irradiances received on the central panel were measured. The albedo factor AF, defined as the front plus back irradiance divided by the front one is presented versus the solar time. We obtained values of AF ranging from 1.5 to 1.8 for this specific application.

Based on the model we have calculated the total yearly energy on a large field of bifacial panels (45 kW) to be installed in the outskirts of Madrid, and it will be connected to the utility network. The bifacially received irradiance is calculated for different tilt angles and for different separation between panels. For a separation of 30 cm we obtain about 60% extra power from the back, as presented in fig. 16.

When the neighbouring structure is more complex, including walls as well as ground, we can obtain almost 80% of energy from the back in some structures.

Another promising application is static concentrators, for which we forecasted very low cost (Luque 1984) and in large productions might reach the U.S.A. Department of Energy target cost of about $0.70/W ($1980); however, no full-size prototype of this application has been yet successfully manufactured.

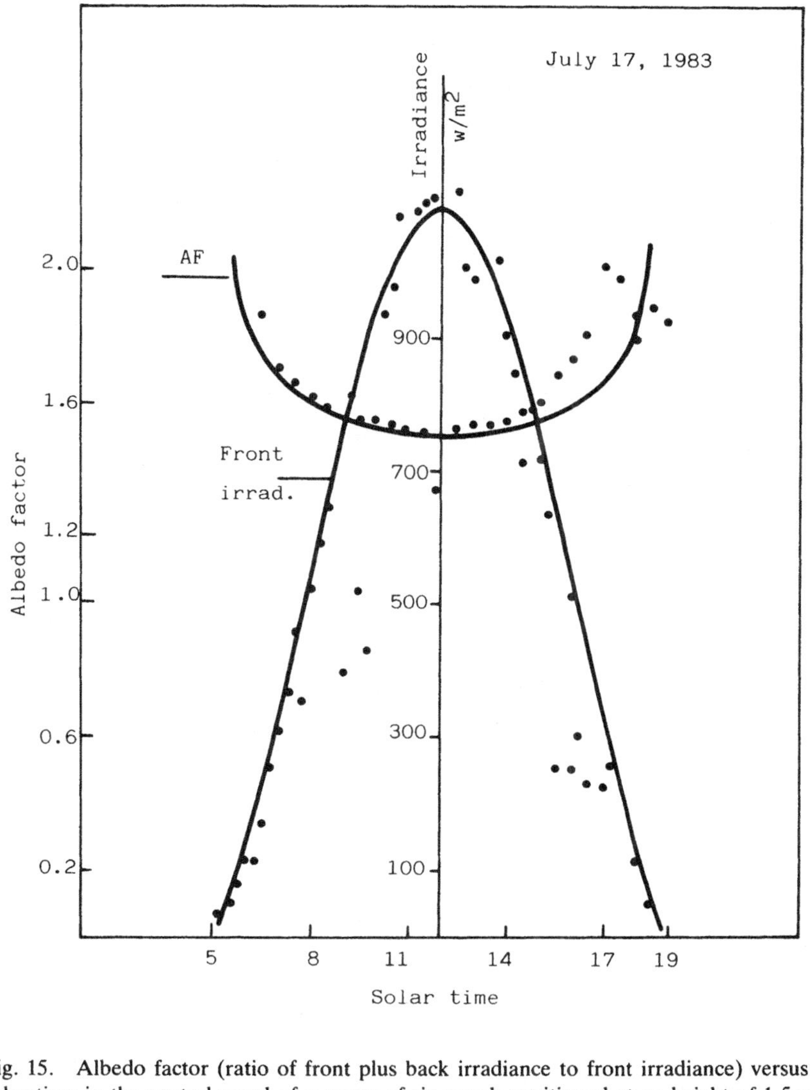

Fig. 15. Albedo factor (ratio of front plus back irradiance to front irradiance) versus the solar time in the central panel of a group of six panels positioned at an height of 1.5 m (at the center) above the ground, an inclination of 30° and a separation of 15 cm between the panels. The dots are measurements. The solid lines are calculated according to the Liu–Jordan model of irradiance, after the clearliness factor of the day was measured. Panel size $110 \times 46\,cm^2$.

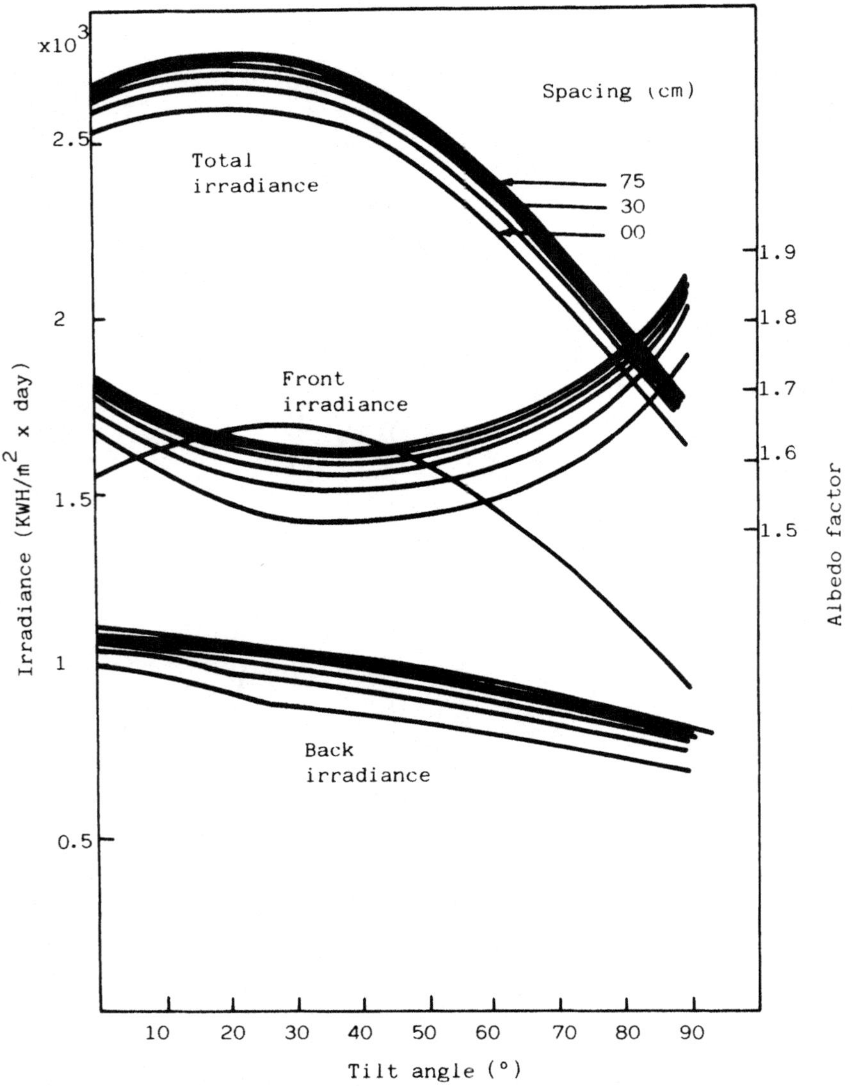

Fig. 16. Energy collected in the front, in the back and in both sides of a bifacial panel in a large array with the ground painted white, for different inclination angles and separations between panels. Panel size is $130 \times 56\,cm^2$, and the highest point is at 2.00 m above the ground.

9. Conclusions

We have presented models of bifacial BSF cells for various levels of injection, suitable for calculations in some cases and able to give an intuitive picture of the phenomena governing the behaviour of these cells.

These models stress the important role of the high–low junction in supporting a voltage, so that high base resistivity bifacial solar cells can be used without voltage loss. Despite the fact that this voltage is produced only under high injection, this level of injection usually occurs near open-circuit in bifacial cells of base resistivity above 5 Ω cm which are the only ones expected to produce, because of their high diffusion length, good bifacial cells, as it has been shown both theoretically and experimentally.

This voltage build-up is more important than the reduction of surface recombination which in bifacial cells (or in interdigitated back contact cells) can be produced by proper passivation of the base surface. This introduces an important technological consequence that forces us to make a leakage-free shallow high–low junction with carefully deposited contacts there.

The short-circuit current of a front-illuminated low-injection bifacial cell has a tendency of being higher than the back-illuminated one. In low injection there is superposition for bifacial illumination. Under very high injection the short-circuit current is somewhat higher under p^+-illumination, no matter where the p–n junction is located (i.e. the base type). Finally, under bifacial illumination the current can be somewhat higher than the one expected from strict superposition, at least when operating at low concentration (1 + 0.5 AM1 suns). The reason for this behaviour is well-explained on the basis of the Ebers–Moll formulation. Cells with base resistivity below 50 Ω cm are in low injection in short-circuit conditions.

In open-circuit, front-illuminated cells have slightly higher open-circuit voltage if they are in low injection. For high-injection cells the voltages are slightly higher for p-illumination, no matter where the p–n junction is located. Cells with resistivities above 5 Ω cm are already in high injection for open-circuit conditions.

As a result, practical cells with resistivities between 5 and 50 Ω cm are in low-injection operation in short-circuit and in high-injection operation in open-circuit. This produces, in some cases, a crossing of the front- and back-illumination cells for n^+–p–p^+ cells, showing a nice experimental demonstration of a violation of the superposition of the dark *I–V* curve and the photogenerated current, as very often assumed.

The curve shape presents a two-exponential aspect, due in low injection to the space-charge recombination, and in high injection to base recombination. The second exponential (the one with $qV/2kT$) in the case of high injection has been measured to be almost an order of magnitude higher.

The curve shape is also modified by the vertical and horizontal series resistance. The vertical one in high-injection cells is intrinsically higher than in low-injection ones. However, this effect is of little importance, as shown theoretically and experimentally, even in concentrator cells. The horizontal resistance, due to the double grid, is basically doubled with respect to that of a monofacial cell. Even though this is a drawback, bifacial cells with efficiencies of 16.5% at 7 AM1 suns have been measured, showing that obtaining good bifacial cells is not a problem.

In practice, the conventional CZ monocrystalline or even the HEM multicrystal fill the requisities to obtain good bifacial cells. This is clearly seen when we look at the symmetry of bifacial cells, which in good cells without AR coating is affected almost solely by the material lifetime.

The viability of bifacial cells is further proven by the fact that they are being industrially produced with good efficiencies. As a matter of fact the open-circuit voltages obtained in these cells are almost as good as the best values theoretically achievable for this parameter.

The applications for bifacial panels make the use of this approach attractive for increasing power collection up to 60% or more and for reducing silicon consumption by using static concentrators able to collect direct or diffuse light. Bifacial cells are considered to be, potentially, a powerful cost reducer in photovoltaic applications.

10. Summary

Bifacial cells are useful to collect the light from the surroundings that amounts up to 80% of the light collected directly from the sun, or to fabricate static concentrators able to approach the theoretical upper limit to effective concentration of about 8 in a clear climate.

Bifacial cells are essentially back surface field cells, but special rules for manufacturing them must be followed to obtain high efficiency under either front or back illumination. These rules are derived and explained very well from theoretical models.

The behaviour of laboratory or industrially made bifacial cells is very satisfactory and the best cells present open-circuit voltages close to the theoretical limits.

Acknowledgement

The development of bifacial cells has been one of the prior tasks of our Instituto de Energía Solar at the Universidad Politecnica Madrid, and the main activity of the newly created firm Isofotón. I want to mention here the scientists and engineers that made possible the development of this product. Among the first, Professor G. Sala, Professor J.M. Ruíz, Professor G.L. Araújo, Professor A. Cuevas and Professor E. Lorenzo; and among the second, J.C. Jimeno and J.C. Miñano of our Institute and Dr. J. Eguren and Miss J. Calleja of Isofotón; also Mrs. M.A. Navarro, Mrs. P. Roman and Mrs. M. Rodrigo and Mr. J. Piñero of our Instituto and Mr. J. Ruíz of Isofotón, as well as the services, as translator, of Miss M.L.P. de la Dehesa.

References

Araújo, G.L., and E. Sánchez, 1982, Sol. Cells **5**, 377–386.

Araújo, G.L., A. Luque, E. Sanchez, J.M. Calleja, S. Lopez Romero and J. Castaño, 1982a, Proc. 16th IEEE Photovoltaic Spec. Conf., San Diego (IEEE, New York) p. 559.

Araújo, G.L., E. Sánchez and M. Marti, 1982b, Sol. Cells **5**, 149.

Araújo, G.L., J. Castaño, A. Luque and E. Sánchez, 1983, IEEE Trans. Electron Devices **ED-30**, 1727.

Borodina, N.M., T.M. Golovner et al., 1976, Geliotekhnika **14**, No. 2, 293.

Calleja, M.J., A. Luque and S. López Romero, 1984a, Comparison of high and low injection cells, Sol. Cells **12**, 295.

Calleja, M.J., A. Luque, J.C. Jimeno and G. Sala, 1984b, Proc. 5th EC Photovoltaic Solar Energy Conference, p. 1138.

Chambouleyron, I., and Y. Chevalier, 1978, Proc. 1st EC Photovoltaic Solar Energy Conf., Luxembourg 1977 (Reidel, Dordrecht) p. 793.

Chiang, S.Y., B.C. Carbajal and G.F. Wakefield, 1978, IEEE Trans. Electron Devices **ED-25**, 337.

Cuevas, A., 1980, Tesis Doctoral, Universidad Politecnica de Madrid, ETSI Telecommunication, Madrid.

Cuevas, A., A. Luque and J.M. Ruiz, 1979a, Proc. Int. IEEE Electron Devices Meeting (IEEE, New York) p. 314.

Cuevas, A., J. Sangrador, A. Luque, J.M. Ruiz and G. Sala, 1979b, Proc. 2nd EC Photovoltaic Solar Energy Conf., Berlin (Reidel, Dordrecht) p. 808.

Cuevas, A., A. Luque, J. Eguren and J. del Alamo, 1982, Sol. Energy **29**, 419.

Cuevas, A., G.L. Araújo and J.M. Ruiz, 1983, Proc. 5th EC Photovoltaic Solar Energy Conf., Athens (Reidel, Dordrecht) p. 114.

Dash, W.C., and R. Newman, 1955, Phys. Rev. **99**, 1151.

del Alamo, J., J. Eguren and A. Luque, 1981, Solid State Electron. **24**, 415.

Dharival, S.R., and A.P. Kulshreshta, 1981, Solid State Electron. **24**, 1161.

Dunbar, P.M., and J.R. Houser, 1973, Solid State Electron. **16**, R61.

Eguren, J., J. del Alamo, A. Cuevas and A. Luque, 1981, Proc. 15th IEEE Photovoltaic Spec. Conf., Kissimee (IEEE, New York) p. 1343.

Fossum, J.G., and D.S. Lee, 1982, Solid State Electron. **25**, 741.
Fossum, J.G., and A. Shibib, 1981, IEEE Trans. Electron Devices **ED-28**, 1018.
Gray, D.E., et al., 1972, American Institute of Physics, Handbook (McGraw-Hill, New York).
Green, M.A., 1977, Solid State Electron. **20**, 265.
Hovel, H.J., 1975, Semiconductors and Semimetals, Vol. II, Solar Cells (Academic Press, New York).
Iles, P.A., and S.I. Soclof, 1975, Proc. 11th IEEE Photovoltaic Spec. Conf., Scottsdale (IEEE, New York).
Lammert, M.D., and J.R. Swartz, 1977, IEEE Trans. Electron Devices **ED-24**, 337.
Lanyon, H.P.D., and R.A. Tuft, 1979, IEEE Trans. Electron Devices **ED-26**, 1014.
Li, S.S., 1978, Solid State Electron. **21**, 1109.
Li, S.S., and W.R. Thurber, 1977, Solid State Electron. **20**, 609.
Lindholm, F.A., and J.G. Fossum, 1981, IEEE Electron Device Lett. **EDL-2**, 230.
Lindholm, F.A., J.G. Fossum and E.L. Burgess, 1979, IEEE Trans. Electron Devices **ED-26**, 165.
Luque, A., 1984, Sol. Cells, **12**, 14.
Luque, A., and G.L. Araújo, 1984, Internal Report UPH/IES/LS/8384, Polytechnical University of Madrid.
Luque, A., and J. Eguren, 1982, Solid State Electron. **25**, 797.
Luque, A., A. Cuevas and J. Eguren, 1978a, Solid State Electron. **21**, 793.
Luque, A., J.M. Ruiz, A. Cuevas, J. Eguren and J.M. Gomez-Agost, 1978b, Proc. 1st EC Photovoltaic Solar Energy Conf., Luxembourg 1977 (Reidel, Dordrecht) p. 269.
Luque, A., J. Eguren and J. del Alamo, 1978c, Rev. Phys. Appl. (France) **13**, 629.
Luque, A., J.M. Ruiz, A. Cuevas, J. Eguren and J.M. Gomez-Agost, 1978d, Proc. ISES Int. Conf., New Delhi (Pergamon Press, New York) p. 729.
Luque, A., A. Cuevas, J.M. Ruiz, 1980, Sol. Cells **2**, 151.
Luque, A., A. Gras and J. Cárabe, 1984a, Proc. 5th EC Photovoltaic Solar Energy Conf., Athens (Reidel, Dordrecht) p. 167.
Luque, A., J.M. Ruiz and S. Lopez-Romero, 1984b, ISES Arab. Section Conf., Kuwait, Dec. 1983, p. 226.
Madelkorn, J., and J.H. Lammeck, 1972, Proc. 9th IEEE Photovoltaic Spec. Conf. (IEEE, New York) p. 66.
Michel, J., 1978, 13th IEEE Photovoltaic Spec. Conf., Washington DC (IEEE, New York) p. 859.
Miñano, J.C., and A. Luque, 1983, Appl. Opt. **16**, 2437.
Roberts, E.M., and E.G.L. Wilkinson, 1971, J. Mater. Sci. **6**, 189.
Ruiz, J.M., M. Cid, A. Cuevas and A. Luque, 1982, Proc. 4th EC Photovoltaic Solar Energy Conf., Stresa (Reidel, Dordrecht) p. 511.
Ruiz, J.M., A. Cuevas and M. Cid, 1983, IEEE Proc. Melecon &LQT;83, Athens (IEEE, New York) paper D9.03.
Sah, C.T., R.N. Noyce and W. Schokley, 1957, Proc. IRE **45**, 1228.
Sheng, W.W., 1975, IEEE Trans. Electron Devices **ED-22**, 25.
Slotboom, J.C., and H.C. de Graaf, 1976, Solid State Electron. **19**, 857.
Tanenbaum, M., and D.E. Thomas, 1956, Bell Syst. Tech. J. **35**, 1.
Thomas, A.P., M.P. Thekaekara and K. Boer, eds, 1976, Proc. Am. Sect. ISES Joint Conf., Cape Canaveral, p. 336.
Thurber, W.R., R.L. Mattis, Y.M. Liu and J.J. Filliben, 1980a, J. Electrochem. Soc. **127**, 1807.

Thurber, W.R., R.C. Mattis, Y.M. Liu and J.J. Filliben, 1980b, J. Electrochem. Soc. **127**, 2291.
Van Overstraeten, R.J., H.J. De Man and R.P. Martens, 1973, IEEE Trans. Electron Devices **ED-20**, 290.
Van Vliet, R.M., 1982, Solid State Electron. **25**, 797.
Volfson, A.A., and V.K. Subashiev, 1967, Sov. Phys. Semicond. **1**, 327.
Von Ross, O., and B. Anspaugh, 1978, Proc. 13th IEEE Photovoltaic Spec. Conf., Washington DC (IEEE, New York) p. 1119.
Weaver, H.T., and R.D. Nasby, 1981, IEEE Trans. Electron Devices **ED-28**, 465.
Winston, R., 1982, J. Opt. Soc. Am. **72**, 1564.
Wolf, M., 1960, Proc. IRE **48**, 1246.

SUBJECT INDEX

Printed in Dunstable, United Kingdom

91338444R00239